BOSTON STUDIES IN THE PHILOSOPHY OF SCIENCE

VOLUME 140

BRIDGING THE GAP

PHILOSOPHY, MATHEMATICS, AND PHYSICS

Lectures on the Foundations of Science

Edited by

GIOVANNA CORSI

Department of Philosophy, University of Florence, Italy

MARIA LUISA DALLA CHIARA

Department of Philosophy, University of Florence, Italy

and

GIAN CARLO GHIRARDI

Department of Theoretical Physics, University of Trieste, Italy

SPRINGER SCIENCE+BUSINESS MEDIA, B.V.

Library of Congress Cataloging-in-Publication Data

Bridging the gap : philosophy, mathematics, and physics : lectures on
  the foundations of science / edited by Giovanna Corsi, Maria Luisa
  Dalla Chiara, Gian Carlo Ghirardi.
      p.   cm. -- (Boston studies in the philosophy of science ; v.
  140)
  Includes bibliographical references.
  ISBN 978-0-7923-1761-6   ISBN 978-94-011-2496-6 (eBook)
  DOI 10.1007/978-94-011-2496-6
  1. Science--Philosophy.   I. Corsi, Giovanna.   II. Dalla Chiara,
  Maria Luisa.   III. Ghirardi, G. C.   IV. Series.
  Q175.B7854   1992
  501--dc20   92-12617

ISBN 978-0-7923-1761-6

TABLE OF CONTENTS

PREFACE

The lectures collected in this volume were given at the International School of Philosophy of Science, held in Trieste (October 1989). The school was devoted to a wide spectrum of subjects, from logic to mathematics and physics. The basic aim of the organizers was to contribute to overcoming a kind of gap that has often been recognized in research into the philosophy of science. There is still today a considerable distance between the kind of questions that are being explored in the framework of general epistemology and the actual problems that arise from technical work on the foundations of the different scientific theories. Such gap phenomena have been particularly serious in the case of the experimental sciences. In contrast, as is well known, philosophy of mathematics has a long tradition that has linked foundational to philosophical enquiries, via the use of logical tools. In spite of this, strangely enough, philosophy of mathematics is sometimes, and artificially, excluded from the field of epistemology, as if philosophy of science ought particularly to mean philosophy of the empirical sciences. As a consequence of such separations, a number of debates about the foundations of mathematics seem to proceed according to a somewhat naive and old-fashioned image of physics, whereas foundational investigations about physics sometimes propose an oversimplified image of the world of mathematics, stressing, for instance, the merely 'linguistic' and 'formal' role of mathematical theories.

One of the aims of the Trieste school was to present a set of topics that should be part of the basic culture of any student or scholar working in philosophy of science today. A characteristic feature of the lectures is the attempt to teach, at the same time, *some* science and *some* philosophy of science, and to show how general epistemological questions arise directly from foundational work in the different scientific theories.

The volume is divided into two parts: 1. Logic, Mathematics and Information; 2. Physics and Probability.

The research for a theory of deduction that would turn out to be at the same time intuitively natural, theoretically satisfactory and adequate with respect to the aims of different scientific theories has been a constant *Leitmotiv* of logical studies. In our days, such research has been particularly stimulated by investigations into the problem of *automatic reasoning* in computer science. Celluci's paper describes advantages, disadvantages and possible refinements of a classical system of deduction, the calculus of *natural deduction*, first proposed by Gentzen.

In modern theories of meaning and truth, a basic role is played by *possible worlds semantics*, also called *Kripke semantics*. In spite of its metaphysical appearance, the notion of *possible world* has turned out to be quite a flexible and adequate abstract tool for conceptual clarification in different fields, from computer science to the logical analysis of physical theories. Bellissima's lecture deals with topics that constitute the 'core' of propositional normal modal logic, providing a wealth of information for anyone working on extensions of classical logic, such as multimodal logics, tense logics, temporal logics of programs. Börger discusses – with new tools, in part arising from recent research into PROLOG – a classical problem of logic, the *Entscheidungsproblem*: to what extent can correct reasoning be mechanically decided? The question is connected with the possibility of a 'technological' realization of a dream pursued by Leibniz: that of solving any dispute in a purely mechanical way ('calculemus!').

Prawitz's lecture analyses the merits and the weaknesses of Hilbert's program, focusing some basic ambiguities in Hilbert's writings, as to the problem: to what extent is the transfinite part of mathematics completely trustworthy? Feferman notices a 'general malaise' about the logical approach to the foundations of mathematics, after the failure of the global views about the nature of mathematics (logicism, formalism, platonism and constructivism). He proposes a new attitude in this field: a research for 'working foundations', without regard to any fixed philosophical standpoint. A philosophical discussion of the mind-body problem, in the framework of information theory, is developed in the lectures by Tahir Shah and Longo.

The lectures about physics are designed to give a concise but comprehensive view of the main epistemological problems arising in connection with the two basic scientific revolutions of our century: relativity and quantum mechanics. As is well known, even though the elaboration of the theory of relativity has required radical changes in the conceptions of space and time, it is the peculiar structure of the quantum mechanical picture of nature which has given rise to a more lively and still ongoing debate.

Budinich introduces readers to the basic principles of special relativity, pointing out the 'paradoxical' consequences of the theory. He also discusses some general questions such as the role of symmetry in physics.

The contribution by Weber focuses on the most important conceptual and formal points of the quantum framework, with particular reference to quantum entanglement and to an exhaustive analysis of the celebrated incompleteness argument of Einstein, Podolsky and Rosen.

Ghirardi brings out the conceptual difficulties which one meets in trying to build a coherent worldview compatible with a macro-objectivist position and based on the quantum picture of nature. Further, he presents an attempt to overcome the difficulties of the formalism, attempt which has recently aroused the interest of both scientists and philosophers.

The contributions by Cassinelli and Lahti, as well that by Cattaneo, present a detailed discussion of the role of probability and of the logical structure of quantum mechanics. As such, these contributions, which also contain new results, should be extremely useful for students and interested readers with a good mathematical preparation but not sufficiently familiar with quantum theory.

The contribution by Dalla Chiara and Toraldo di Francia gives an exposition of the epistemological and conceptual problems which arise in connection with the identity of the constituents of a composite system in quantum mechanics. Finally, van Fraassen deals with some basic problems of the foundations of probability theory. In particular, he discusses how to describe *opinion change* after the acquisition of new *evidence* that may be totally or only partially reliable.

The volume will be of particular interest to those postgraduate students in logic, mathematics or physics who are working on interdisciplinary researches about the foundations of science as developed in contemporary debates. It will also be of interest to all readers of modern epistemological literature.

The papers by S. Feferman, M.L. Dalla Chiara–G. Toraldo di Francia and B. van Fraassen are reprinted (with slight modifications in each case) from: 'Working Foundations', *Synthese*, **62** (1985) 229–254; 'Individuals, Kinds and Names in Physics', *Logica e Filosofia della Scienze, oggi*, E. Agazzi, M. Mondadori and S. Tugnoli Pattaro (eds.), CLUEB, Bologna, 1986; *Laws and Symmetry*, Clarendon Press, Oxford, 1989.

Florence, March 1992   GIOVANNA CORSI
                                           MARIA LUISA DALLA CHIARA
                                              GIAN CARLO GHIRARDI

ACKNOWLEDGEMENTS

The International School of Philosophy of Science, Trieste, 2–13 October, 1989, was organized and sponsored by the following institutions. The editors wish to thank them for their contribution to the success of the school:

International Centre for Theoretical Physics (Trieste),
Istituto Gramsci del Friuli Venezia-Giulia,
Laboratorio dell'Immaginario Scientifico della Fondazione Internazionale di Trieste per il Progresso e la Libertà della Scienza,
Laboratorio Interdisciplinare per Scienze Naturali ed Umanistiche della Scuola Internazionale Superiore di Studi Avanzati di Trieste,
Società Italiana di Logica e Folosofia della Scienza,
Università degli Studi di Trieste.

G.C., M.L.D.C. AND G.C.G.

PART I

LOGIC, MATHEMATICS AND INFORMATION

CARLO CELLUCCI

THEORY OF DEDUCTION

1. INTRODUCTION

It is often said that logic is the theory of deduction. While such a character-ization may appear too narrow because, in our century, several new chapters have been added to logic, such as the theory of truth, definition, computation, communication, action etc., the theory of deduction remains a basic part of logic. The foundations of such a theory were laid down by Aristotle through his theory of syllogism. The modern approach to the subject is due to Gentzen (1934–5) who proposed two different analyses of deduction:

(1) *the calculus of sequents*;
(2) *natural deduction*.

Gentzen's work has been developed in various directions, as shown by the following figure:

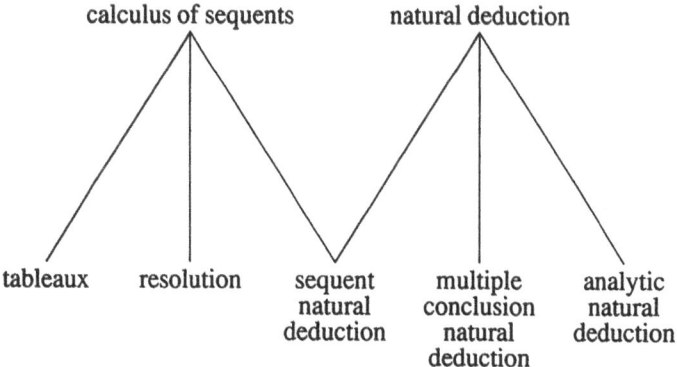

At present no comprehensive survey of all such directions exists. (Some of them are dealt with in Sundholm (1983).) Space limitations prevent us from providing such a survey here: only natural deduction will be dealt with in some

3

G. Corsi et al. (eds), Bridging the Gap: Philosophy, Mathematics, and Physics, 3–37.
© 1993 Kluwer Academic Publishers.

detail. This choice is primarily a matter of taste but, as we shall see, it also has objective motivations.

Natural deduction was not Gentzen's favorite choice: he preferred the calculus of sequents (see below for his motivations). It seems fair to say, however, that the calculus of sequents has both advantages and disadvantages. The advantages consist in the fact that: (1) proof search procedures in the cut-free part of the calculus of sequents are easily formulated; (2) properties of cut-free derivations, such as the subformula property, are easily grasped while this is not generally the case with natural deduction. The disadvantages consist in the fact that: (1) the formulation of the calculus of sequents is neither simple nor elegant; (2) derivations in the calculus of sequents are not easy to handle; and (3) the procedure for transforming derivations in the calculus of sequents into cut-free derivations is a very complicated one and involves a lot of rebuilding. The advantages do not seem to compensate for the disadvantages of the calculus.

2. FIRST-ORDER LANGUAGES

We consider *first-order languages* whose symbols include individual variables $x_0, x_1, \ldots$, individual parameters $a_0, a_1, \ldots$, individual constants $k_0, k_1, \ldots$, n-ary function constants $f_0^n, f_1^n, \ldots$ ($n \geq 1$), n-ary predicate constants $P_0^n, P_1^n, \ldots$ ($n \geq 0$), logical symbols $\bot, \wedge, \vee, \rightarrow, \forall, \exists$, auxiliary symbols (,) and , (comma). We use letters $x, y, \ldots$ to denote individual variables, letters $a, b, \ldots$ to denote individual parameters, letters $k, l, \ldots$ to denote individual constants, letters $f, q, \ldots$ to denote function constants, letters $P, Q, \ldots$ to denote predicate constants.

Terms are defined as follows: (i) individual parameters and individual constants are terms; (ii) if f is an n-ary function constant and $t_1, \ldots, t_n$ are terms, then $f(t_1, \ldots, t_n)$ is a term.

Formulas are defined as follows: (i) if P is an n-ary predicate constant and $t_1, \ldots, t_n$ are terms, then $P(t_1, \ldots, t_n)$ is a formula; (ii) $\bot$ is a formula; (iii) if φ and ψ are formulas, then so are $(\varphi \wedge \psi)$, $(\varphi \vee \psi)$, $(\varphi \rightarrow \psi)$; (iv) if $\varphi(a)$ is a formula containing at least one occurrence of the individual parameter a, and x is an individual variable not occurring in φ, then $\forall x \varphi(x)$ and $\exists x \varphi(x)$ are formulas, where $\varphi(x)$ is the expression obtained from $\varphi(a)$ by replacing at least one occurrence of a by x.

We write $ft_1 \ldots t_n$ for $f(t_1, \ldots, t_n)$ and $Pt_1 \ldots t_n$ for $P(t_1, \ldots, t_n)$. We usually omit outermost parentheses in formulas of one of the forms $(\varphi \wedge \psi)$,

$(\varphi \vee \psi)$, $(\varphi \rightarrow \psi)$. We use the notation $\rho \left[\begin{array}{c} \sigma \\ \tau \end{array} \right]$ to denote the result of replacing every occurrence of σ in ρ by τ, where ρ, σ, τ are arbitrary expressions.

The *degree* of a formula φ is the number of occurrences of logical symbols in φ, other than $\perp$.

The *subformulas* of a formula φ are defined as follows: (i) φ is a subformula of φ; (ii) if $\psi \wedge \chi$, $\psi \vee \chi$ or $\psi \rightarrow \chi$ is a subformula of φ, then so are ψ and χ; (iii) if $\forall x \psi(x)$ or $\exists x \psi(x)$ is a subformula of φ, then so is $\psi \left[\begin{array}{c} x \\ t \end{array} \right]$ for each term t.

The *positive [negative] subformulas* of a formula φ are defined as follows: (i) φ is a positive subformula of φ; (ii) if $\psi \wedge \chi$ or $\psi \vee \chi$ is a positive [negative] subformula of φ, then so are ψ and χ; (iii) if $\psi \rightarrow \chi$ is a positive [negative] subformula of φ, then ψ is a negative [positive] subformula of φ and χ is a positive [negative] subformula of φ; (iv) if $\forall x \psi(x)$ or $\exists x \psi(x)$ is a positive [negative] subformula of φ, then so is $\psi \left[\begin{array}{c} x \\ t \end{array} \right]$ for each t.

We have formulated first-order languages so as not to include $\neg$ as a primitive symbol. This is no limitation because, given $\perp$ and $\rightarrow$, one may introduce $\neg \varphi$ by the definition $\neg \varphi \equiv \varphi \rightarrow \perp$. Alternatively one may also include $\perp$ as a primitive symbol, as in Gentzen (1934–5).

3. INFERENCE RULES

The *inference rules* consist of an introduction (I) rule and an elimination (E) rule for each logical symbol except $\perp$:

I-rules   *E-rules*

$(\wedge \text{I})$ $\dfrac{\varphi \quad \psi}{\varphi \wedge \psi}$ $(\wedge \text{E})$ $\dfrac{\varphi \wedge \psi}{\varphi}$, $\dfrac{\varphi \wedge \psi}{\psi}$

$(\vee \text{I})$ $\dfrac{\varphi}{\varphi \vee \psi}$, $\dfrac{\psi}{\varphi \vee \psi}$ $(\vee \text{E})$ $\dfrac{\varphi \vee \psi \quad \overset{[\varphi]}{\chi} \quad \overset{[\psi]}{\chi}}{\chi}$

$$(\to \text{I}) \quad \frac{\begin{array}{c}[\varphi]\\ \psi\end{array}}{\varphi \to \psi} \qquad\qquad (\to \text{E}) \quad \frac{\varphi \quad \varphi \to \psi}{\psi}$$

$$(\forall \text{I}) \quad \frac{\varphi(a)}{\forall x \varphi(x)} \qquad\qquad (\forall \text{E}) \quad \frac{\forall x \varphi(x)}{\varphi(t)}$$

$$(\exists \text{I}) \quad \frac{\varphi(t)}{\exists x \varphi(x)} \qquad\qquad (\exists \text{E}) \quad \frac{\exists x \varphi(x) \quad \begin{array}{c}[\varphi(a)]\\ \psi\end{array}}{\psi} \quad .$$

In an inference obtained by an E-rule the premise containing the logical symbol which is eliminated in the conclusion is called the *major* premise and the other premises (if any) are called the *minor* premises of that inference.

In a ($\forall$I) or ($\exists$E) inference $\varphi(a)$ denotes $\varphi(x) \left[\begin{array}{c} x \\ a \end{array} \right]$; the individual parameter a is called the *proper parameter* of that inference. In a ($\forall$E) or ($\exists$I) inference $\varphi(t)$ denotes $\varphi(x) \left[\begin{array}{c} x \\ t \end{array} \right]$.

The above rules were stated by Gentzen (1934–5). If, as in Gentzen (1934–5), the language is formulated so as to include $\neg$ as a primitive symbol, one may add the rules:

$$(\neg \text{I}) \quad \frac{\begin{array}{c}[\neg\varphi]\\ \bot\end{array}}{\varphi} \qquad\qquad (\neg \text{E}) \quad \frac{\varphi \quad \neg\varphi}{\bot} \quad .$$

On the other hand, if $\neg$ is introduced by the definition $\neg\varphi \equiv \varphi \to \bot$, such rules are special cases of ($\to$ I) and ($\to$ E), respectively.

4. DERIVATIONS

By a *derivation* we mean a tree of formulas in which every non-topmost formula is obtained from the formulas standing immediately above it by one of the inference rules, subject to certain restrictions to be stated below. The topmost formulas are arbitrary formulas and are called the *assumptions* of the derivation. The downmost formula is called the *conclusion* of the derivation.

Introducing derivations as trees of formulas we are somewhat deviating from the analogy with proofs familiar from mathematical practice. Such a deviation is acknowledged by Gentzen (1934–5) when he points out that in proofs: (1) we necessarily have a linear sequence of propositions because of the linear arrangement of our utterances; and (2) we may repeatedly use a result once it has been obtained, whereas in derivations in tree form one is allowed only a single use of a derived formula. While there is no difficulty in introducing derivations as linear sequences of formulas (indeed this is the standard practice in elementary presentations of natural deduction such as Lemmon (1965)), in view of the structural analysis of derivations envisaged below the tree form is more convenient.

($\vee$E), ($\rightarrow$ I), ($\exists$E) allow assumptions of the form indicated within square brackets to be *discharged*. Discharge is not compulsory: any number of assumptions of the given form (possibly zero) may be discharged (on the relevance of allowing any number of assumptions to be discharged see Leivant (1979)). A formula φ in a derivation $\mathcal{D}$ is said to *depend* on the assumptions standing above φ in $\mathcal{D}$, which have not been discharged by any ($\vee$E), ($\rightarrow$ I), ($\exists$E) inference standing above φ in $\mathcal{D}$. The *open* assumptions of a derivation $\mathcal{D}$ are the assumptions on which the conclusion of $\mathcal{D}$ depends.

We write

$$[\varphi]$$
$$\mathcal{D}$$

to indicate that $[\varphi]$ is a list of occurrences of φ, the elements of which are open assumptions of $\mathcal{D}$. We write

$$\mathcal{D}$$
$$\varphi$$

to indicate that φ is the conclusion of $\mathcal{D}$ (so φ is a part of $\mathcal{D}$ itself). Given two derivations,

$$\mathcal{D}_1 \quad \text{and} \quad \begin{array}{c} [\varphi] \\ \mathcal{D}_2 \\ \psi \end{array},$$
$$\varphi$$

we write

$$\mathcal{D}_1$$
$$[\varphi]$$
$$\mathcal{D}_2$$
$$\psi$$

to denote the tree of formulas obtained from $\mathcal{D}$ by writing $\mathcal{D}_1$ above each topmost formula in $\mathcal{D}_2$ which is in $[\varphi]$. (If $[\varphi] = \emptyset$, then the above notation simply denotes $\mathcal{D}_2$). Finally, we write

$$\mathcal{D}\begin{bmatrix} a \\ t \end{bmatrix}.$$

to denote the result of replacing each formula φ in $\mathcal{D}$ by $\varphi\begin{bmatrix} a \\ t \end{bmatrix}$.

5. RESTRICTIONS ON RULES

The inference rules are subject to the following restrictions:

R1.   In a (∀I) inference the proper parameter a must not occur in the conclusion $\forall x\varphi(x)$ of that inference or in any assumption on which the conclusion depends.

R2.   In a (∃E) inference the proper parameter a must not occur in the major premise $\exists x\varphi(x)$, in the minor premise ψ or in any assumption on which the minor premise depends other than those in $[\varphi(a)]$.

The need for such restrictions is shown by the following examples where an incorrect conclusion is derived from given assumptions (as can be easily seen by interpreting, say, Pa as 'a is an even number' and Qa as 'a is an odd number').

$$(1) \qquad (\exists E) \quad \dfrac{\exists x Px \quad (\forall I)\ \dfrac{\overset{(1)}{Pa}}{\forall x Px}}{\forall x Px}\ (1)\ .$$

The indicated (∀I) inference violates R1 because the proper parameter a occurs in the assumption Pa on which the premise Pa depends.

$$(2) \qquad (\exists E) \quad \dfrac{\exists x Px \quad \overset{(1)}{Pa}}{\dfrac{Pa}{\forall x Px}}\ (1)\ .$$

The indicated (∃E) inference violates R2 because the proper parameter a occurs in the minor premise Pa.

$$\frac{\begin{array}{cc}(1) & (2)\\ Pa & Qa\end{array}}{Pa \wedge Qa}$$

$$(3)\quad \frac{\exists x Px \quad (\exists E)\ \dfrac{\exists x Qx \quad \dfrac{\exists x(Px \wedge Qx)}{\exists x(Px \wedge Qx)}\ (2)}{\exists x(Px \wedge Qx)}}{\exists x(Px \wedge Qx)}\ (1)$$

The indicated (∃E) inference violates R2 because the proper parameter a occurs in the assumption Pa on which the minor premise $\exists x(Px \wedge Qx)$ depends (distinct from the discharged assumption Qa).

6. MINIMAL, INTUITIONISTIC AND CLASSICAL LOGIC

The rules (∧I), (∧E), (∨I), (∨E), (→ I), (→ E), (∀I), (∀E), (∃I), (∃E) yield *minimal logic* **NM**. Now **NM** is a very weak system in which such basic laws as $\neg\varphi \to (\varphi \to \psi)$ and $\varphi \vee \neg\varphi$ are not provable (in the sense: there is no derivation in **NM** with no open assumptions whose conclusion is one of such formulas). In order to prove $\neg\varphi \to (\varphi \to \psi)$ we need the rule of *ex falso quodlibet*,

$$(EF)\quad \frac{\bot}{\varphi}\,.$$

Adding (EF) to **NM** yields *intuitionistic logic* **NJ**. The system **NJ** was stated by Gentzen (1934–5), except that ¬ was included as a primitive symbol.

In order to prove $\varphi \vee \neg\varphi$ we have several options. We confine ourselves to the following (see Curry, 1963, Tennant, 1978 for other possible options):

(i)   trivially, to adopt the axiom schema of *excluded middle*,

$$(EM)\quad \varphi \vee \neg\varphi\,;$$

(ii)   to adopt the rule of *dilemma*,

$$(D)\quad \frac{\begin{array}{cc}[\varphi] & [\neg\varphi]\\ \psi & \psi\end{array}}{\psi}\,;$$

(iii)   to adopt the rule of *classical reductio ad adsurdum*,

$$(CR)\quad \frac{\begin{array}{c}[\neg\varphi]\\ \bot\end{array}}{\varphi}\,;$$

(iv)　　to adopt the rule of *consequentia mirabilis*,

$$(CM) \quad \frac{\genfrac{}{}{0pt}{}{[\neg\varphi]}{\varphi}}{\varphi} \ .$$

It can be easily seen that (EM), (D), (CR), (CM) are equivalent in **NJ**. This can be shown as follows.

(EM) ⇒ (D):

$$(\vee E) \ \frac{(EM) \quad \overset{\overset{\textstyle[\varphi]}{\textstyle\mathcal{D}_1}}{\psi} \quad \overset{\overset{\textstyle[\varphi]}{\textstyle\mathcal{D}_2}}{\psi}}{\psi} \ (1), (2)$$

(D) ⇒ (CR):

$$(D) \ \frac{\overset{(1)}{\varphi} \quad (EF) \ \frac{\overset{\overset{\textstyle[\neg\varphi]}{\textstyle\mathcal{D}_1}}{\bot}}{\varphi}}{\varphi} \ (1), (2)$$

(CR) ⇒ (CM):

$$\begin{array}{c} (\neg E) \\ (CR) \end{array} \ \frac{\overset{\overset{\textstyle[\neg\varphi]}{\textstyle\mathcal{D}_1}}{\varphi} \quad \overset{(2)}{\neg\varphi}}{\frac{\bot}{\varphi} \ (1)}$$

(CM) ⇒ (EM):

$$\begin{array}{c} (\vee I) \\ (\neg E) \end{array} \ \frac{(\neg I) \ \dfrac{\dfrac{\overset{(1)}{\varphi}}{\varphi \vee \neg\varphi} \quad \overset{(2)}{\neg(\varphi \vee \neg\varphi)}}{\bot}}{\begin{array}{c}(\vee I)\\(CM)\end{array} \dfrac{\dfrac{\bot}{\neg\varphi} \ (1)}{\dfrac{\varphi \vee \neg\varphi}{\varphi \vee \neg\varphi}} \ (2)}$$

Adding one of the rules (EM), (D), (CR), (CM) to intuitionistic logic **NJ** yields *classical logic* **NK**. More specifically, we write **NK(EM)**, **NK(D)**, **NK(CR)**, **NK(CM)** to denote the version of **NK** obtained by adding (EM), (D), (CR), (CM) respectively to **NJ**.

The system **NK**(EM) was stated by Gentzen (1934–5), except that $\neg$ was introduced as a primitive symbol. Gentzen (1934–5) also envisaged alternative formulations of **NK**, such as the one obtained by adopting the rule of *double negation*:

$$\text{(DN)} \quad \frac{\neg\neg\varphi}{\varphi} \, .$$

(DN) is strictly related to (CR), indeed it may even considered to be a variant of the latter. Clearly (DN) and (CR) are equivalent in **NM**:

(DN) $\Rightarrow$ (CR):

$$
\begin{array}{c}
(1) \\
[\neg\varphi] \\
\mathcal{D}_1 \\
\end{array}
$$

$$
\begin{array}{c}
(\neg\text{I}) \\
(\text{DN})
\end{array}
\quad
\dfrac{\dfrac{\bot}{\neg\neg\varphi} \,\, (1)}{\varphi}
$$

(CR) $\Rightarrow$ (DN):

$$
(\neg\text{E}) \quad
\dfrac{
\begin{array}{cc}
(1) & \mathcal{D}_1 \\
\neg\varphi & \neg\neg\varphi
\end{array}
}{
(\text{CR}) \,\, \dfrac{\bot}{\varphi} \, (1)
} \, .
$$

Note that, while (EM), (D), (CR), (CM) are equivalent in **NJ**, they are not all equivalent in **NM**. Specifically (EM), (D), (CM) are equivalent in **NM**. This follows from the fact that in **NM**, as we have already shown, (EM) $\Rightarrow$ (D) and (CM) $\Rightarrow$ (EM), and moreover:

(D) $\Rightarrow$ (CM):

$$
(\text{D}) \quad
\dfrac{
\begin{array}{cc}
& (2) \\
& [\neg\varphi] \\
(1) & \mathcal{D}_1 \\
\varphi & \varphi
\end{array}
}{\varphi} \,\, (1), (2) \, .
$$

However (CR) is stronger than each of (EM), (D), (CM). For, adding (CR) to **NM** yields classical logic **NK**, because (EF) is the special case of (CR) where no assumption is discharged. On the other hand adding (EM), (D) or (CM) to **NM** does not yield classical logic **NK** because $\neg\varphi \to (\varphi \to \psi)$ is not provable in the resulting systems. Thus there exists an asymmetry between (CR), on the one hand, and (EM), (D), (CM) on the other hand.

7. LOGICAL REDUCTIONS

There exists a certain symmetry between the I-rules and the E-rules, which may be expressed by saying that the corresponding I-rules and E-rules are inverses of each other in the following sense: the conclusion of an inference obtained by an E-rule does not establish anything more than that must have already been established if the major premise of that inference was the conclusion of an inference obtained by an I-rule (this is called the *inversion principle* in Prawitz (1965, 1971).

For example, if the premise $\varphi_1 \wedge \varphi_2$ of a $(\wedge E)$ inference was the conclusion of a $(\wedge I)$ inference, then the conclusion φ_i $(i = 1, 2)$ of the $(\wedge E)$ inference must already occur as a premise of the $(\wedge I)$ inference. Similarly, if the premise $\forall x \varphi(x)$ of a $(\forall E)$ inference was the conclusion of a $(\forall I)$ inference, then the conclusion $\varphi(t)$ of the $(\forall E)$ inference may be obtained from the derivation ending with the premise $\varphi(a)$ of the $(\forall I)$ inference simply by replacing each occurrence of a by t.

This suggests that an inference obtained by an E-rule whose major premise is the conclusion of an inference obtained by an I-rule is redundant, being an unnecessary detour in a derivation. Such detours are quite common in mathematical practice where, say, having shown $\varphi(t)$, we interrupt the main proof and establish a lemma $\forall x(\varphi(x) \rightarrow \psi(x))$ by showing $\varphi(a) \rightarrow \psi(a)$, then we use the lemma to obtain $\varphi(t) \rightarrow \psi(t)$ and finally go back to the main proof concluding $\psi(t)$. This use of a lemma is clearly an unnecessary detour because we may use the proof of $\varphi(a) \rightarrow \psi(a)$ to obtain $\varphi(t) \rightarrow \psi(t)$ simply by replacing each occurrence of a in it by t.

Detours of this kind can be eliminated by the following *logical reduction* rules:

$\wedge$-*reduction*

$$
\begin{array}{cc}
\begin{array}{c}
\mathcal{D}_1 \ \ \mathcal{D}_2 \\
\hline
\varphi_1 \ \ \varphi_2 \\
\hline
\varphi_1 \wedge \varphi_2 \\
\hline
\varphi_i \\
\mathcal{D}_3
\end{array}
&
\begin{array}{c}
\mathcal{D}_i \\
\varphi_i \\
\mathcal{D}_3
\end{array}
\end{array}
$$

(∧I) (∧E) $\longmapsto$ $(i = 1, 2)$

∨-*reduction*

$$
\text{(∨E)} \quad \cfrac{\text{(∨I)} \cfrac{\mathcal{D}_0}{\varphi_1 \vee \varphi_2} \quad \begin{matrix}[\varphi_1]\\ \mathcal{D}_1 \\ \psi\end{matrix} \quad \begin{matrix}[\varphi_2]\\ \mathcal{D}_2 \\ \psi\end{matrix}}{\psi} \\ \mathcal{D}_3
\qquad \mapsto \qquad
\begin{matrix}\mathcal{D}_0 \\ [\varphi_i] \\ \mathcal{D}_i \\ \psi \\ \mathcal{D}_3\end{matrix} \quad (i = 1, 2)
$$

→-*reduction*

$$
\text{(→E)} \quad \cfrac{\cfrac{\mathcal{D}_1}{\varphi} \quad \text{(→I)} \cfrac{\begin{matrix}[\varphi]\\ \mathcal{D}_2 \\ \psi\end{matrix}}{\varphi \to \psi}}{\psi} \\ \mathcal{D}_3
\qquad \mapsto \qquad
\begin{matrix}\mathcal{D}_1 \\ [\varphi] \\ \mathcal{D}_2 \\ \psi \\ \mathcal{D}_3\end{matrix}
$$

∀-*reduction*

$$
\begin{matrix}\text{(∀I)}\\ \text{(∀E)}\end{matrix} \quad \cfrac{\cfrac{\mathcal{D}_1}{\varphi(a)}}{\cfrac{\forall x \varphi(x)}{\varphi(t)}} \\ \mathcal{D}_2
\qquad \mapsto \qquad
\begin{matrix}\mathcal{D}_1 \begin{bmatrix} a \\ t \end{bmatrix} \\ \varphi(t) \\ \mathcal{D}_2\end{matrix}
$$

∃-*reduction*

$$
\begin{matrix}\text{(∃I)}\\ \text{(∃E)}\end{matrix} \quad \cfrac{\cfrac{\mathcal{D}_1}{\varphi(t)}}{\cfrac{\exists x \varphi(x)}{\psi}} \quad \begin{matrix}[\varphi(a)]\\ \mathcal{D}_2 \\ \psi\end{matrix}}{\psi} \\ \mathcal{D}_3
\qquad \mapsto \qquad
\begin{matrix}\mathcal{D}_1 \\ [\varphi(t)] \\ \mathcal{D}_2 \begin{bmatrix} a \\ t \end{bmatrix} \\ \psi \\ \mathcal{D}_3\end{matrix}
$$

8. COMMUTATIVE REDUCTIONS

The logical reduction rules, while eliminating a detour, may create new ones. The latter, however, have a complexity lower than that of the eliminated detour in all cases except (∨E) and (∃E). For example in the ∨-reduction

$$
\text{(∨E)} \cfrac{\text{(∨I)}\cfrac{\mathcal{D}_0}{\varphi_i} \cfrac{\text{(∧I)}\cfrac{[\varphi_1] \; [\varphi_1]}{\begin{matrix}\mathcal{D}_1 \; \mathcal{D}_1' \\ \rho \quad \sigma\end{matrix}}{\rho \wedge \sigma} \quad \text{(∧I)}\cfrac{[\varphi_2] \; [\varphi_2]}{\begin{matrix}\mathcal{D}_2 \; \mathcal{D}_2' \\ \rho \quad \sigma\end{matrix}}{\rho \wedge \sigma}}{\text{(∧E)}\cfrac{\rho \wedge \sigma}{\rho}}
\; \mapsto \;
\text{(∧I)}\cfrac{\begin{matrix}\mathcal{D}_0 \quad \mathcal{D}_0 \\ [\varphi_i] \; [\varphi_i] \\ \mathcal{D}_i \quad \mathcal{D}_i' \\ \rho \quad \sigma\end{matrix}}{\text{(∧E)}\cfrac{\rho \wedge \sigma}{\rho}} \; (i = 1, 2),
$$

the degree of $\rho \wedge \sigma$ may be higher than that of $\varphi_1 \vee \varphi_2$. In order to avoid this problem we introduce the following *commutative reduction rules:*

$\vee$-*commutative reduction*

$$
(\vee E)\dfrac{\begin{array}{ccc}\mathcal{D}_0 & \mathcal{D}_1 & \mathcal{D}_2 \\ \varphi \vee \psi & \rho & \rho\end{array}}{\text{E-rule }\dfrac{\rho}{\dfrac{\sigma}{\mathcal{D}_4}}\quad \mathcal{D}_3}
\;\mapsto\;
(\vee E)\dfrac{\begin{array}{c}\mathcal{D}_0 \\ \varphi \vee \psi\end{array}\quad \text{E-rule}\dfrac{\rho\quad\mathcal{D}_3}{\sigma}\quad \text{E-rule}\dfrac{\rho\quad\mathcal{D}_3}{\sigma}}{\dfrac{\sigma}{\mathcal{D}_4}}
$$

with $\mathcal{D}_1$ over the first E-rule and $\mathcal{D}_2$ over the second.

$\exists$-*commutative reduction*

$$
(\exists E)\dfrac{\begin{array}{cc}\mathcal{D}_0 & \mathcal{D}_1 \\ \exists x\varphi(x) & \rho\end{array}}{\text{E-rule }\dfrac{\rho}{\dfrac{\sigma}{\mathcal{D}_3}}\quad \mathcal{D}_2}
\;\mapsto\;
(\exists E)\dfrac{\begin{array}{c}\mathcal{D}_0 \\ \exists x\varphi(x)\end{array}\quad \text{E-rule}\dfrac{\begin{array}{cc}\mathcal{D}_1 \\ \rho\end{array}\quad \mathcal{D}_2}{\sigma}}{\dfrac{\sigma}{\mathcal{D}_3}}
$$

Thus the aim of a commutative reduction is not to eliminate any detour but only to move a $(\vee E)$ or $(\exists E)$ inference past any inference obtained by an E-rule.

Both logical and commutative reduction rules were stated by Prawitz (1971) who provided the first major development in the theory of deduction since Gentzen (1934–5).

9. THE LEMMA ON PROPER PARAMETERS

The above reduction rules raise some problems. A first problem is that the result of applying a reduction to a given derivation need not be a derivation. In order to see this we must introduce some notions.

A *connection* in a derivation $\mathcal{D}$ between two formulas φ and ψ is a sequence $\varphi_1, \ldots \varphi_n$ of formulas of $\mathcal{D}$ such that $\varphi_1 \equiv \varphi$, $\varphi_n \equiv \psi$ and one of the following conditions holds for $1 \leq i < n$: (1) φ_i stands immediately above φ_{i+1}, or vice versa; (2) φ_i and φ_{i+1} are the two premises of an $(\rightarrow E)$ inference; (3) φ_i is the conclusion of an $(\rightarrow I)$ inference and φ_{i+1} is an assumption discharged by that inference, or vice versa; (4) φ_i is the major premise of a $(\vee E)$ or $(\exists E)$ inference and φ_{i+1} is an assumption discharged by that inference, or vice versa.

Two occurrences of an individual parameter a in a derivation $\mathcal{D}$ are said to be *linked* in $\mathcal{D}$ if there is a connection between the two formulas containing these two occurrences of a such that a occurs in every formula of that connection.

Consider then the following examples of reductions.

$$(1) \, (\to E) \; \frac{Pa \wedge \forall x Px \quad (\to I) \; \dfrac{(\forall I)\dfrac{\forall x Px}{\dfrac{Pa}{\forall y Py}}}{\forall x Px \to \forall y Py}(2)}{\forall y Py} \quad \mapsto \quad (\forall I) \dfrac{\dfrac{(1)}{Pa \wedge \forall x Px}}{\dfrac{\forall x Px}{\dfrac{Pa}{\forall y Py}}}$$

where the left side has $(1) \; \dfrac{Pa \wedge \forall x Px}{\forall x Px}$ and $(\forall I)$ over the (2) assumption $\forall x Px$.

The ($\forall$I) inference in the derivation on the right violates R1 because its proper parameter a occurs in an assumption on which its premise depends. This originates from the fact that the two occurrences of a in the derivation on the left are not linked.

$$(2) \, (\forall E) \; \frac{(\forall I)\dfrac{(\forall I)\dfrac{\dfrac{(1)}{\forall x Px}}{\dfrac{Pa}{\forall y Py}}}{\dfrac{Pa}{\forall z Pz}}}{Pk} \quad \mapsto \quad (\forall I)\dfrac{\dfrac{(1)}{\forall x Px}}{\dfrac{Pk}{\forall y Py}}$$

with conclusion Pk on the right.

The ($\forall$I) inference shown in the derivation on the right is incorrect because k is an individual constant. This originates from the fact that the two occurrences of a shown in the inference on the left are not linked.

The problem concerning reduction illustrated by the above examples can be solved introducing a standardization on proper parameters.

A derivation $\mathcal{D}$ is said to satisfy the *proper parameter condition* if every occurrence in $\mathcal{D}$ of the proper parameter a of an ($\forall$I) [($\exists$E)] inference is linked to an occurrence of a in the premise of that ($\forall$I) [in an assumption discharged by that ($\exists$E)] inference.

Clearly a derivation $\mathcal{D}$ satisfying the proper parameter condition has the following properties: (1) the proper parameter of a ($\forall$I) inference in $\mathcal{D}$ occurs only in formulas standing above the conclusion of that inference; (2) the proper parameter of an ($\exists$E) inference in $\mathcal{D}$ occurs only in formulas standing above the minor premise of that inference; (3) every proper parameter of $\mathcal{D}$ is the proper parameter of a single ($\forall$I) or ($\exists$E) inference.

The following result can be easily established (see Prawitz [1965]):

LEMMA ON PROPER PARAMETERS. *Every derivation $\mathcal{D}$ can be uniformly transformed into a derivation $\mathcal{D}'$ satisfying the proper parameter condition, where $\mathcal{D}'$ has the same open assumptions and the same conclusion as $\mathcal{D}$ and is obtained from $\mathcal{D}$ by replacing the proper parameters of $\mathcal{D}$ in a suitable way.*

This solves the problem concerning reductions. In what follows reductions will be applied only to derivations satisfying the proper parameter condition. In view of the previous result this does not involve any loss of generality.

10. NORMALIZATION THEOREM FOR NM

We are now in a position to solve the problem of eliminating all detours from a derivation in NM.

Generally a derivation is said to be *normal* (with respect to a given class of reductions) if no reduction (of that class) can be applied to it. Normal derivations have a special character which was expressed by Gentzen (1934–5), although only w.r.t. the calculus of sequents, by saying that a normal derivation makes no detours: no concepts enter into the derivation except those occurring in its conclusion, and their use is therefore essential to the achievement of that conclusion.

In particular, a derivation in NM is *normal* (with respect to logical and commutative reductions) if no (logical or commutative) reduction can be applied to it.

Normal derivations in NM have the following:

SUBFORMULA PROPERTY. *Every formula occurring in a normal derivation $\mathcal{D}$ in NM with open assumptions Γ and conclusion φ is a subformula of some formula in Γ or of φ.*

The following result can easily be established:

NORMALIZATION THEOREM FOR NM. *Every derivation $\mathcal{D}$ in NM can be transformed into a normal derivation $\mathcal{D}'$ whose open assumptions are among those of $\mathcal{D}$ and whose conclusion is the same as that of $\mathcal{D}$, by an appropriate sequence of logical or commutative reductions.*

11. EF-REDUCTIONS

Further redundancies in derivations, of a kind different from the detours arising from the I-rules and the E-rules, are yielded by the rules (EF), (D), (CR) and (CM). For example, in the case of (EF) and (CR), this stems from the fact that such rules allow to introduce formulas φ of arbitrary complexity. In order to eliminate such additional redundacies it is convenient to state the following restriction:

R3.   In any inference obtained by one of the rules (EF), (D), (CR), (CM) the
       formula φ must be different from $\bot$.

Such a restriction (which avoids considering a number of trivial cases in establishing properties of normal derivations) is inessential, as shown by the
following transformations.

$$
(1)\ (EF)\quad
\begin{array}{c}
\mathcal{D}_1 \\
\bot \\
\hline
\bot \\
\mathcal{D}_2
\end{array}
\quad\mapsto\quad
\begin{array}{c}
\mathcal{D}_1 \\
\bot \\
\mathcal{D}_2
\end{array}
$$

$$
(2)\ (D)\quad
\begin{array}{c}
(1)\quad (2) \\
[\bot]\ \ [\neg\bot] \\
\mathcal{D}_1\ \ \ \mathcal{D}_2 \\
\psi\ \ \ \ \psi \\
\hline
\psi \\
\mathcal{D}_3
\end{array}\ (1),(2)
\quad\mapsto\quad
\begin{array}{c}
(1) \\
(\rightarrow I)\ \dfrac{\bot}{[\neg\bot]}\ (1) \\
\mathcal{D}_2 \\
\psi \\
\mathcal{D}_3
\end{array}
$$

$$
(3)\ (CR)\ or\ (CM)\quad
\begin{array}{c}
(1) \\
[\neg\bot] \\
\mathcal{D}_1 \\
\bot \\
\hline
\bot \\
\mathcal{D}_2
\end{array}\ (1)
\quad\mapsto\quad
\begin{array}{c}
(1) \\
(\rightarrow I)\ \dfrac{\bot}{[\neg\bot]}\ (1) \\
\mathcal{D}_1 \\
\bot \\
\mathcal{D}_2
\end{array}
$$

First we consider the case of the redundancies yielded by (EF). In order to
eliminate such redundancies we introduce the following EF-*reduction rules*:

EF∧-*reduction*

$$
(EF)\quad
\begin{array}{c}
\mathcal{D}_1 \\
\bot \\
\hline
\varphi\wedge\psi \\
\mathcal{D}_2
\end{array}
\quad\mapsto\quad
\begin{array}{c}
(EF)\ \dfrac{\mathcal{D}_1}{\dfrac{\bot}{\varphi}}\quad (EF)\ \dfrac{\mathcal{D}_1}{\dfrac{\bot}{\psi}} \\
(\wedge I)\ \dfrac{\qquad\qquad}{\varphi\wedge\psi} \\
\mathcal{D}_2
\end{array}
$$

EF∨-*reduction*

$$
(EF) \; \frac{\begin{array}{c} \mathcal{D}_1 \\ \bot \end{array}}{\begin{array}{c} \varphi \vee \psi \\ \mathcal{D}_2 \end{array}} \quad \mapsto \quad \begin{array}{c} (EF) \\ (\vee I) \end{array} \frac{\dfrac{\begin{array}{c} \mathcal{D}_1 \\ \bot \end{array}}{\varphi}}{\begin{array}{c} \varphi \vee \psi \\ \mathcal{D}_2 \end{array}}
$$

EF→-*reduction*

$$
(EF) \; \frac{\begin{array}{c} \mathcal{D}_1 \\ \bot \end{array}}{\begin{array}{c} \varphi \to \psi \\ \mathcal{D}_2 \end{array}} \quad \mapsto \quad \begin{array}{c} (EF) \\ (\to I) \end{array} \frac{\dfrac{\begin{array}{c} \mathcal{D}_1 \\ \bot \end{array}}{\psi}}{\begin{array}{c} \varphi \to \psi \\ \mathcal{D}_2 \end{array}}
$$

EF∀-*reduction*

$$
(EF) \; \frac{\begin{array}{c} \mathcal{D}_1 \\ \bot \end{array}}{\begin{array}{c} \forall x \varphi(x) \\ \mathcal{D}_2 \end{array}} \quad \mapsto \quad (\forall I) \frac{(EF)\dfrac{\begin{array}{c} \mathcal{D}_1 \\ \bot \end{array}}{\varphi(a)}}{\begin{array}{c} \forall x \varphi(x) \\ \mathcal{D}_2 \end{array}}
$$

EF∃-*reduction*

$$
(EF) \; \frac{\begin{array}{c} \mathcal{D}_1 \\ \bot \end{array}}{\begin{array}{c} \exists x \varphi(x) \\ \mathcal{D}_2 \end{array}} \quad \mapsto \quad (\exists I) \frac{(EF)\dfrac{\begin{array}{c} \mathcal{D}_1 \\ \bot \end{array}}{\varphi(a)}}{\begin{array}{c} \exists x \varphi(x) \\ \mathcal{D}_2 \end{array}}
$$

12. NORMALIZATION THEOREM FOR NJ

We are now in a position to solve the problem of eliminating all redundancies from a derivation in **NJ**.

A derivation in **NJ** is *normal* (with respect to logical, commutative and EF-reductions) if no (logical, commutative or EF-) reduction can be applied to it. Like normal derivations in **NM**, also normal derivations in **NJ** have the subformula property. Further properties of normal derivations in **NJ** are discussed in Prawitz (1965, 1971) and Troelstra and Van Dalen (1988).

The following result can be easily established (see Prawitz, 1965, 1971, and Troelstra-Van Dalen, 1988).

NORMALIZATION THEOREM FOR **NJ**. *Every derivation* $\mathcal{D}$ *in* **NJ** *can be* *transformed into a normal derivation $\mathcal{D}'$ whose open assumptions are among those of $\mathcal{D}$ and whose conclusion is the same as that of $\mathcal{D}$, by an appropriate sequence of logical, commutative or EF-reductions.*

13. NORMALIZATION THEOREM FOR $\mathbf{NK(D)_P}$

Next we consider the case of redundancies yielded by (D). In order to eliminate all such redundancies we introduce the following D-*reduction rules.*

D$\wedge$-*reduction*

$$
\begin{array}{cc}
(1) & (2) \\
[\varphi \wedge \psi] & [\neg(\varphi \wedge \psi)] \\
\mathcal{D}_1 & \mathcal{D}_2 \\
(D)\dfrac{\rho \qquad \rho}{\rho} \,(1),(2) \\
\mathcal{D}_3
\end{array}
\;\longmapsto\;
\begin{array}{c}
(3) \\
\dfrac{\varphi \wedge \psi}{} \quad (4) \\
\dfrac{(1)\;(2)}{} \quad \dfrac{\varphi \qquad \neg\varphi}{} \\
\dfrac{\varphi \quad \psi}{[\varphi \wedge \psi]} \quad \dfrac{\bot}{[\neg(\varphi \wedge \psi)]}\,(3) \\
\mathcal{D}_1 \qquad \mathcal{D}_2 \\
(D)\dfrac{\rho \qquad\qquad \rho}{} \,(1),(4)
\end{array}
\;\;
\begin{array}{c}
(5) \\
\dfrac{\varphi \wedge \psi}{} \quad (6) \\
\dfrac{\psi \qquad \neg\psi}{} \\
\dfrac{\bot}{[\neg(\varphi \wedge \psi)]}\,(5) \\
\mathcal{D}_2 \\
\dfrac{\rho}{}\,(2),(6) \\
\rho \\
\mathcal{D}_3
\end{array}
$$

D$\vee$-*reduction*

$$
\begin{array}{cc}
(1) & (2) \\
[\varphi \vee \psi] & [\neg(\varphi \vee \psi)] \\
\mathcal{D}_1 & \mathcal{D}_2 \\
(D)\dfrac{\rho \qquad \rho}{\rho} \,(1),(2) \\
\mathcal{D}_3
\end{array}
\;\longmapsto\;
\begin{array}{c}
(4)\;(6)\;\;(5)\;(7) \\
(3) \quad \dfrac{\varphi \;\; \neg\varphi \;\; \psi \;\; \neg\psi}{} \\
(2) \;\; \dfrac{\varphi \vee \psi \quad \bot \qquad \bot}{}\,(4),(5) \\
(1) \;\; \dfrac{\varphi}{[\varphi \vee \psi]} \quad \dfrac{\bot}{[\neg(\varphi \vee \psi)]}\,(3) \\
\dfrac{\psi}{[\varphi \vee \psi]} \quad \mathcal{D}_1 \qquad \mathcal{D}_2 \\
\mathcal{D}_1 \quad (D)\dfrac{\rho \qquad\qquad \rho}{}\,(2),(6) \\
(D)\dfrac{\rho \qquad\qquad\qquad \rho}{\rho}\,(1),(7) \\
\mathcal{D}_3
\end{array}
$$

D$\rightarrow$-*reduction*

$$
\begin{array}{cc}
(1) & (2) \\
[\varphi \rightarrow \psi] & [\neg(\varphi \rightarrow \psi)] \\
\mathcal{D}_1 & \mathcal{D}_2 \\
(D)\dfrac{\rho \qquad \rho}{\rho} \,(1),(2) \\
\mathcal{D}_3
\end{array}
\;\longmapsto\;
\begin{array}{c}
(2) \quad (3) \\
\dfrac{\varphi \quad \varphi \rightarrow \psi}{} \quad (4) \\
\dfrac{\psi \qquad \neg\psi}{} \\
(1) \;\; \dfrac{\bot}{[\neg(\varphi \rightarrow \psi)]}\,(3) \\
\dfrac{\psi}{[\varphi \rightarrow \psi]} \quad \mathcal{D}_2 \\
\mathcal{D}_1 \quad (D)\dfrac{\rho \qquad\qquad \rho}{\rho}\,(1),(4) \\
\mathcal{D}_3
\end{array}
\;\;
\begin{array}{c}
(5)\;(6) \\
\dfrac{\varphi \quad \neg\varphi}{} \\
\dfrac{\bot}{\psi} \\
\dfrac{}{[\varphi \rightarrow \psi]}\,(5) \\
\mathcal{D}_1 \\
\dfrac{\rho}{}\,(2),(6)
\end{array}
$$

In addition to the above D-reduction rules we must consider also the following D-*commutative reduction rule.*

D-*commutative reduction*

$$
\text{(D)} \quad \frac{\begin{array}{cc} (1) & (2) \\ [\varphi] & [\neg\varphi] \\ \mathcal{D}_1 & \mathcal{D}_2 \\ \rho & \rho \end{array}}{\begin{array}{c} \rho \\ \mathcal{D}_3 \\ \psi \end{array}} \ (1),(2) \quad \mapsto \quad \text{(D)} \quad \frac{\begin{array}{cc} (1) & (2) \\ [\varphi] & [\neg\varphi] \\ \mathcal{D}_1 & \mathcal{D}_2 \\ \rho & \rho \\ \mathcal{D}_3 & \mathcal{D}_3 \\ \psi & \psi \end{array}}{\psi} \ (1),(2)
$$

Using such a transformation all (D) inferences can be moved downwards, so that any derivation $\mathcal{D}$ in $\mathbf{NK(D)_P}$ whose conclusion is ψ can be transformed into a derivation $\mathcal{D}'$ such that all (D) inferences in $\mathcal{D}'$ have ψ as conclusion.

14. NORMALIZATION THEOREM FOR $\mathbf{NK(D)_P}$

The above D- and D-commutative reductions allow us to solve the problem of eliminating all redundancies from a derivation in the propositional part of $\mathbf{NK(D)}$, say $\mathbf{NK(D)_P}$, which consists of the rules $(\wedge I)$, $(\wedge E)$, $(\vee I)$, $(\vee E)$, $(\rightarrow I)$, $(\rightarrow E)$, (EF), (D).

A derivation in $\mathbf{NK(D)_P}$ is *normal* (with respect to logical, commutative, EF-, D- and D-commutative reductions) if no (logical, commutative, EF-, D- or D-commutative) reduction can be applied to it.

Normal derivations in $\mathbf{NK(D)_P}$ have the following:

WEAK SUBFORMULA PROPERTY. *A normal derivation $\mathcal{D}$ in $\mathbf{NK(D)_P}$ with open assumptions Γ and conclusion φ may contain only subformulas of some formula in Γ or of φ, assumptions of the form $\neg P$ discharged by some (D) inference and occurrences of $\perp$ standing immediately below such assumptions.*

The following result can be easily established (see Tennant, 1978).

NORMALIZATION THEOREM FOR $\mathbf{NK(D)_P}$. *Every derivation $\mathcal{D}$ in $\mathbf{NK(D)_P}$ can be transformed into a normal derivation $\mathcal{D}'$ whose open assumptions are among those of $\mathcal{D}$ and whose conclusion is the same as that of $\mathcal{D}$ by an appropriate sequence of logical, commutative, EF, D- or D-commutative reductions.*

It should be noted that this result does not extend to $\mathbf{NK(D)}$. For, on the one hand, by the latter result if in **NK(D)**$_P$ one restricts (D) to atomic φ, then the resulting system is complete. On the other hand, if in **NK(D)** one restricts (D) to atomic φ, then the resulting system can be seen to be incomplete.

15. CR-REDUCTIONS

A partial solution to the problem of eliminating all redundancies yielded by (CR) can be given by considering a system **C** poorer than **NK(CR)**. In order to introduce such a system we modify our notion of first-order language not including ∨ and ∃ as primitive symbols and introducing $\varphi \lor \psi$ and $\exists x \varphi(x)$ by the definitions $\varphi \lor \psi \equiv \neg\varphi \rightarrow \psi$ and $\exists x \varphi(x) \equiv \neg\forall x \neg\varphi(x)$. Then **C** consists of the rules (∧I), (∧E), (→ I), (→ E), (∀I), (∀E), (CR). The system **C** was introduced by Prawitz (1965).

In order to eliminate all unnecessary detours yielded by (CR) we introduce the following CR-*reduction rules*.

CR∧-*reduction*

CR→-*reduction*

CR∀-*reduction*

$$\frac{\dfrac{\overset{(1)}{\forall x\varphi(x)}}{\varphi(a)} \quad \overset{(2)}{\neg\varphi(a)}}{\dfrac{\bot}{[\neg\forall x\varphi(x)]}}\ (1)$$

$$
\begin{array}{c}
(1)\\
[\neg\forall x\varphi(x)]\\
\mathcal{D}_1\\
(\mathrm{CR})\ \dfrac{\bot}{\forall x\varphi(x)}\ (1)\\
\mathcal{D}_2
\end{array}
\quad\mapsto\quad
\begin{array}{c}
\mathcal{D}_1\\
(\mathrm{CR})\ \dfrac{\bot}{\varphi(a)}\ (2)\\
\forall x\varphi(x)\\
\mathcal{D}_2
\end{array}
$$

Note here that the restriction to first-order languages not including ∨ or ∃ as primitive symbols is essential: no CR∨- or CR∃-reduction exists. Indeed the system obtained from **NK(CR)** by imposing the restriction that the conclusion of any (CR) inference must be atomic is incomplete.

16. NORMALIZATION THEOREM FOR **C**

We may now solve the problem of eliminating all redundancies from a derivation in **C**.

A derivation in **C** is *normal* (with respect to logical and CR-reductions) if no (logical or CR) reduction can be applied to it.

Normal derivations in **C** have the following:

WEAK SUBFORMULA PROPERTY. *A normal derivation $\mathcal{D}$ in* **C** *with open assumptions Γ and conclusion φ may contain only subformulas of some formula in Γ or of φ, assumptions discharged by some* (CR) *inference and occurrences of $\bot$ standing immediately below such assumptions.*

Further properties of normal derivations in **C** are discussed in Prawitz (1965) and Cellucci (1978). The following result can be easily established (see Prawitz 1965, and Cellucci 1978).

NORMALIZATION THEOREM FOR **C**. *Every derivation $\mathcal{D}$ in* **C** *can be transformed into a normal derivation $\mathcal{D}'$ whose open assumptions are among those of $\mathcal{D}$ and whose conclusion is the same as that of $\mathcal{D}$, by an appropriate sequence of logical or CR-reductions.*

Gentzen (1934–5) believed that, while a normalization theorem could be established for **NM** and **NJ**, no such result could be established for (his formulation, **NK(EM)**, of) **NK** because the axiom schema (EM) occupies a special position w.r.t. the properties essential for the validity of such a theorem. Apparently this was his main reason for preferring the calculus of sequents to natural deduction. By establishing the normalization theorem for C Prawitz (1965) showed that this belief was unfounded, at least as far as the particular version of **NK** embodied in C is concerned.

The question then arises whether a normalization theorem can be established for the full system **NK(CR)**, not only for C. Tennant (1980) claimed to have solved this problem but, as Schroeder-Heister (1981) pointed out, his proof is incorrect. A correct solution was provided by Seldin (1990).

17. (CM)-REDUCTION RULES

A solution to the problem of eliminating all redundancies yielded by (CM) can be provided in the case of **NK(CM)**. In order to solve such problem we introduce the following (CM)-*reduction rules.*

($\neg$E), (CM)-*reduction*

If no open assumption of $\mathcal{D}_1$ is discharged in $\mathcal{D}_2$,

$$
(\neg\text{E})\ \dfrac{\overset{\substack{(1)\\ [\neg\varphi]\\ \mathcal{D}_1}}{\varphi}\quad \overset{(1)}{\neg\varphi}}{\underset{\substack{\mathcal{D}_2}}{\bot}}
\qquad
(\text{CM})\ \dfrac{\varphi}{\underset{\mathcal{D}_3}{\varphi}}\ (1)
\quad\mapsto\quad
(\text{CM})\ \dfrac{\overset{\substack{(1)\\ [\neg\varphi]\\ \mathcal{D}_1}}{\varphi}}{\underset{\mathcal{D}_3}{\varphi}}\ (1)
$$

In particular,

$$
(\neg\text{E})\ \dfrac{\overset{\mathcal{D}_1}{\varphi}\quad \overset{(1)}{\neg\varphi}}{\underset{\mathcal{D}_2}{\bot}}
\qquad
(\text{CM})\ \dfrac{\varphi}{\underset{\mathcal{D}_3}{\varphi}}\ (1)
\quad\mapsto\quad
\overset{\mathcal{D}_1}{\underset{\substack{\varphi\\ \mathcal{D}_3}}{\varphi}}
$$

(CM), (R)-*reduction*

For any inference rule (R) different from (CM) and (∀I),

$$
\text{(CM)} \ \dfrac{\begin{array}{c}(1)\\ [\neg\varphi]\\ \mathcal{D}_1\\ \varphi\end{array}}{\ \varphi\ }\ (1)
$$
$$
\text{(R)} \ \dfrac{\qquad\qquad \mathcal{D}_2}{\begin{array}{c}\chi\\ \mathcal{D}_3\end{array}}
$$
$$
\longmapsto
$$
$$
\text{(R)}\ \dfrac{\dfrac{\dfrac{(1)}{\varphi\ (\mathcal{D}_2)}\ \dfrac{(2)}{\qquad}}{\dfrac{\dfrac{\chi\qquad\neg\chi}{\bot}\ (1)}{[\neg\varphi]}}}{\varphi}\qquad\qquad\mathcal{D}_2
$$
$$
\text{(CM)}\ \dfrac{\chi}{\begin{array}{c}\chi\\ \mathcal{D}_3\end{array}}\ (2)
$$

if $\chi \not\equiv \bot$,

$$
\longmapsto
$$

$$
\text{(R)}\ \dfrac{\dfrac{(1)}{\varphi\ \mathcal{D}_2}}{\dfrac{\dfrac{\bot}{[\neg\varphi]}\ (1)}{\mathcal{D}_1}}
$$
$$
\text{(R)}\ \dfrac{\varphi\qquad\qquad\mathcal{D}_2}{\begin{array}{c}\bot\\ \mathcal{D}_3\end{array}}
$$

if $\chi \equiv \bot$

(CM), (CM)-*reduction*

$$
\text{(CM)}\ \dfrac{\begin{array}{cc}(1)&(2)\\ [\neg\varphi]&[\neg\varphi]\\ &\mathcal{D}_1\\ &\varphi\end{array}}{\varphi}\ (1)
$$
$$
\text{(CM)}\ \dfrac{\varphi}{\begin{array}{c}\varphi\\ \mathcal{D}_2\end{array}}\ (2)
$$
$$
\longmapsto
$$
$$
\text{(CM)}\ \dfrac{\begin{array}{cc}(1)&(1)\\ [\neg\varphi]&[\neg\varphi]\\ &\mathcal{D}_1\\ &\varphi\end{array}}{\begin{array}{c}\varphi\\ \mathcal{D}_2\end{array}}\ (1)
$$

Note that all (CM)-reductions except (¬E), (CM)-reduction move a (CM) inference down a given derivation. This fact allows to show that (CM)-reductions can always be used to push (CM) inferences down to the end of any derivation.

18. NORMALIZATION THEOREM FOR NK(CM)

We may now solve the problem of eliminating all redundancies from a derivation in NK(CM).

A derivation in NK(CM) is *normal* (with respect to logical, commutative and CM-reductions) if no (logical, commutative or CM-) reduction can be applied to it.

Every (CM) inference in a normal derivation $\mathcal{D}$) in NK(CM) is either the last inference in $\mathcal{D}$ or stands immediately above a ($\forall$I) inference. Using this fact, the following result can easily be established (see Seldin 1990).

STANDARDIZATION LEMMA. *Every normal derivation $\mathcal{D}$ in NK(CM) with open assumptions Γ and conclusion φ can be transformed into a normal derivation $\mathcal{D}'$ in NJ with open assumptions $\Gamma' \cup \{\neg\varphi'\}$ and conclusion φ', where Γ' and φ' are obtained from Γ and φ, respectively, by replacing given occurrences of certain components $\forall x\varphi(x)$ which are either negative subformulas of formulas in Γ or positive subformulas of φ by $\forall x(\neg\varphi(x) \rightarrow \varphi(x))$. Indeed $\mathcal{D}'$ can be obtained from $\mathcal{D}$ by the following transformation:*

$$
\begin{array}{c}
(1) \\
[\neg\varphi(a)] \\
\mathcal{D}_1 \\
\varphi(a)
\end{array}
\qquad\qquad
\begin{array}{c}
(1) \\
[\neg\varphi(a)] \\
\mathcal{D}_1 \\
\varphi(a)
\end{array}
$$

$$
\begin{array}{cc}
(\text{CM}) & \dfrac{\varphi(a)}{\cfrac{\varphi(a)}{\forall x\varphi(x)}}\,(1) \\
(\forall\text{I}) & \\
& \mathcal{D}_2
\end{array}
\quad\mapsto\quad
\begin{array}{cc}
(\rightarrow\text{I}) & \dfrac{\cfrac{\neg\varphi(a) \rightarrow \varphi(a)}{\forall x(\neg\varphi(x) \rightarrow \varphi(x))}}{}\,(1) \\
(\forall\text{I}) & \\
& \mathcal{D}_2'
\end{array}
$$

where $\mathcal{D}_2'$ is obtained from $\mathcal{D}_2$ by replacing occurrences of $\forall x\varphi(x)$ by $\forall x(\neg\varphi(x) \rightarrow \varphi(x))$.

Since normal derivations in NJ have the subformula property, we obtain the following:

WEAK SUBFORMULA PROPERTY FOR NK(CM). *Each formula occurring in the normal derivation $\mathcal{D}'$ provided by the above result is a subformula of some formula in Γ' or of $\neg\varphi'$.*

By a rather involved argument one can establish the following result (see Seldin 1990).

NORMALIZATION THEOREM FOR NK(CM). *Every derivation $\mathcal{D}$ in NK(CM) can be transformed into a normal derivation $\mathcal{D}'$ whose open as-*

sumptions are among those of $\mathcal{D}$ and whose conclusion is the same as that of $\mathcal{D}$, by an appropriate sequence of logical, commutative or CM-reductions.

19. IDENTITY OF PROOFS

Strictly speaking, derivations in tree form do not represent proofs, at least proofs of the kind familiar from mathematical practice because, as has already been pointed out, the latter are essentially linear sequences of propositions. Since, however, derivations in tree form can easily be transformed into derivations in linear form, one may loosely say that derivations represent proofs (modulo the transformation to linear form).

The problem then arises: when a derivation $\mathcal{D}$ is transformed into a normal derivation $\mathcal{D}'$, do $\mathcal{D}$ and $\mathcal{D}'$ represent the same proof? Apparently transforming $\mathcal{D}$ into $\mathcal{D}'$ only removes unnecessary parts of $\mathcal{D}$, hence it would seem reasonable to answer the above question in the affirmative.

This overlooks, however, the fact that essential information may be lost in the passage from $\mathcal{D}$ to $\mathcal{D}'$. For example, let $\mathcal{D}$ be of the form

$$
\begin{array}{ll}
& \mathcal{D}_1 \\
& \varphi(a) \\
(\forall\text{I}) & \overline{\forall x \varphi(x)} \\
(\forall\text{E}) & \overline{\varphi(t)}
\end{array}
$$

and let $\mathcal{D}'$ be obtained from $\mathcal{D}$ by a single $\forall$-reduction, whence $\mathcal{D}'$ is

$$
\mathcal{D}_1 \left[\begin{array}{c} a \\ t \end{array} \right]
$$
$$
\varphi(t) \, .
$$

Then in the passage from $\mathcal{D}$ to $\mathcal{D}'$ one loses the information that $\forall x \varphi(x)$ holds, not only its special case $\varphi(t)$. While, as $\forall$-reduction shows, such an information is not strictly required to obtain the conclusion $\varphi(t)$, it is the use of just this information that allowed the conclusion $\varphi(t)$ to be originally obtained. Hence the information in question, while in principle dispensable, may be crucial heuristically.

Thus it seems fair to say that, when a derivation $\mathcal{D}$ is transformed into a normal derivation $\mathcal{D}'$, the latter does not represent the same proof as $\mathcal{D}$ but only a proof which extracts from the proof represented by $\mathcal{D}$ just the information which is strictly required for its particular conclusion, disregarding additional information which may otherwise be essential heuristically.

This argument is relevant to the following conjecture concerning the problem of identity of proofs (ascribed to Martin-Löf in Prawitz 1971).

CONJECTURE. *Two derivations represent the same proof if and only if they can be transformed into the same normal derivation.*

In view of the above discussion the 'if' part of the conjecture appears to be false. But the 'only if' part can also be challenged: as noted by Troelstra (1975), if it is possible that different formulas express the same proposition, then it is possible that different normal derivations represent the same proof. A different opinion is expressed by Prawitz (1971), who seems to assume that different normal derivations should represent different derivations. (For further discussion see Martin-Löf 1975, Feferman 1985, Troelstra 1975, and Cellucci 1980).

20. DEFECTS OF NATURAL DEDUCTION

Like the calculus of sequents, natural deduction has both advantages and disadvantages. The advantages consist in the fact that: (1) the inference rules, with the only exception of $(\vee E)$ and $(\exists E)$, are simple and elegant; (2) derivations are relatively easy to handle, in any case easier to handle than derivations in the calculus of sequents; and (3) the procedure for transforming derivations into normal derivations is also simple and elegant. The disadvantages consist in the fact that: (1) the rules of **NJ** or **NK** are actually rules for minimal logic extended with new principles – such as (EF), (D), (CR), (CM) – which do not fit into the pattern of the other rules; (2) because of the form of $(\vee E)$ and $(\exists E)$ the treatment of disjunction and existential quantification is problematic, even for minimal logic, because the rules are not symmetrical; (3) since **NJ** and **NK** are really systems for minimal logic, extended with new principles which do not fit into the pattern of other rules, only derivations of minimal laws are easy to handle in **NJ** or **NK**: the derivation of essentially intuitionistic or classical laws can be very complicated.

It is not easy to avoid these defects while remaining within the context of standard natural deduction. For example, consider the problem of finding a more symmetrical formulation of the rules for $\vee$ (see Weir 1986). A possibility would be to state the following alternative rules:

$$(\vee I') \quad \frac{\overset{[\neg\varphi]}{\psi}}{\varphi \vee \psi}, \frac{\overset{[\neg\psi]}{\varphi}}{\varphi \vee \psi} \qquad\qquad (\vee E') \quad \frac{\varphi \vee \psi \ \neg\varphi}{\psi}, \frac{\varphi \vee \psi \ \neg\psi}{\varphi}$$

Let **S** be the system consisting of the rules $(\wedge I)$, $(\wedge E)$, $(\vee I')$, $(\vee E')$, $(\rightarrow I)$, $(\rightarrow E)$, $(\forall I)$, $(\forall E)$, $(\exists I)$, $(\exists E)$. The new rules make the need of assuming (EM) superfluous. For, in S we have the derivation:

$$(\text{VI}') \quad \frac{\overset{\displaystyle (1)}{\neg\varphi}}{\varphi \vee \neg\varphi} \, (1)$$

The standard $(\vee\text{E})$ is derivable in S as follows:

$$(\vee\text{E}') \quad \frac{\mathcal{D}_0}{\varphi \vee \psi} \quad \frac{\overset{\displaystyle (1)}{[\varphi]} \quad (2)}{\underset{\displaystyle \mathcal{D}_2}{\psi}} \quad \frac{\chi \quad \neg\chi}{\dfrac{\bot}{\neg\varphi}} \, (1)}{\psi}$$

$$(\text{VI}') \quad (\vee\text{E}') \quad \frac{\overset{\displaystyle (3)}{\neg\chi}}{\chi \vee \neg\chi} \, (3) \qquad \frac{\overset{\displaystyle \mathcal{D}_2 \quad (2)}{\chi \quad \neg\chi}}{\dfrac{\bot}{\neg\neg\chi}} \, (2)}{\chi}$$

However $\vee$-reduction is a problem. Its first two cases are straightforward:

$\vee$-*reduction*

$$(\text{VI}') \quad (\vee\text{E}') \quad \frac{\frac{\overset{\displaystyle\overset{(1)}{[\neg\varphi]}}{\mathcal{D}_1}}{\varphi \vee \psi} \, (1) \quad \overset{\mathcal{D}_2}{\neg\varphi}}{\underset{\displaystyle \mathcal{D}_3}{\psi}} \quad \longmapsto \quad \begin{array}{c} \mathcal{D}_2 \\ [\neg\varphi] \\ \mathcal{D}_1 \\ \psi \\ \mathcal{D}_3 \end{array}$$

$$(\text{VI}') \quad (\vee\text{E}') \quad \frac{\frac{\overset{\displaystyle\overset{(1)}{[\neg\psi]}}{\mathcal{D}_1}}{\varphi \vee \psi} \, (1) \quad \overset{\mathcal{D}_2}{\neg\psi}}{\underset{\displaystyle \mathcal{D}_3}{\varphi}} \quad \longmapsto \quad \begin{array}{c} \mathcal{D}_2 \\ [\neg\psi] \\ \mathcal{D}_1 \\ \varphi \\ \mathcal{D}_3 \end{array}$$

The remaining two cases, however, are problematic:

$$\text{(VI')} \quad \cfrac{\cfrac{\begin{array}{c}(1)\\ [\neg\varphi]\\ \mathcal{D}_1 \\ \psi \end{array}}{\varphi \lor \psi}\,(1) \quad \begin{array}{c}\mathcal{D}_2\\ \neg\psi\end{array}}{\begin{array}{c}\varphi \\ \mathcal{D}_3\end{array}} \quad \longmapsto \quad \mathbf{?}$$

$$\text{(VI')} \quad \cfrac{\cfrac{\begin{array}{c}(1)\\ [\neg\psi]\\ \mathcal{D}_1 \\ \varphi \end{array}}{\varphi \lor \psi}\,(1) \quad \begin{array}{c}\mathcal{D}_2\\ \neg\varphi\end{array}}{\begin{array}{c}\psi \\ \mathcal{D}_3\end{array}} \quad \longmapsto \quad \mathbf{?}$$

The best we can do is:

and thus we enter into an infinite loop. This shows that we need a new approach.

If we want to make the rules for $\lor$ more symmetrical, an obvious solution is to state (VE) in the form

$$\text{(VE)} \quad \frac{\varphi \lor \psi}{\varphi \quad \psi}\,,$$

but this involves that derivations are no longer tree-like structures. Such an approach, which is explored in *multiple conclusion natural deduction*, however, raises a number of difficulties (see Kneale 1956, Kneale and Kneale 1962, Shoesmith and Smiley 1978, and Ungar, forthcoming).

If we want to keep derivations in tree form, we need something like

$$\text{(VE)} \quad \frac{\varphi \lor \psi}{\varphi, \psi}\,.$$

This involves considering finite lists of formulas instead of formulas. Such an approach is explored in *sequent natural deduction* (see Boričić 1985, Cellucci 1988, 1991, 1992).

21. THE SYSTEM $\mathbf{ND}_\varepsilon$

The only problem which can be solved while remaining within the context of standard (i.e. non-sequent) natural deduction is that of giving a more symmetrical formulation to the rules for $\exists$.

To this effect we modify first-order languages by: (1) adding function parameters $\varepsilon_0, \varepsilon_1, \ldots$; (2) dropping the logical symbol $\lor$; (3) adding the auxiliary symbols [,] and ; (semicolon). In addition to terms we introduce ε-*terms*, i.e. expressions of the form $\varepsilon_i[\exists x\varphi(x); \Gamma]$ where Γ is a finite list of formulas (this involves defining formulas and ε-terms simultaneously). The inference rules are expanded by adding the following new E-rule:

$$(\exists E_\varepsilon) \quad \frac{\exists x\varphi(x)}{\varphi(\varepsilon[\exists x\varphi(x); \Gamma])} \; .$$

In an $(\exists E_\varepsilon)$ inference Γ is the list of all assumptions on which the premise $\exists x\varphi(x)$ depends. The ε-term $\varepsilon[\exists x\varphi(x); \Gamma]$ is called the *proper ε-term* of that inference. We call $\exists x Px$ the *formula* part of $\varepsilon[\exists x\varphi(x); \Gamma]$ and Γ its *assumption* part.

In addition to R1–R3 we state the following new restriction:

R4.   In a $(\to I)$ inference in which some assumption φ is discharged, for any ε-term $\varepsilon[\exists x\chi(x); \Delta]$ occurring in the conclusion $\varphi \to \psi$ or in any assumption on which the conclusion depends, φ must not be in the assumption part Δ.

That such a restriction is essential for an intuitionistic system is shown by the following examples where an intuitionistically invalid conclusion is obtained violating R4.

(1) Let $\varepsilon \equiv \varepsilon[\exists x Px; \exists x Px]$.

$$(\to I) \quad \frac{\dfrac{\overset{(1)}{\exists x Px}}{P\varepsilon}}{\dfrac{\exists x Px \to P\varepsilon}{\exists y(\exists x Px \to Py)}} \, (1)$$

The indicated $(\to I)$ inference violates R4 because the ε-term ε occurring in the conclusion contains the discharged assumption $\exists x Px$ in its assumption part.

(2) Let $\varepsilon \equiv \varepsilon[\exists x Q x; \; P, P \to \exists x Q x]$.

$$\dfrac{\dfrac{(\to \mathrm{I}) \dfrac{\dfrac{\dfrac{(1) \quad (2)}{P \quad P \to \exists x Q x}}{\dfrac{\exists x Q x}{Q\varepsilon}}}{P \to Q\varepsilon} \, (1)}{\dfrac{\exists x (P \to Q x)}{(P \to \exists x Q x) \to \exists x (P \to Q x)}} \, (2)}{}$$

The indicated ($\to$ I) inference violates R4 because the ε-term ε occurring in the conclusion contains the discharged assumption P in its assumption part.

Let **ND** consist of the rules ($\wedge$I), ($\wedge$E), ($\to$ I), ($\to$ E), ($\forall$I), ($\forall$E), ($\exists$I), ($\exists$E), (EF), i.e. **ND** is obtained from **NJ** by omitting the rules ($\vee$I), ($\vee$E). Let **ND**$_\varepsilon$ consist of the rules ($\wedge$I), ($\wedge$E), ($\to$ I), ($\to$ E), ($\forall$I), ($\forall$E), ($\exists$I), ($\exists$E$_\varepsilon$), (EF). Let **ND**$_\varepsilon^+$ be obtained from **ND**$_\varepsilon$ by adding the rule ($\exists$E), i.e. **ND**$_\varepsilon^+$ contains two E-rules for $\exists$, ($\exists$E) and ($\exists$E$_\varepsilon$). (**ND**$_\varepsilon$ and **ND**$_\varepsilon^+$ were introduced by Leivant 1973.)

22. NORMALIZATION THEOREM FOR **ND**$_\varepsilon$

A derivation $\mathcal{D}$ in **ND**$_\varepsilon$ or **ND**$_\varepsilon^+$ with open assumptions Γ and conclusion φ is said to be *pure* if Γ and φ contain no ε-term.

The following result can easily be established.

LEMMA. *Every pure derivation $\mathcal{D}$ in* **ND**$_\varepsilon^+$ *can be transformed into a derivation $\mathcal{D}'$ in* **ND** *whose open assumptions are among those of $\mathcal{D}$ and whose conclusion is the same as that of $\mathcal{D}$.*

Since every derivation in **ND**$_\varepsilon$ is a derivation in **ND**$_\varepsilon^+$ this establishes:

CONSERVATIVITY OF **ND**$_\varepsilon$ OVER **ND**. *Every pure derivation $\mathcal{D}$ in* **ND**$_\varepsilon$ *can be transformed into a derivation $\mathcal{D}'$ in* **ND** *whose open assumptions are among those of $\mathcal{D}$ and whose conclusion is the same as that of $\mathcal{D}$.*

To establish the converse we need the following:

LEMMA. *Evey pure derivation $\mathcal{D}$ in* **ND**$_\varepsilon^+$ *can be transformed into a pure derivation $\mathcal{D}'$ in* **ND**$_\varepsilon$ *whose open assumptions are among those of $\mathcal{D}$ and whose conclusion is the same as that of $\mathcal{D}$.*

Since every derivation in **ND** is a pure derivation in **ND**$_\varepsilon^+$ this established:

CONSERVATIVITY OF **ND** OVER **ND**$_\varepsilon$. *Every derivation $\mathcal{D}$ in* **ND** *can be transformed into a pure derivation $\mathcal{D}'$ in* **ND**$_\varepsilon$ *whose open assumptions are among those of $\mathcal{D}$ and whose conclusion is the same as that of $\mathcal{D}$.*

Logical reduction rules are modified by replacing $\exists$-reduction by:

$\exists\varepsilon$-reduction

$$
(\exists E_\varepsilon) \quad
\cfrac{(\exists I) \cfrac{\begin{array}{c}\Gamma\\\mathcal{D}_1\\\varphi(t)\end{array}}{\exists x\varphi(x)}}{\varphi(\varepsilon[\exists x\varphi(x);\,\Gamma])} \\
\qquad\qquad\qquad \mathcal{D}_2
\qquad\mapsto\qquad
\mathcal{D}_2\left[\begin{array}{c}\Gamma\\\mathcal{D}_1\\\varphi(t)\\\cfrac{\varepsilon[\exists x\varphi(x);\,\Gamma]}{t}\end{array}\right]
$$

A derivation $\mathcal{D}$ in **ND**$_\varepsilon$ is *normal* (w.r.t. the modified logical and EF-reductions) if no (modified logical or EF-) reduction can be applied to it.

Normal derivations in **ND**$_\varepsilon$ have the following:

SUBFORMULA PROPERTY. *Formulas occurring in a normal derivation $\mathcal{D}$ in* **ND**$_\varepsilon$ *are subformulas of some open assumption of $\mathcal{D}$ or of the conclusion of $\mathcal{D}$.*

The following result can easily be established (see Leivant 1973).

NORMALIZATION THEOREM OF **ND**$_\varepsilon$. *Every pure derivation $\mathcal{D}$ in* **ND**$_\varepsilon$ *can be uniformly transformed into a pure normal derivation $\mathcal{D}'$ whose open assumptions are among those of $\mathcal{D}$ and whose conclusion is the same as that of $\mathcal{D}$, by an appropriate sequence of modified logical or EF-reductions.*

23. NECESSITY OF RESTRICTIONS

It should be noted that R4 is essential not only for intuitionistic validity but also for the validity of the normalization theorem. This is shown by the following example where $\varepsilon \equiv \varepsilon[\exists y Py;\ Pa]$:

$$
\begin{array}{c}
(\exists I)\ (\exists E_\varepsilon)\ \cfrac{\cfrac{\cfrac{(1)}{Pa}}{\exists y Py}\quad \cfrac{(2)}{P\varepsilon \to Q}}{\cfrac{\cfrac{Q}{Pa \to Q}\ (1)}{\forall x(Px \to Q)}}
\quad\mapsto\quad
\cfrac{\cfrac{(1)\quad(2)}{\cfrac{Pa\quad Pa\to Q}{\cfrac{Q}{Pa\to Q}\ (1)}}}{\forall x(Px \to Q)}
\end{array}
$$

The indicated ($\forall$I) inference on the right violates restriction R1 because its proper parameter a occurs in the assumption $Pa \to Q$ on which its premise depends. This originates from the fact that the ε-term ε occurring in the assumption $P\varepsilon \to Q$ on which the conclusion of the indicated ($\to$ I) inference on the left depends contains the discharged assumption Pa in its assumption part. This problem is avoided by R4.

It should be also noted that, whenevery a derivation $\mathcal{D}$ has two subderivations with the same open assumptions and the same conclusion, $\exists E_\varepsilon$ reduction must be applied to them simultaneously. This is shown by the following example where $\varepsilon \equiv \varepsilon[\exists x Px;\ Pa]$:

$$
\begin{array}{ccc}
\begin{array}{c}
(1) \\
(\exists\mathrm{I})\ \dfrac{Pa}{\exists x Px} \\
(\exists\mathrm{E}_\varepsilon)\ \dfrac{}{P\varepsilon}
\end{array}
&
\begin{array}{c}
(1) \\
(\exists\mathrm{I})\ \dfrac{Pa}{\exists x Px} \\
(\exists\mathrm{E}_\varepsilon)\ \dfrac{}{P\varepsilon}
\end{array}
&
\\[2em]
\multicolumn{2}{c}{\dfrac{P\varepsilon \wedge P\varepsilon}{\exists x (Px \wedge Px)}}
\end{array}
\quad\mapsto\quad
\begin{array}{cc}
\begin{array}{c}
(1) \\
Pa
\end{array}
&
\begin{array}{c}
(1) \\
(\exists\mathrm{I})\ \dfrac{Pa}{\exists x Px} \\
(\exists\mathrm{E}_\varepsilon)\ \dfrac{}{Pa}
\end{array}
\\[2em]
\multicolumn{2}{c}{\dfrac{Pa \wedge Pa}{\exists x (Px \wedge Px)}}
\end{array}
$$

Obviously the ($\exists E_\varepsilon$) inference indicated on the right is incorrect because $a \not\equiv \varepsilon$.

24. THE SYSTEM **BK**

The idea of introducing restrictions on discharge rules in intuitionistic systems including a rule like ($\exists E_\varepsilon$) was explicitly stated by Smirnov (1971, 1978) and Leivant (1973) (see Mints (1982) for a different approach). Such an idea, however, is implicit in certain classical natural deduction systems including an existential instantiation rule, such as those of Kalish, Montague and Mar (1964), Klenk (1983) or Fine (1985).

For example, let us consider the system **BK** of Klenk (1983), which is due to Belnap. Unlike the natural deduction systems considered so far, derivations in **BK** are in linear form. The rules of **BK** are standard except for universal generalization and existential instantiation. Universal generalization (UG) takes the form:

$$
\frac{\begin{array}{|l} \to \text{flag } a \\ \ \ \vdots \\ \ \ \varphi(a) \end{array}}{\forall x \varphi(x)}
$$

Thus (UG) requires that one starts off a new subordinate derivation in which the proper parameter a is 'flagged'. Existential instantiation (EI) takes the form:

$$\frac{\exists x \varphi(x)}{\varphi(a)} \quad \text{flag } a$$

Thus, unlike (UG), the rule (EI) does not require that one starts off any new subordinate derivation but only that the proper parameter a is flagged to the right.

Flagging is subject to the following simple restrictions, which are related to the restrictions on the rules of $\mathbf{ND}_\varepsilon$.

R1'.   A flagged parameter may not occur, either in a formula or as a flagged parameter, above the line in which it gets flagged or below the subordinate derivation in which it gets flagged.

R2'.   A flagged parameter may not appear in the open assumptions or in the conclusion of a derivation.

To see the working of such restrictions in preventing from proving intuitionistically invalid formulas we consider the following examples.

	1	$\exists x Px$	Ass.
(1)	2	Pa	EI 1, flag a
	$\Rightarrow$ 3	$\exists x Px \rightarrow Pa$	CP 1 – 2
	4	$\exists y(\exists x Px \rightarrow Py)$	EG 3

Here the CP inference on line 3 violates restriction R1' because $\exists x Px \rightarrow Pa$ contains the flagged parameter a.

	1	$P \rightarrow \exists x Qx$	Ass.
	2	P	Ass.
(2)	3	$\exists x Qx$	MP 1, 2
	4	Qa	EI 3, flag a
	$\Rightarrow$ 5	$P \rightarrow Qa$	CP 2 – 4
	6	$\exists y(P \rightarrow Qy)$	EG 5

Here again the CP inference on line 5 violates restriction R1' because $P \rightarrow Qa$ contains the flagged parameter a.

Clearly R1' disallows such fallacious inferences in a way that is somehow related to the working of restriction R4 on $(\rightarrow I)$.

While the restrictions on the rules of $\mathbf{BK}$ are related to the restrictions on the rules on $\mathbf{ND}_\varepsilon$, they are by no means the same. This can be seen, for instance, with reference to $(\forall I)$. Whereas in $\mathbf{ND}_\varepsilon$ the restriction R1 on $(\forall I)$ is different in kind from the restriction R4 on $(\rightarrow I)$, in $\mathbf{BK}$ these two restrictions are merged into the single restriction R1'. What are disparate restrictions in $\mathbf{ND}_\varepsilon$ are aspects of a single restriction in $\mathbf{BK}$.

25. CONCLUSION

There is more to natural deduction than we have been able to deal with here. Important topics that have been omitted for space limitations are complexity of derivations (see Statman 1974, 1978, Mints 1982, Solov'ev 1982, Schwichtenberg, 1982, Pereira, 1982, Cellucci 1985), the relationship between natural deduction and the calculus of sequents (see Prawitz 1965, Zucker 1974, Pottinger, 1977), the relationship between natural deduction and typed lambda calculus (see Howard 1980, Hindley and Seldin 1986, Girard, Lafont and Taylor 1989, Fortune, Leivant and O'Donnell 1983), second-order natural deduction (see Girard 1987, Prawitz 1971, 1981), validity notions (see Prawitz 1971, 1973, 1981), the use of natural deduction in interactive theorem proving (see Paulson 1986, 1987, 1989, Felty and Miller 1988).

It remain, however, a fact that standard natural deduction presents problems which cannot easily be solved by remaining within its simple framework. To solve such problems it is more convenient to adopt a different approach such as the one provided by sequent natural deduction (Boričić 1985, Cellucci 1988, 1991, 1992); but that is another story.

Dipartimento di Studi Filosofici ed Epistemologici,
Università di Roma 'La Sapienza'.

REFERENCES

Boričić, B.R.: 1985, 'On sequence-conclusion natural deduction systems', *J. Phil. Logic* **14**, 359–377.
Cellucci, C.: 1978, *Teoria della dimostrazione*, Boringhieri, Torino, 1978.
Cellucci, C.: 1980, 'Proof theory and theory of meaning', in *Italian Studies in the Philosophy of Science* (ed. M.L. Dalla Chiara), Reidel, Dordrecht, pp. 13–29.
Cellucci, C.: 1985, 'Proof theory and complexity', *Synthese* **62**, 173–189.
Cellucci, C.: 1988, 'Efficient natural deduction', in *Temi e prospettive della logica e della filosofia della scienza contemporanee* (ed. C. Cellucci and G. Sambin), Vol. I, CLUEB, Bologna, pp. 29–57.
Cellucci, C.: 1991, 'Sequent natural deduction and intuitionistic logic', in *Nuovi problemi della logica e della filosofia della scienza*, Vol. II (ed. G. Corsi and G. Sambin), CLUEB, Bologna, 1991, pp. 259–266.
Cellucci, C.: 1992, 'Existential instantiation and normalization in sequent natural deduction', *Annals of Mathematical Logic* **58**, in press.
Curry, H.B.: 1963, *Foundations of Mathematical Logic*, Dover, New York (second ed. 1977).
Feferman, S.: 1975, Review of Prawitz (1971), *The Journal of Symbolic Logic* **40**, 232–234.
Felty, A. and Miller, D.: 1988, 'Specifying theorem provers in a higher order proof system for

mechanizing mathematics', in *Ninth Conference on Automated Deduction* (ed. E. Lusk and R. Overbeek), Springer-Verlag, Berlin, pp. 61–80.

Fine, K.: 1985, *Reasoning with Arbitrary Objects*, Blackwell, Oxford.

Fortune, S., Leivant, D. and O'Donnell, M.: 1983, 'The expressiveness of simple and second-order type structures', *Journal of the Association for Computing Machinery* **30**, 151–185.

Gentzen, G.: 1934–5, 'Untersuchungen über die logische Schliessen', *Mathematische Zeitschrift* **39**, 176–210, 405–431; Engl. transl. in *The collected papers of Gerhard Gentzen* (ed. M.E. Szabo), North-Holland, Amsterdam, pp. 68–131.

Girard, J.-Y.: 1987, *Proof Theory and Logical Complexity*, Bibliopolis, Napoli.

Girard, J.-Y., Lafont, Y. and Taylor, P.: 1989, *Proofs and Types*, Cambridge University Press, Cambridge.

Howard, W.A.: 1980, 'The formulae-as-types notion of construction', in *To H.B. Curry: Essays on Combinatory Logic, Lambda Calculus and Formalism* (ed. J.R. Hindley and J.P. Seldin), Academic Press, London, pp. 479–490.

Hindley, J.R. and Seldin, J.P.: 1986, *Introduction to Combinators and λ-Calculus*, Cambridge University Press, Cambridge.

Kalish, D., Montague, R. and Mar, G.: 1964, *Logic. Techniques of Formal Reasoning*, Harcourt Brace Jovanovich, New York (second ed. 1980).

Klenk, V.: 1983, *Understanding Symbolic Logic*, Prentice Hall, Englewood Cliffs (second ed. 1989).

Kneale, W.: 1956, 'The province of logic', in *Contemporary British Philosophy* (ed. H.D. Lewis), 3rd series, Allen & Unwin, London, pp. 237–261.

Kneale, W. and Kneale, M.: 1962, *The Development of Logic*, Oxford University Press, Oxford.

Leivant, D.: 1973, 'Existential instantiation in a system of natural deduction for intuitionistic arithmetic', Stichting Mathematisch Centrum, Amsterdam, Report ZW 13/73.

Leivant, D.: 1979, 'Assumption classes in natural deduction', *Zeitschrift für mathematische Logik und Grundlagen der Mathematik* **25**, 1–4.

Lemmon, E.J.: 1965, *Beginning Logic*, Nelson, London.

Martin-Löf, P.: 1975, 'About models for intuitionistic type theories and the notion of definitional equality', in *Proceedings of the Third Scandinavian Logic Symposium* (ed. S. Kanger), North-Holland, Amsterdam, pp. 81–109.

Mints, G.E.: 1977, 'Heyting predicate calculus with epsilon symbol', *Journal of Soviet Mathematics* **8**, 317–323.

Mints, G.E.: 1982, 'Primitive recursive estimate of strong normalization for predicate calculus', *Journal of Soviet Mathematics* **20**, 2334–2336.

Paulson, L.C.: 1986, 'Natural deduction as higher-order resolution', *J. Logic Programming* **3**, 237–258.

Paulson, L.C.: 1987, *Logic and Computation. Interactive Proof with Cambridge LCF*, Cambridge University Press, Cambridge.

Paulson, L.C.: 1989, 'The foundation of a generic theorem prover', *J. Autom. Reasoning* **5**, 363–397.

Pereira, L.C.: 1982, *On the Estimation of the Length of Normal Derivations*, Philosophical Studies Vol. 4, Stockholm.

Pottinger, G.: 1977, 'Normalization as a homomorphic image of cut-elimination', *Annals of Mathematical Logic* **12**, 323–357.

Prawitz, D.: 1965, *Natural Deduction. A Proof-Theoretical Study*, Almqvist & Wiksell, Stockholm.

Prawitz, D.: 1971, 'Ideas and results in proof theory', in *Proceedings of the Second Scandinavian Logic Symposium* (ed. J.E. Fenstad), North-Holland, Amsterdam, pp. 235–307.

Prawitz, D.: 1973, 'Towards a foundation of a general proof theory', in *Logic, Methodology and*

Philosophy of Science IV (ed. P. Suppes, L. Henkin, A. Joja and Gr.C. Moisil), North-Holland, Amsterdam, pp. 225–250.

Prawitz, D.: 1981, 'Validity and normalizability of proofs in 1st and 2nd order classical and intuitionistic logic', in *Atti del Congresso Nazionale di Logica* (ed. S. Bernini), Bibliopolis, Napoli, pp. 11-36.

Schroeder-Heister, P.J.: 1981, *Untersuchungen zur Regellogischen Deutung von Aussagen-verknüpfungen*, Dissertation, Bonn.

Schwichtenberg, H.: 1982, 'Complexity of normalization in the pure typed lambda-calculus', in *The L.E.J. Brouwer Centenary Symposium* (ed. A.S. Troelstra and D. van Dalen), North-Holland, Amsterdam, pp. 453–457.

Seldin, J.P.: 1986, 'On the proof theory of the intermediate logic MH', *The Journal of Symbolic Logic* **51**, 626–647.

Seldin, J.P.: 1990, 'Normalization and excluded middle. I', *Studia Logica* **48**, 193–217.

Shoesmith, D.J. and Smiley, T.J.: 1978, *Multiple-Conclusion Logic*, Cambridge University Press, Cambridge.

Smirnov, V.A.: 1971, 'Elimination des termes ε dans la logique intuitionniste', *Revue Internationale de Philosophie* **98**, 512–519.

Smirnov, V.A.: 1978, 'Theory of quantification and ε-calculi', in *Essays on Mathematical and Philosophical Logic* (ed. J. Hintikka, I. Niiniluoto and E. Saarinen), Reidel, Dordrecht, pp. 41–82.

Solov'ev, S.V.: 1982, 'Growth of length of sequential derivation transformed into natural one', *Journal of Soviet Mathematics* **20**, 2367–2369.

Statman, R.: 1974, *Structural Complexity of Proofs*, Dissertation, Stanford University, Stanford.

Statman, R.: 1978, 'Bounds for proof-search and speed-up in the predicate calculus', *Annals of Mathematical Logic* **15**, 225–287.

Sundholm, G.: 1983, 'Systems of deduction', in *Handbook of Philosophical Logic* (ed. D. Gabbay and F. Guenthner), Vol. I, Reidel, Dordrecht, pp. 133–188.

Tennant, N.: 1978, *Natural Logic*, Edinburgh University Press, Edinburgh (second ed. 1990).

Tennant, N.: 1980, 'A proof-theoretical approach to entailment', *Journal of Philosophical Logic* **9**, 185–209.

Troelstra, A.S.: 1975, 'Non-extensional equality', *Fundamenta Mathematicae* **82**, 307–322.

Troelstra, A.S. and Van Dalen, D.: 1988, *Constructivity in Mathematics. An Introduction*, Vol. II, North-Holland, Amsterdam.

Ungar, A.M.: forthcoming, 'Normalization and Cut Elimination in First-Order Logic'.

Weir, A.: 1986, 'Classical harmony', *Notre Dame J. Formal Logic* **27**, 459–482.

Zucker, J.: 1974, 'The correspondence between cut-elimination and normalization', *Annals of Mathematical Logic* **7**, 1-112, 113–155.

FABIO BELLISSIMA

AN INTRODUCTION TO MODAL SEMANTICS

1. INTRODUCTION AND BASIC NOTIONS

Introduction

The present note constitutes an introductory course in propositional modal logic. It does not require specific prerequisites, but only a standard knowledge of classical propositional calculus (denoted by PC in the following). Its main goal consists in making the reader familiar with both the traditional 'possible-worlds semantics', and the more recent concept of 'general frame'.

In conformity with this goal, our note presents the following features:

- syntactic notions and syntactic results are as few as possible;
- only the logic **K** is given individual attention. Other logics are used only as examples;
- many of the exercises are simple theorems, often used in more complex proofs. Because of this fact, all the exercises are solved;
- algebraic semantics for modal logic is not mentioned, although the concept of general frame arises from (or at least is strongly linked to) the algebraic concept of dual space of modal algebra. In fact, thanks to the duality theory and the subsequent development of the general semantics, the algebraic language may be viewed as an alternative and equivalent approach to the problem.

The paper is divided into four sections and some subsections, as follows:

0. Introduction and basic notions (language and formulas; models; frames; logics). 1. General frames and filtrations. 2. Relations between structures (relations between $KF(L)$, $KM(L)$ and $GF(L)$; subframes and submodels; p-morphisms, disjoint unions). 3. Canonical structures and completeness.

Language and Formulas

A propositional normal modal logic can be regarded as an 'enrichment' of the propositional calculus PC. As is well known, PC has the following two properties, a syntactic one and a semantic one:

G. Corsi et al. (eds), Bridging the Gap: Philosophy, Mathematics, and Physics, 39–69.
© *1993 Kluwer Academic Publishers.*

(a)   the set of theorems of PC is maximal among the sets which are proper
       subsets of the set of all propositional formulas and are closed under Modus
       Ponens and Substitution Rule;
(b)   each truth-function is expressible by means of Boolean connectives.

From (a), it follows that the word 'enrichment', used above, cannot mean
the addition of new axioms, expressed in the language of PC, but must involve
the addition of (at least) a new linguistic symbol; in fact, our new language
contains a new symbol, denoted by $\Box$. Moreover, a consequence of (b) is that
this new symbol cannot be interpreted as a truth-functional operator (we will
come back to this after introducing models).

DEFINITION 0.0. The propositional modal language that we consider is a
six-tuple $L_M = \langle \mathbf{P}, \wedge, \neg, \perp, \Box \rangle$ where $\mathbf{P} = \{p_i, i \in \omega\}$ is the set of *propositional letters*, $\wedge$ and $\neg$ are the usual conjunction and negation connectives, and
$\perp$ and $\Box$ are symbols which are read as 'falsum' and 'box'. The connectives
$\vee$, $\rightarrow$ and $\equiv$ are defined as usual, $\top$ (the 'truth') is defined as $\neg\perp$ and $\Diamond$
(the 'diamond') is $\neg\Box\neg$. We will often use the italic letters p, q, r as variables
through the set of propositional letters.

DEFINITION 0.1. The set *Wff* of *well formed formulas* of L_M (*formulas*
in the following) is defined as follows:

   $\perp \in$ *Wff* and $p_i \in$ *Wff*, for each $i \in \omega$,
   if $\varphi \in$ *Wff* then $\neg\varphi \in$ *Wff*,
   if $\varphi, \psi \in$ *Wff* then $(\varphi \wedge \psi) \in$ *Wff*,
   if $\varphi \in$ *Wff* then $\Box\varphi \in$ *Wff*,

nothing else is a formula.

We use $\Box^n$ instead of $\Box \ldots n$-times$\ldots \Box$, and $\varphi(p_0, \ldots, p_n)$ to denote a formula φ whose set of propositional letters forms a subset of $\{p_0, \ldots, p_n\}$. By
$\varphi(p_0/\psi_0, \ldots, p_n/\psi_n)$ we denote the formula obtained from the formula φ by
replacing each occurrence of p_i with the formula ψ_i, for every $i \leq n$. For
each cardinal number $\alpha \leq \omega$ we denote by *Wff*$_\alpha$ the set of all α-*formulas*,
i.e. the formulas not containing propositional letters p_i for $i \geq \alpha$. Obviously,
Wff$_\omega$ = *Wff*. The definition of a *subformula* is standard.
   (Throughout the paper, the letters φ, ψ, ξ and σ are used to denote formulas,
while α and β denote ordinal numbers.)

Models

As we have observed, in order that $\Box$ be undefinable in terms of the operators of PC, it must be the case that $\Box$ is not truth-functional; therefore, the notion of 'truth' of an L_M-formula, and also the concept of L_M-model, must change. Usually, in the classical propositional case, 'assignments' are used instead of 'models'. But, of course, we can define a PC-model to be a subset X of the set P of the propositional letters, and define the 'truth of φ in X' as follows: $X \models p$ iff $p \in X$, $X \models \neg\varphi$ iff $X \not\models \varphi$, and $X \models \varphi \wedge \psi$ iff $X \models \varphi$ and $X \models \psi$. (In such a way, the completeness theorem assumes the familiar form: $PC \vdash \varphi$ iff $X \models \varphi$ for each model X.) Using PC-models of this form we can easily present the concept of L_M-model as a generalization of that of PC-model. In fact: let us imagine that $\Box\varphi$ means 'it is necessary that φ', and think of each PC-model X as a 'state of affairs', or a 'possible word'. We intuitively require, in order to consider $\Box\varphi$ as true in X, that φ be true in each word X' that we, being in X, are able to conceive. Analogously, if we read $\Box\varphi$ as 'φ forever' and interpret X as an 'instant', then $\Box\varphi$ is true at X if φ is true at every Y which follows X. In both cases, we refer to a set of PC-models and to a binary relation among them (respectively 'Y is conceivable by X' and 'Y follows X'). The following Definitions 0.2 and 0.3 may be viewed as a formalization of this intuitive idea. In fact, as we shall see, the only (inessential) difference consists in considering, instead of a set of PC-models, a generic set W and a function V from P into $\mathcal{P}(W)$; obviously, there is a one-one correspondence between pairs W and V of this form, and sets of PC-models; for instance, the set of PC-models corresponding to a pair $\langle W, V \rangle$ is $\{\{p \,:\, w \in V(p)\} \,:\, w \in W\}$.

DEFINITION 0.2. A *Kripke model (model)* is a triple $\mathbf{M} = \langle W, R, V \rangle$ where

> W is a non empty set,
> R is a binary relation on W.
> V is a function from the set $\mathbf{P}$ of propositional variables
> into the set $\mathcal{P}(W)$ of all subsets of W.

W is called the *domain* of $\mathbf{M}$, and its elements are called *points* and denoted by $w, v, \ldots$, etc. V is called *valuation*. For each $w \in W$, we set $V^{-1}(w) = \{p \,:\, w \in V(p)\}$.

DEFINITION 0.3. The concept of *truth* of a formula φ at a point w of a model $\mathbf{M} = \langle W, R, V \rangle$ (in symbols: $\mathbf{M} \models \varphi[w]$ or simply $w \models \varphi$) is defined as follows:

$w \not\models \bot$,

$w \models p_i$ iff $w \in V(p_i)$

$w \models \neg\varphi$ iff $w \not\models \varphi$

$w \models \varphi \wedge \psi$ iff $w \models \varphi$ and $w \models \psi$,

$w \models \Box\varphi$ iff $v \models \varphi$ for each $v \in W$ such that wRv,

(and hence, from the definition of $\Diamond$, $w \models \Diamond\varphi$ iff there exists $v \in W$ such that wRv and $v \models \varphi$).

For each $\varphi \in Wff$ we set

$$V^+(\varphi) = \{w \in W \ : \ w \models \varphi\} \ .$$

A formula φ is *true* in a model M (in symbols $M \models \varphi$) if $M \models \varphi[w]$ for each $w \in W$, i.e. if $V^+(\varphi) = W$. Let Σ be a set of formulas. M is a *model for* Σ if, for each $\varphi \in \Sigma$, $M \models \varphi$ (in symbols, $M \models \Sigma$).

Exercise 0.4. Define V^+ by induction.

Solution. Let $V^+(\bot) = \emptyset$, $V^+(p) = V(p)$, $V^+(\neg\varphi) = W - V^+(\varphi)$ and $V^+(\varphi \wedge \psi) = V^+(\varphi) \cap V^+(\psi)$. As regards the inductive step for $\Box$, $V^+(\Box\varphi) = \{w \in W \ : \text{for each } v \in W, \text{if } wRv \text{ then } v \in V^+(\varphi)\}$.

Because of the importance of the set $V^+(\Box\varphi)$, it is useful to define the following function from $\mathcal{P}(W)$ into $\mathcal{P}(W)$:

$$\tau(X) = \{w \in W \ : \ \text{for each } v \in W, \text{if } wRv \text{ then } v \in X\} \ .$$

Thus we obtain that $V^+(\Box\varphi) = \tau(V^+(\varphi))$.

Frames

Let us consider PC once again. The formulas of PC in which we are more interested are those formulas which are true 'for each assignment'; and since assignments are the only factor on which the truth of a PC-formula depends, we obtain a unique set of significant formulas, that is, the set of tautologies. This is not so in the modal case; in fact the truth of a formula depends on assignments as well as on the relations among these assignments. This gives rise to a multitude of sets of formulas which are independent of valuations (i.e. closed under the rule of substitution) and depending on properties of the relations: we shall have in fact many modal logics. For the same reason, a new class of semantic structures naturally arises, i.e. structures obtained from models by deleting valuations. What remains, called a frame, is the semantic structure mostly referred to in modal logic.

DEFINITION 0.5. A *Kripke frame (frame)* is a pair $\mathbf{F} = \langle W, R \rangle$ where W is a non-empty set and R is a binary relation on W. A *valuation* over $\mathbf{F}$ is a function V from $\mathbf{P}$ into $\mathcal{P}(W)$; a model $\langle W, R, V \rangle$ (also denoted by $\langle \mathbf{F}, V \rangle$) is called a *model* on $\mathbf{F}$. A formula φ is *true* in a frame $\mathbf{F}$ ($\mathbf{F} \models \varphi$ in symbols) if $\langle \mathbf{F}, V \rangle \models \varphi$ for each V over $\mathbf{F}$. $\mathbf{F}$ is a *frame* for a set of formulas Σ ($\mathbf{F} \models \Sigma$ in symbols) if $\mathbf{F} \models \varphi$ for each $\varphi \in \Sigma$.

The easy result that follows is often used in semantic proofs.

THEOREM 0.6. *Let $p \in \mathbf{P}$ and $\psi \in Wff$, and let V, V' be two valuations on $\langle W, R \rangle$ such that $V(p) = V'^{+}(\psi)$. Then, for each $\varphi \in Wff$ and each $w \in W$, $\langle W, R, V \rangle \models \varphi[w]$ iff $\langle W, R, V' \rangle \models \varphi(p/\psi)\,[w]$.*

Proof. By induction on the complexity of φ.

DEFINITION 0.7. Let $w, v \in W$. We write $wR^n v$ if there exist $u_0, \ldots, u_n \in W$ such that $u_0 = w$, $u_n = v$ and $u_i R u_{i+1}$. Moreover, we denote by R^* the transitive closure of R. For each $n \in \omega$ we define

$$S_n(w) = \{v \ : \ wR^n v\}$$

and

$$S^*(w) = \{v \ : \ wR^* v\}.$$

(In particular, $S_0(w) = \{w\}$ and $S_1(w) = \{v \ : \ wRv\}$; if R is transitive, then $S^*(w) = S_1(w)$). We say that a point w is *terminal* if $S_1(w) = \emptyset$ (and therefore $S_n(w)$ and $S^*(w)$ are empty).)

DEFINITION 0.8. A modal formula φ is a *modal translation* of φ^*, where φ^* is a formula expressed in a predicative language with identity and a binary relation symbol R, if, for each frame $\mathbf{F}$, $\mathbf{F} \models \varphi$ iff $\mathbf{F}$ (regarded as a classical predicative model) satisfies φ^*.

Modal translations of the most significant relational properties, together with the more general problem of characterizing the class of those properties which can be modally translated, constitute an important topic of research in contemporary modal logic. We give some examples of relational properties which can be modally translated; in Ex. 2.13 we list some relations which can not be modally translated.

Example 0.9. The reflexivity property ($\varphi^* = \forall x(xRx)$) is modally translated by the formula $\varphi = \Box p \rightarrow p$. This formula is usually denoted by 'T' (names of modal formulas are often esoteric).

Proof. We can immediately verify that if R is reflexive then $\langle W, R \rangle \models \Box p \rightarrow p$. Vice versa, suppose that there exists $w \in W$ such that not wRw. Let v be such that $V(p) = W - \{w\}$. Then $\langle W, R, V \rangle \models \Box p[w]$ and $\langle W, R, V \rangle \not\models p[w]$, and therefore $\langle W, R, V \rangle \not\models \Box p \rightarrow p[w]$ and $\langle W, R \rangle \not\models \Box p \rightarrow p$ (in fact, a formula φ is false in a frame $\mathbf{F}$ iff there exist a valuation V and a point w such that $\langle \mathbf{F}, V \rangle \not\models \varphi[w]$).

Example 0.10. Transitivity $(\forall x \, \forall y \, \forall z (xRy \wedge yRz \rightarrow xRz))$ is modally translated by the formula $\Box p \rightarrow \Box\Box p$ (the name of this formula is '4').

Proof. As regards the non-trivial direction, suppose that R is not transitive; then there exist $x, y, z \in W$ such that xRy, yRz and not xRz. Let V be such that $V(p) = W - \{z\}$. Again we obtain that $\langle W, R, V \rangle \models \Box p[x]$ and $\langle W, R, V \rangle \not\models \Box\Box p[x]$.

Example 0.11. $R = \emptyset$ (i.e. R satisfies $\forall x \, \forall y (\neg(xRy))$) iff $\langle W, R \rangle \models \Box \bot$.

Example 0.12. Seriality $(\forall x \, \exists y (xRy))$ is modally translated by $\neg\Box\bot$ (the name of this formula is '**D**').

Definition 0.8 refers to frames and not to models; in fact, the first two examples above show that it is necessary that valuations be used freely. We shall come back to this subject in Ex. 2.14. For the moment, we show that those examples fail when substituting frames by models.

Exercise 0.13. (i) Find a model $\mathbf{M}$ whose relation is not reflexive and such that $\mathbf{M} \models \Box p \rightarrow p$.

(ii) Find a model $\mathbf{M}$ whose relation is not transitive and such that $\mathbf{M} \models \Box p \rightarrow \Box\Box p$.

Solution. In order to satisfy both $\Box p \rightarrow p$ and $\Box p \rightarrow \Box\Box p$ it is sufficient that $V(p) = W$, independently of R. Moreover, as regards (ii) it is also sufficient to have $V(p) = \emptyset$; in this case in fact, if $\langle W, R, V \rangle \models \Box p[w]$ then w is terminal, and therefore $\langle W, R, V \rangle \models \Box\Box p[w]$. As regards (i), if $V(p) = \emptyset$ then $\langle W, R, V \rangle \models \Box p \rightarrow p$ iff R is a serial relation.

Logics

The first modal logics (or modal calculi) were born in a syntactic ambit, with the operator $\Box$ interpreted as 'to be necessary'. Those logics were obtained by adjoining to PC axioms and rules whose 'meaning' in terms of necessity seemed plausible, and then by deriving theorems from the axioms. In such a context, one of the principal problems was that of the so-called irreducible modalities. More recently, the starting point in defining new logics has been mostly semantic: the axioms considered have been the modal translations of some of the most significant relational properties.

But how, in general, does one define, a modal logic? Once again, the general definition has a semantic origin. In fact (see definitions below), a modal logic is an extension of a minimal logic (named **K** after Kripke) whose set of theorems is exactly the set of the formulas which are true in all models (or, equivalently, in all frames). Logics defined in such a way are called normal, and these logics are the only ones in which we are interested here (for definitions of non-normal logics see Segerberg (1971)).

DEFINITION 0.14. A *propositional normal modal logic* L is a subset of *Wff* containing:

0.14.0. all the propositional tautologies,
0.14.1. the formula $\Box(p \wedge q) \equiv (\Box p \wedge \Box q)$;

and closed under the following rules:

0.14.2. Modus ponens (if $\varphi \in L$ and $\varphi \rightarrow \psi \in L$ then $\psi \in L$),
0.14.3. The rule of substitution (if $\varphi \in L$, $p \in \mathbf{P}$ and $\psi \in Wff$ then $\varphi(p/\psi) \in L$),
0.14.4. Necessitation (if $\varphi \in L$ then $\Box\varphi \in L$).

A logic L is *consistent* if it is a proper subset of *Wff* (or, equivalently, if $\bot \notin L$). Observe that, from 0.14.0 and 0.14.3 it follows that if φ is a propositional tautology then, for each $\psi_0, \ldots, \psi_n \in Wff$, the formula $\varphi(p_0/\psi_0, \ldots, p_n/\psi_n)$ belongs to each logic. It is customary to call these formulas *tautologies*, too.

The minimal modal logic, i.e. the minimal set of formulas satisfying 0.14.0–0.14.4, is named **K**. We shall show in Theorems 0.16 and 3.3 that $\varphi \in \mathbf{K}$ iff $\mathbf{M} \models \varphi$, for each $\mathbf{M}$.

If $\mathbf{M} \models L$, or $\mathbf{F} \models L$, we say that $\mathbf{M}$ is an L-*model* [$\mathbf{F}$ an L-*frame*].

DEFINITION 0.15. Let $\Gamma \subseteq Wff$ and $\varphi \in Wff$. Then $\Gamma \vdash \varphi$ (in words, φ is a *theorem* of Γ) if there exists a proof of φ starting from $\Gamma \cup \{0.14.0, 0.14.1\}$

as set of axioms and $\{0.14.2–0.14.4\}$ as set of derivability rules. In detail:

$\Gamma \vdash \varphi$ if there exists a finite sequence $\{\psi_i \ : \ i \leq n\} \subseteq Wff$ such that, for each i, one of the following (i)–(iv) occurs:

(i) $\psi_i \in \Gamma$,
(ii)   there exist $i', i'' < i$ such that $\psi_{i''} = \psi_{i'} \rightarrow \psi_i$,
(iii) $\psi_i = \Box\psi_j$, for $j < i$,
(iv) $\psi_i = \psi_j(p/\xi)$, for $j < i$.

If L is a logic and $\Gamma \subseteq Wff$ then Γ *axiomatizes* L if $L = \{\varphi \ : \ \Gamma \vdash \varphi\}$. As an instance, **K** is axiomatized by $\Gamma = X \cup \{\Box(p \wedge q) \equiv (\Box p \wedge \Box q)\}$, where X is any set of axioms for PC.

THEOREM 0.16. *For each* **M**, **M** $\models$ **K** *(and therefore, for each* **F**, **F** $\models$ **K**).

Proof. As is well known, in propositional calculi the rule of substitution can be replaced by axiom schemata. So it is sufficient to show that **M** $\models \varphi$, where **M** is any model and φ is any axiom schema, and also that the set $\{\varphi \ : \ \mathbf{M} \models \varphi\}$ is closed under modus ponens and necessitation rule. All of these steps are trivial.

In Theorem 3.3 we shall show that the converse of this result is true, i.e., if **M** $\models \varphi$ for every **M** then $\varphi \in$ **K**.

As is customary in modal logic, we identify a logic L with any set of its axioms; therefore we indifferently write $L \vdash \varphi$ and $\varphi \in L$. We observed that significant modal formulas have individual names; it is customary to denote by $\mathbf{KX}_1 \ldots \mathbf{X}_n$ the logic axiomatized by $\mathbf{K} \cup \{\mathbf{X}_1, \ldots, \mathbf{X}_n\}$. (There are several exceptions: for instance, **KT** and **KT4**, i.e., the logics axiomatized by $K \cup \{\mathbf{T}\}$ and $K \cup \{\mathbf{T}, 4\}$ respectively, are more frequently denoted by **T** and **S4**).

If $\mathbf{K} \vdash \varphi \equiv \psi$ we simply say that φ and ψ are *equivalent*; in fact, for each L, $L \vdash \varphi \equiv \psi$.

THEOREM 0.17. *Each formula φ is equivalent to a formula in normal form, i.e., of the form*

$$\bigvee_{i \in I} \left(\psi_i \wedge \bigwedge_{s \in S} \Diamond\xi_{i,s} \wedge \bigwedge_{k \in K} \Box\sigma_{i,k} \right),$$

where $\psi_i, \xi_{i,s}, \sigma_{i,k}$ are formulas and, for each $i \in I$, ψ_i is a PC-formula.

Proof. The claim trivially follows from the normal disjunctive form theorem for PC, by considering each subformula of φ having the form $\Box\varphi'$ and $\Diamond\varphi'$ as a propositional variable, and considering the equivalences $\neg\Box\varphi' \equiv \Diamond\neg\varphi'$ and $\neg\Diamond\varphi' \equiv \Box\neg\varphi'$.

DEFINITION 0.18. Let L be a consistent logic, $\alpha \leq \omega$ and $\Delta \subseteq Wff_\alpha$; Δ is an (L, α)-*maximal consistent set* if $L \cap Wff_\alpha \subseteq \Delta$ and, for each $\varphi \in Wff_\alpha$, exactly one among φ and $\neg\varphi$ belongs to Δ.

We omit the proof of the following

THEOREM 0.19. *Let W_L^α be the set of all (L, α)-maximal consistent sets for a given consistent L and a given $\alpha \leq \omega$. Then:*

(i) W_L^α *is not empty,*
(ii) $L \cap Wff_\alpha = \cap\{\Delta : \Delta \in W_L^\alpha\}$, *i.e., for each $\varphi \in Wff_\alpha$, $L \vdash \varphi$ iff φ belongs to each (L, α)-maximal consistent set.*
(iii)   *for each $\varphi \in Wff_\alpha$ and each $\Delta \in W_L^\alpha$, $\Box\varphi \in \Delta$ iff for every Δ' such that $\{\psi : \Box\psi \in \Delta\} \subseteq \Delta'$, $\varphi \in \Delta'$.*

Exercise 0.20. Show that, for each consistent L, $|W_L^\omega| = 2^{\aleph_0}$.
Solution. For each $X \subseteq \mathbf{P}$ there exists at least a Δ such that $\Delta \cap \mathbf{P} = X$.

2. GENERAL FRAMES AND FILTRATIONS

General Frames

DEFINITION 1.0. A *general frame* is a triple $\langle W, R, T \rangle$, where $\langle W, R \rangle$ is a frame and T is a subset of $\mathcal{P}(W)$ which satisfies the following conditions:

1.0.0. $\emptyset \in T$,
1.0.1. T is closed under intersection and complementation,
1.0.2. T is closed under the operator τ (see Ex. 0.4).

We say in this case that T is a τ-*field of sets* on $\langle W, R \rangle$.

A model $\mathbf{M} = \langle W, R, V \rangle$ is called a *model on* $\langle W, R, T \rangle$ if, for each $p \in \mathbf{P}$, $V(p) \in T$. In such a case V is said to be a *valuation on* $\langle W, R, T \rangle$.

A formula φ is *true* in a general frame $\langle W, R, T \rangle$ (in symbols $\langle W, R, T \rangle \models \varphi$) if $\langle W, R, V \rangle \models \varphi$ for each V on $\langle W, R, T \rangle$.

Important. For every non-empty set W and every binary relation R on W, the set $T = \mathcal{P}(W)$ satisfies 1.0.0–1.0.2, that is, $\langle W, R, \mathcal{P}(W) \rangle$ is a general frame. Each valuation on $\langle W, R \rangle$ is a valuation on $\langle W, R, \mathcal{P}(W) \rangle$, and so, for each φ, $\langle W, R \rangle \models \varphi$ iff $\langle W, R, \mathcal{P}(W) \rangle \models \varphi$. Therefore, in the following, we regard the class of frames as a subclass of the class of general frames. In other words:

a general frame $\langle W, R, T \rangle$ is a frame iff $T = \mathcal{P}(W)$.

We write $\langle W, R \rangle$ instead of $\langle W, R, \mathcal{P}(W) \rangle$ and use the letter $\mathbf{F}$ to denote general frames, too.

Given a logic L, we denote by $KM(L)$ the class of all L-models (i.e., $KM(L) = \{\mathbf{M} : \mathbf{M} \models L\}$), by $GF(L)$ the class $\{\mathbf{F} : \mathbf{F} \models L\}$ of all L-general frames and by $KF(L)$ the class of all L-frames. Under the above-mentioned identification of frames with certain particular general frames, we have that $KF(L) \subseteq GF(L)$.

The real 'meaning' of the definition of general frames is revealed by the following

Exercise 1.1. (i) Show that conditions 1.0.0–1.0.2 are necessary and sufficient in order to satisfy the following property: if $\langle W, R, V \rangle$ is a model on $\langle W, R, T \rangle$, then, for each $\varphi \in \mathit{Wff}$, $V^+(\varphi) \in T$.

(ii) Show that for each model $\langle W, R, V \rangle$ and each $\alpha \leq \omega$, the set $T_{V,\alpha} = \{V^+(\varphi) : \varphi \in \mathit{Wff}_\alpha\}$ is a τ-field of sets over $\langle W, R \rangle$.

Solution. The claims follow from Ex. 0.4.

DEFINITION 1.2. Let $\alpha \leq \omega$, and let $\mathbf{F}$ be a general frame, $\mathbf{M}$ a model and w a point of $\mathbf{M}$. Then:

$Th_\alpha(\mathbf{F}) = \{\varphi \in \mathit{Wff}_\alpha : \mathbf{F} \models \varphi\}$,
$Th_\alpha(\mathbf{M}) = \{\varphi \in \mathit{Wff}_\alpha : \mathbf{M} \models \varphi\}$ and
$Th_\alpha(w, \mathbf{M}) = \{\varphi \in \mathit{Wff}_\alpha : \mathbf{M} \models \varphi[w]\}$.
$Th_\alpha(w, \mathbf{F}) = \{\varphi \in \mathit{Wff}_\alpha : \langle \mathbf{F}, V \rangle \models \varphi[w] \text{ for each } V \text{ on } \mathbf{F}\}$.

We shall write Th instead of Th_ω. From the various definitions of truth it follows that, for each $\alpha \leq \omega$, $Th_\alpha(\mathbf{M}) = \cap\{Th_\alpha(w, \mathbf{M}) : w \in W\}$ and $Th_\alpha(\mathbf{F}) = \cap\{Th_\alpha(\mathbf{M}) : \mathbf{M} \text{ is a model on } \mathbf{F}\} = \cap\{Th_\alpha(w, \mathbf{F}) : w \in W\}$.

THEOREM 1.3. *Let* $\mathbf{M} = \langle W, R, V \rangle$ *be a model on* $\mathbf{F} = \langle W, R, T \rangle$ *and* $w \in W$. *Then:*

(i) $Th(\mathbf{F})$ *is a logic.*
(ii) $Th(\mathbf{M})$ *is not, in general, a logic;*
(iii) $Th(w, \mathbf{F})$ *is not, in general, a logic;*
(iv) $Th(w, \mathbf{M})$ *is never a logic.*

Proof. (i) We observe immediately that $Th(\mathbf{F})$ satisfies conditions 0.14.0–2. Moreover it satisfies 0.14.3. Suppose that $\varphi(p/\psi) \notin Th(\mathbf{F})$. Then there exist a V on $\mathbf{F}$ and a $w \in W$ such that $\langle \mathbf{F}, V \rangle \not\models \varphi(p/\psi)[w]$. Let us consider $V^+(\psi)$; by Ex. 1.1, we have that $V^+(\psi) \in T$, and therefore there exists a valuation V' on $\mathbf{F}$ such that $V'(p) = V^+(\psi)$. Then, by Theorem 0.6 we obtain that $\langle \mathbf{F}, V' \rangle \not\models \varphi[w]$ and so $\varphi \notin Th(\mathbf{F})$. Finally, we show that, for each $\mathbf{M}$, $Th(\mathbf{M})$ satisfies 0.14.4, from which it immediately follows that $Th(\mathbf{F})$ also satisfies 0.14.4. In fact, if $\Box\varphi \notin Th(\mathbf{M})$ then $\Box\varphi \notin Th(w, \mathbf{M})$ for some $w \in W$, and hence $\varphi \notin Th(v, \mathbf{M})$ for some $v \in W$ such that wRv. So $\varphi \notin Th(\mathbf{M})$.

(ii) $Th(\mathbf{M})$ satisfies conditions 0.14.0–2; but, in general, it is not closed under the rule of substitution. Suppose, for instance, that $V(p) = W$ for a certain $p \in \mathbf{P}$. Let $\varphi = p$; then $\varphi \in Th(\mathbf{M})$ but $\varphi(p/\neg p)$ (i.e. $\neg p$) does not belong to $Th(\mathbf{M})$. We have shown in (i) that $Th(\mathbf{M})$ satisfies 0.14.4.

(iii) $Th(w, \mathbf{F})$ satisfies 0.14.0–3 but, in general, does not satisfy 0.14.4.

(iv) $Th(w, \mathbf{M})$ satisfies 0.14.0–2, but it is never closed under the substitution rule.

An example of a model $\mathbf{M}$ such that $Th(\mathbf{M})$ is a logic is given by the canonical model of a logic (see Definition 3.0); another example is in Ex. 3.4. Moreover, examples of points w such that $Th(w, \mathbf{F})$ is a logic are given in Ex. 1.11.

Exercise 1.4. Show that if $\mathbf{M}$ is finite then $Th(\mathbf{M})$ is not a logic.

    Solution. Assume $W = \{w_0, \ldots, w_n\}$ and consider the formula $\varphi = \bigvee\{p_i^* : i \leq n\}$, where p_i^* is p_i if $p_i \in V^{-1}(w_i)$, $\neg p_i$ otherwise. Thus $\mathbf{M} \models \varphi$ holds, but φ cannot belong to any consistent logic.

An obvious consequence of Theorem 1.3 is the following

REMARK 1.5. *Let Γ be a set of axioms for a logic L. $\mathbf{M} \models \Gamma$ does not imply that $\mathbf{M} \models L$, whereas $\mathbf{F} \models \Gamma$ implies that $\mathbf{F} \models L$.*

REMARK 1.6. *Let $\langle W, R \rangle$ be a frame, T a τ-field of sets over $\langle W, R \rangle$ and V a valuation on $\langle W, R, T \rangle$. Then:*

(i)   For each $\alpha \leq \omega$, $Th_\alpha(\langle W, R \rangle) \subseteq Th_\alpha(\langle W, R, T \rangle) \subseteq Th_\alpha(\langle W, R, V \rangle)$;

(ii) $Th_0(\langle W, R \rangle) = Th_0(\langle W, R, T \rangle) = Th_0(\langle W, R, V \rangle)$.

    Proof. (i): obvious from the definitions.

(ii) By definition, $\varphi \in Wff_0$, i.e. φ is a 0-formula, iff no propositional variable appears in φ. Hence, Wff_0 can be inductively defined as follows: $\bot \in Wff_0$; if $\varphi, \psi \in Wff_0$ then $\neg\varphi, \Box\varphi, \varphi \wedge \psi \in Wff_0$; nothing else. Obviously, the truth of a 0-formula in a model does not depend upon the valuation.

Let $\langle W, R \rangle$ be a frame. As shown in Ex. 1.1 (ii), for each V on $\langle W, R \rangle$ and each $\alpha \leq \omega$, $T_{V,\alpha} = \{V^+(\varphi) : \varphi \in Wff_\alpha\}$ is a τ-field of sets over $\langle W, R \rangle$. But, as observed, if $\varphi \in Wff_0$, then V is inessential in order to determine $\{w : \langle W, R, V \rangle \models \varphi[w]\}$ (i.e. $V^+(\varphi)$); in fact, $V^+(\varphi)$ is equal to $\{w \in W : F \models \varphi[w]\}$. To avoid heavy notation, we still denote this set by $V^+(\varphi)$, even if it depends only upon $\langle W, R \rangle$. But we denote $T_{V,0}$ by T_0, i.e. we let $T_0 = \{V^+(\varphi) : \varphi \in Wff_0\}$. By Ex. 1.1 (ii), T_0 is a τ-field of sets over $\langle W, R \rangle$. Moreover:

Exercise 1.7. Show that, for each $\langle W, R \rangle$, T_0 is the minimal τ-field of sets over $\langle W, R \rangle$ (i.e. $T_0 = \cap\{T' : T'$ is a τ-field of sets over $\langle W, R \rangle\}$).
  Solution. The claim follows from the inductive definition of Wff_0 given in Remark 1.6 (ii) and from Ex. 1.1.

Exercise 1.8. Find necessary and sufficient conditions on a frame $\langle W, R \rangle$ in order that $T = \{\emptyset, W\}$ be a τ-field of sets.
  Solution. From Ex. 1.7 it follows that $\{\emptyset, W\}$ is a τ- field of sets iff $\{\emptyset, W\} = T_0$. This happens iff either (i) $\tau(\emptyset) = \emptyset$ or (ii) $\tau(\emptyset) = W$. Moreover, (i) holds iff, for each $w \in W$, $S_1(w) \neq \emptyset$, and (ii) holds iff for each $w \in W$, $S_1(w) = \emptyset$ (i.e. iff $R = \emptyset$).

Exercise 1.9. Find, for each $n < \omega$, a frame $\langle W, R \rangle$, with $|W| = n$, which admits exactly one τ-field of sets.
  Solution. For each $\langle W, R \rangle$, $\mathcal{P}(W)$ and T_0 are τ-fields of sets; the former is the largest one and (see Ex. 1.7) the latter is the smallest one. Therefore $\langle W, R \rangle$ admits exactly one τ-field of sets iff $T_0 = \mathcal{P}(W)$, i.e., since W is finite, iff for each $w \in W$ there exists a 0-formula φ such that $V^+(\varphi) = \{w\}$. Let $W = \{m : m < n\}$ and $R = \{\langle m, m' \rangle : m' < m\}$. For each $m < n$ let us consider the 0-formula $\Box^m \bot$. Obviously, $w \models \Box^m \bot$ iff $S_m(w) = \emptyset$; thus, in our case, $V^+(\Box^m \bot) = \{0, \ldots, m - 1\}$. Let us define the following 0-formula:

$$\text{lev}(m) = \Box^{m+1} \bot \wedge \neg\Box^m \bot .$$

For each $m < n$, $V^+(\text{lev}(m)) = \{m\}$.

Exercise 1.10. Show that $Th(w, \mathbf{M})$ is closed under the necessitation rule iff, for each $v \in S_1(w)$, $Th(w, \mathbf{M}) = Th(v, \mathbf{M})$.
Solution. Trivial.

Exercise 1.11. Find sufficient conditions on $S_1(w)$ in order that $Th(w, \mathbf{F})$ be a logic.
  Solution. If either $S_1(w) = \emptyset$ (i.e. w is terminal) or $S_1(w) = \{w\}$, then $Th(w, \mathbf{F})$ is a logic. In the former case, $\Box\varphi \in Th(w, \mathbf{F})$ for each φ, and, in the latter case, $\Box\varphi \in Th(w, \mathbf{F})$ iff $\varphi \in Th(w, \mathbf{F})$; in both cases $Th(w, \mathbf{F})$ is closed under the necessitation rule.

Filtrations

DEFINITION 1.12. Given a model $\mathbf{M} = \langle W, R, V \rangle$ and a set Σ of formulas closed under subformulas, we say that a model $\mathbf{M}' = \langle W_\Sigma, R', V_\Sigma \rangle$ is a *filtration of* $\mathbf{M}$ *through* Σ if:

1.12.0. $W_\Sigma = \{[w]_\equiv \ : \ w \in W\}$, where $\equiv$ is the following equivalence relation on W: $w \equiv w'$ if, for each $\varphi \in \Sigma$, $\mathbf{M} \models \varphi[w]$ iff $\mathbf{M} \models \varphi[w']$.
1.12.1. wRv implies $[w]_\equiv R'[v]_\equiv$.
1.12.2. $[w]_\equiv R'[v]_\equiv$ implies that, for each $\Box\varphi \in \Sigma$, if $\mathbf{M} \models \Box\varphi[w]$ then $\mathbf{M} \models \varphi[v]$.
1.12.3. $V_\Sigma(p_i) = \{[w]_\equiv \ : \ w \in V(p_i)\}$ if $p_i \in \Sigma$, $V_\Sigma(p_i) = \emptyset$ otherwise.

  From condition 1.12.0 it follows that

1.12.4. $|W_\Sigma| \leq 2^{|\Sigma|}$.

  Observe that, while 1.12.0 and 1.12.3 respectively define W_Σ and V_Σ, 1.12.1 and 1.12.2 are conditions that a certain R' must satisfy. We show that there exists at least one such R', i.e.:

THEOREM 1.13. *For each* $\mathbf{M}$ *and each* Σ *closed under subformulas, there exists at least one* R' *such that* $\mathbf{M} = \langle W_\Sigma, R', V_\Sigma \rangle$ *is a filtration of* $\mathbf{M}$ *through* Σ.
  Proof. Let us define R' as follows:

1.13.0. $[w]_\equiv R'[v]_\equiv$ iff wRv for some $w' \in [w]_\equiv$ and $v' \in [v]_\equiv$.

Obviously R' satisfies 1.12.2. As regards 1.12.3, suppose that $[w]_\equiv R'[v]_\equiv$, $\Box\varphi \in \Sigma$ and $\mathbf{M} \models \Box\varphi[w]$; $[w]_\equiv R'[v]_\equiv$ implies that $w'Rv'$ for some $w' \in [w]_\equiv$ and $v' \in [v]_\equiv$. Hence, by definition of $\equiv$, $\mathbf{M} \models \Box\varphi[w']$, which implies $\mathbf{M} \models \varphi[v']$ and, since $\varphi \in \Sigma$, $\mathbf{M} \models \varphi[v]$.

The filtration $\mathbf{M}' = \langle W_\Sigma, R', V_\Sigma \rangle$, where R' is defined as in 1.13.0 is called the *finest* filtration of $\mathbf{M}$ through Σ. We shall see that, in general, there are filtrations that are different from the finest filtration.

The importance of filtrations lies in the following theorem.

THEOREM 1.14. (Filtration Theorem). *Let* $\mathbf{M}'$ *be a filtration of* $\mathbf{M}$ *through* Σ. *For each* $\varphi \in \Sigma$ *and each* $w \in W$ *it holds that*

$$\mathbf{M}' \models \varphi[[w]_\equiv] \quad \text{iff} \quad \mathbf{M} \models \varphi[w] \,,$$

and therefore $Th(\mathbf{M}') \cap \Sigma = Th(\mathbf{M}) \cap \Sigma$.
    Proof. By induction on the construction of φ.

Filtrations will be used in Section 3 to prove the finite model property for certain logics. Another important consequence of the filtration thoerem is given in Theorem 2.1.

3. RELATIONS BETWEEN STRUCTURES

Relations between $KF(L)$, $KM(L)$ *and* $GF(L)$

DEFINITION 2.0. A model $\mathbf{M} = \langle W, R, V \rangle$ is α-*distinguishable* (for $\alpha \leq \omega$) if, for each $w, v \in W$, $w \neq v$ implies that $Th_\alpha(w, \mathbf{M}) \neq Th_\alpha(v, \mathbf{M})$. In the case $\alpha = \omega$, $\mathbf{M}$ is said to be *distinguishable* instead of ω-distinguishable. Obviously, if $\mathbf{M}$ is α-distinguishable then it is β-distinguishable for each $\alpha \leq \beta \leq \omega$.

THEOREM 2.1. *For each* $\mathbf{M}$ *and each* $\alpha \leq \omega$ *there exists an* α-*distinguishable* $\mathbf{M}'$ *such that* $Th_\alpha(\mathbf{M}) = Th_\alpha(\mathbf{M}')$.
    Proof. The claim follows from Theorem 1.14, letting $\Sigma = Wff_\alpha$.

THEOREM 2.2. *For each finite general frame* $\langle W, R, T \rangle$ *there exists a finite frame* $\langle W', R' \rangle$ *such that* $Th(\langle W, R, T \rangle) = Th(\langle W', R' \rangle)$.

Proof. Let $\equiv$ be the following equivalence relation on W: $w \equiv v$ iff, for each $X \in T$, $w \in X$ iff $v \in X$. By definition of a valuation on $\langle W, R, T \rangle$, if $w \equiv v$ then $V^{-1}(w) = V^{-1}(v)$ for each V. We define $\langle W', R' \rangle$ as follows: $W' = \{[w]_\equiv \ : \ w \in W\}$, and $[w]_\equiv R'[v]_\equiv$ iff there exist $w' \equiv w$ and $v' \equiv v$ such that $w'Rv'$. Since W is finite, $[w]_\equiv \in T$, for each $w \in W$: in fact, by definition, $[w]_\equiv = \cap\{X \in T \ : \ w \in X\}$, and T is closed under finite intersections. Therefore, the function f, from the set of all valuations on $\langle W, R, T \rangle$ into the set of all valuations on $\langle W', R' \rangle$, and defined as follows:

$$f(V) = V' \, , \quad \text{where } V'(p) = \{[w]_\equiv \ : \ w \in V(p)\} \, ,$$

is readily seen to be a bijection. Finally, it is easy to show, by induction on the construction of φ, that, for each V and w, $\langle W, R, V \rangle \models \varphi[w]$ iff $\langle W', R', V' \rangle \models \varphi[[w]_\equiv]$. This, together with the fact that f is bijective, implies that $Th(\langle W, R, T \rangle) = Th(\langle W', R' \rangle)$.

THEOREM 2.3. *Let $\langle W, R, V \rangle$ be a finite distinguishable model. Then, for each logic L, $\langle W, R, V \rangle \models L$ iff $\langle W, R \rangle \models L$.*

Proof. Obviously $\langle W, R \rangle \models L$ implies that $\langle W, R, V \rangle \models L$. To show the converse, let $W = \{w_0, \ldots, w_m\}$. Since $\langle W, R, V \rangle$ is distinguishable, for each $i, j \leq m$ and $i \neq j$ there exists a formula $\xi_{i,j}$ such that $w_i \models \xi_{i,j}$ and $w_j \not\models \xi_{i,j}$. Therefore, defining $\xi_i = \bigwedge(\xi_{i,j} : j \neq i < m)$, we obtain that $V^+(\xi_i) = \{w_i\}$, for each $i \leq m$. Now suppose that $\langle W, R \rangle \not\models L$ and let $\varphi(p_0, \ldots, p_{\alpha-1}) \in L$ be such that $\langle W, R, V' \rangle \not\models \varphi(p_0, \ldots, p_{\alpha-1}) [w]$, for some V' on $\langle W, R \rangle$ and $w \in W$. For each $\beta < \alpha$ let $\psi_\beta = \bigvee(\xi_i : w_i \in V'(p_\beta))$. By Theorem 0.6 we obtain that $\langle W, R, V \rangle \not\models \varphi(p_0/\psi_0, \ldots, p_{\alpha-1}/\psi_{\alpha-1}) [w]$. But, as L is closed under the rule of substitution, $\varphi(p_0/\psi_0, \ldots, p_{\alpha-1}/\psi_{\alpha-1}) \in L$ and therefore $\langle W, R, V \rangle \not\models L$.

THEOREM 2.4. *For each logic L and each model $\langle W, R, V \rangle$, $\langle W, R, V \rangle \models L$ iff $\langle W, R, T_V \rangle \models L$, where, see Ex. 1.1. (ii), $T_V = T_{V,\omega} = \{X \in W \ : \ X = \{w \ : \ \langle W, R, V \rangle \models \varphi[w]\} \text{ for } \varphi \in \text{Wff}\}$.*

Proof. Obviously $Th(\langle W, R, T_V \rangle) \subseteq Th(\langle W, R, V \rangle)$, and therefore $\langle W, R, T_V \rangle \models L$ implies that $\langle W, R, V \rangle \models L$. As regards the non-trivial direction, suppose $\langle W, R, T_V \rangle \not\models GF(L)$, and let $\varphi(p_0, \ldots, p_{\alpha-1})$ be a theorem of L which is not true in $\langle W, R, T_V \rangle$. Then there exists a valuation V' on $\langle W, R, T_V \rangle$ and a $w \in W$ such that $\langle W, R, V' \rangle \not\models \varphi(p_0, \ldots, p_{\alpha-1}) [w]$. By definition of T_V, for each $i < \alpha$ there exists a formula ψ_i such that $V'(p_i) = V^+(\psi_i)$. So, by Theorem 0.6, $\langle W, R, V \rangle \not\models \varphi(p_0/\psi_0, \ldots, p_{\alpha-1}/\psi_{\alpha-1})$. But L is closed under the rule of substitution and hence $\varphi(p_0/\psi_0, \ldots, p_{\alpha-1}/\psi_{\alpha-1}) \in L$ and

$\langle W, R, V \rangle \notin KM(L)$.

Subframes and Submodels

DEFINITION 2.5. $\mathbf{F}' = \langle W', R', T' \rangle$ is a *subframe* of the general frame $\mathbf{F} = \langle W, R, T \rangle$ ($\mathbf{F}' \subseteq \mathbf{F}$ in symbols) if

2.5.0. $W' \subseteq W$,
2.5.1. $R' = R \cap (W')^2$,
2.5.2. $T' = \{X \cap W' : X \in T\}$.

$\mathbf{F}'$ is a *generated subframe* of $\mathbf{F}$ (in symbols $\mathbf{F}' \subseteq_+ \mathbf{F}$) if $\mathbf{F}' \subseteq \mathbf{F}$ and

2.5.3. if $w \in W'$ and wRv then $v \in W'$.

$\mathbf{M}' = \langle W', R', V' \rangle$ is *submodel [generated submodel]* of $\mathbf{M} = \langle W, R, V \rangle$ (in symbols $\mathbf{M}' \subseteq \mathbf{M}$ [$\mathbf{M}' \subseteq_+ \mathbf{M}$]) if 2.5.0 and 2.5.1 [2.5.0, 2.5.1 and 2.5.3] hold, and

2.5.4. $V'(p) = V(p) \cap W'$.

Let $\mathbf{F}$ be $\langle W, R, T \rangle$ and let $X \subseteq W$ ($X \neq \emptyset$). We denote by $\mathbf{F} \restriction X$ the subframe $\langle X, R \restriction X, T \restriction X \rangle$ and by $\mathbf{F}/X$ the minimal generated subframe of $\mathbf{F}$ including X, i.e. the general frame $\mathbf{F} \restriction W'$, where $W' = X \cup \bigcup (S^*(w) : w \in X)$.

Exercise 2.6. Show that if $\mathbf{F}$ is a frame and $\mathbf{F}' \subseteq \mathbf{F}$ then $\mathbf{F}'$ is a frame.
    Solution. Since $T = \mathcal{P}(W)$, then, by 2.5.2, $T' = \mathcal{P}(W')$, and hence $\mathbf{F}'$ is a frame.

THEOREM 2.7. (Generation Theorem). *Let* $\mathbf{F}' \subseteq_+ \mathbf{F}$ *and* $\mathbf{M}' \subseteq_+ \mathbf{M}$. *Then*

(i) $Th(w, \mathbf{M}') = Th(w, \mathbf{M})$, *for each* $w \in W'$;
(ii) $Th(\mathbf{M}) \subseteq Th(\mathbf{M}')$;
(iii) $Th(\mathbf{F}) \subseteq Th(\mathbf{F}')$.

    Proof. Trivial.

An immediate consequence of Ex. 2.5.b and Theorem 2.7 is the following

COROLLARY 2.8. *For each L, we have that* $KM(L), GF(L)$ *and* $KF(L)$

are closed under generated submodels and generated subframes, respectively.

The previous result fails if referred to submodel and subframes which are not generated, as is shown in the following

Exercise 2.9. Find logics whose classes of frames are closed under subframes, and logics whose classes of frames are not closed under subframes.

Solution. (1) $GF(\mathbf{KT})$ is closed under subframes. (2) $GF(L)$, where $L = \mathbf{K} \cup \{\Box\bot \vee \Diamond\Box\bot\}$, is not closed under subframes. The frame $\mathbf{F} = \langle W, R \rangle$, with $W = \{w_0, w_1\}$ and $R = \{\langle w_1, w_1 \rangle, \langle w_1, w_0 \rangle\}$, is an L-frame; in fact $\mathbf{F} \models \Box\bot[w_0]$ and $\mathbf{F} \models \Diamond\Box\bot[w_1]$. By contrast, $\mathbf{F}' = \mathbf{F} \restriction \{w_1\}$ is not an L-frame, since $\mathbf{F}' \not\models \Box\bot \vee \Diamond\Box\bot[w_1]$.

p-Morphisms

DEFINITION 2.10. $\mathbf{F}' = \langle W', R', T' \rangle$ is a *p-morphic image* of $\mathbf{F} = \langle W, R, T \rangle$ (in symbols $\mathbf{F} \to \mathbf{F}'$) if there exists a function f from W to W' such that:

2.10.0. f is onto,
2.10.1. wRv implies that $f(w) R' f(v)$,
2.10.2. $f(w) R' f(v)$ implies that there exists $u \in W$ such that $f(u) = f(v)$ and wRu,
2.10.3. $X \in T'$ implies that $f^{-1}[X] \in T$.

Such a function f is called a *p-morphism*. Observe that if $\mathbf{F}$ is a frame then 2.10.3 is always satisfied. Moreover, if f is a one-to-one function, then f is called an *isomorphism* (and denoted by $\mathbf{F} \cong \mathbf{F}'$); in such a case 2.10.2 becomes: $f(w) R' f(v)$ implies wRv. A model $\mathbf{M} = \langle W, R, V \rangle$ is a *p-morphic image* of $\mathbf{M}' = \langle W', R', V' \rangle$ (in symbols $\mathbf{M} \to \mathbf{M}'$) if $\langle W, R \rangle \to \langle W', R' \rangle$ and

2.10.4. $w \in V(p)$ iff $f(v) \in V'(p)$, for each $p \in \mathbf{P}$.

THEOREM 2.11. (*p-Morphism Theorem*). *Let f be a p-morphism from* $\mathbf{M} = \langle F, V \rangle$ *onto* $\mathbf{M}' = \langle F', V' \rangle$. *Then*
(i)   *for each $w \in W$, $Th(w, \mathbf{M}) = Th(f(w), \mathbf{M}')$,*

(ii) $Th(\mathbf{M}) = Th(\mathbf{M}')$,

(iii) $Th(\mathbf{F}) \subseteq Th(\mathbf{F}')$.

Proof. (i) By induction on the construction of the formulas. (ii) From (i) and 2.10.0. (iii) For each V' on $\mathbf{F}'$ let V'' be the function from $\mathbf{P}$ into $\mathcal{P}(W)$ such that, for each $p \in \mathbf{P}$, $V''(p) = f^{-1}[V'(p)]$. By definition of a valuation, $V'(p) \in T'$, and hence, by 2.10.3, $V''(p) \in T$, thus obtaining that V'' is a valuation on $\mathbf{F}$ and $\langle \mathbf{F}, V'' \rangle \rightarrow \langle \mathbf{F}', V' \rangle$. Since this happens for each V' on $\mathbf{F}'$, (iii) follows from (ii).

Observe that in (iii) we have $Th(\mathbf{F}) \subseteq Th(\mathbf{F}')$ instead of $Th(\mathbf{F}) = Th(\mathbf{F}')$ because if $w \neq v$, $f(w) = f(v)$ and V is a valuation on $\mathbf{F}$ such that $V^{-1}(w) \neq V^{-1}(v)$ then there is no V' on $\mathbf{F}$ such that $\langle \mathbf{F}, V \rangle \rightarrow \langle \mathbf{F}', V' \rangle$. Obviously, if $\mathbf{F}$ and $\mathbf{F}'$ are isomorphic, then $Th(\mathbf{F}) = Th(\mathbf{F}')$.

COROLLARY 2.12. *For every L, $KM(L)$ and $GF(L)$ are closed under p-morphic images. $KF(L)$ is closed under p-morphic images if it is considered as a subclass of the class of all frames.*

(The fact that $\mathbf{F} \rightarrow \mathbf{F}'$, and $\mathbf{F}$ is a frame, does not imply that $\mathbf{F}'$ is a frame.)

Exercise 2.13. Using Corollary 2.12, show that antireflexivity, antisymmetry and antitransitivity can not be modally translated.

Solution. Consider the frame $\mathbf{F}_{00} = \langle \{w_{00}\}, \{\langle w_{00}, w_{00} \rangle\} \rangle$. $\mathbf{F}_{00}$ is p-morphic image of the frame $\langle \omega, \text{Suc} \rangle$ (the natural numbers ordered by the successor relation). $\langle \omega, \text{Suc} \rangle$ satisfies antireflexivity, antisymmetry and antitransitivity. Suppose, by contradiction, that φ is a modal translation of one of these properties and let L be the logic which is axiomatized by $K \cup \{\varphi\}$; then $\langle \omega, \text{Suc} \rangle \in KF(L)$ and hence, by Corollary 2.12, $\mathbf{F}_{00} \in KF(L)$. But $\mathbf{F}_{00}$ satisfies none of those properties.

Exercise 2.14. Show that for each reflexive frame $\mathbf{F}$ there exists an antireflexive general frame $\mathbf{F}'$ such that $Th(\mathbf{F}) = Th(\mathbf{F}')$.

Proof. Let $\mathbf{F} = \langle W, R \rangle$ be a reflexive frame. We define $\mathbf{F}_{\mathfrak{D}} = \langle W_{\mathfrak{D}}, R_{\mathfrak{D}}, T_{\mathfrak{D}} \rangle$ as follows:

$W_{\mathfrak{D}} = W \times \{0\} \cup W \times \{1\}$ (we write w_0 and w_1 instead of $\langle w, 0 \rangle$ and $\langle w, 1 \rangle$);

$R_{\mathfrak{D}} = \{\langle w_0, w_1 \rangle, \langle w_1, w_0 \rangle : w \in W\} \cup \{\langle w_0, v_0 \rangle, \langle w_1, v_1 \rangle: \langle w, v \rangle \in R \text{ and } w \neq v\}$.

$T_{\mathfrak{D}} = \{X \subseteq W_{\mathfrak{D}} : w_0 \in X \text{ iff } w_1 \in X, \text{ for each } w \in W\}$ (i.e., $T_{\mathfrak{D}} = \mathcal{P}(K)$, where $K = \{\{w_0, w_1\} : w \in W\}$).

It can immediately be verified that $R_{\mathfrak{D}}$ is antireflexive and that $T_{\mathfrak{D}}$ is a τ-field of sets. Let us consider the following function Φ from the set of all valuations on $\mathbf{F}_{\mathfrak{D}}$ into the set of all valuations on $\mathbf{F}$:

$$\Phi(V) = V', \quad \text{where } V'(p) = \{w \in W \; : \; w_0 \in V(p)\} \, .$$

Since, by definition of $T_{\mathfrak{D}}$, $w_0 \in V(p)$ iff $w_1 \in V(p)$, it follows that Φ is a bijection between the set of all valuations on $\mathbf{F}_{\mathfrak{D}}$ and the set of all valuations on $\mathbf{F}$. Let f be the function from $\mathbf{F}_{\mathfrak{D}}$ onto $\mathbf{F}$ such that $f(w_i) = w$, for each $w \in \mathbf{F}$ and $i \in \{0, 1\}$. It is straightforward to show that, for each V on $\mathbf{F}_{\mathfrak{D}}$, f is a p-morphism from $\langle \mathbf{F}_{\mathfrak{D}}, V \rangle$ to $\langle \mathbf{F}, \Phi(V) \rangle$. Thus, by Theorem 2.11 (ii), $Th(\langle \mathbf{F}_{\mathfrak{D}}, V \rangle) = Th(\langle \mathbf{F}, \Phi(V) \rangle)$, and, since Φ is a bijection, $Th(\mathbf{F}_{\mathfrak{D}}) = Th(\mathbf{F})$.

DEFINITION 2.15. A general frame $\mathbf{F}$ [a model $\mathbf{M}$] is g-reducible to a general frame $\mathbf{F}'$ [a model $\mathbf{M}'$] if there exists $\mathbf{F}''[\mathbf{M}'']$ such that $\mathbf{F}'' \subseteq_+ \mathbf{F}$ and $\mathbf{F}'' \rightarrow \mathbf{F}' \, [\mathbf{M}'' \subseteq_+ \mathbf{M} \text{ and } \mathbf{M}'' \rightarrow \mathbf{M}']$.

A immediate consequence of Corollaries 2.8 and 2.12 is given by the following

COROLLARY 2.16. *For each L, $KM(L)$, $GF(L)$ and $KF(L)$ are closed under g-reducibility.*

THEOREM 2.17. *Consider $\mathbf{F}_{00}$ (see Ex. 2.13) and the frame $\mathbf{F}_0 = \langle \{w_0\}, \emptyset \rangle$. For each general frame $\mathbf{F}$, $\mathbf{F}$ is g-reducible either to $\mathbf{F}_0$ or to $\mathbf{F}_{00}$.*
  Proof. If $\mathbf{F}$ contains a terminal point w, then $\mathbf{F} \restriction \{w\}$ is a generated subframe of $\mathbf{F}$, and, obviously, $\mathbf{F} \restriction \{w\} \cong \mathbf{F}_0$. Hence suppose that $\mathbf{F}$ is without terminal points, i.e., R is a serial relation. Then it is straightforward to show that the function f such that $f(w) = w_{00}$, for each $w \in W$, is a p-morphism from $\mathbf{F}$ to $\mathbf{F}_{00}$.

Disjoint Union

DEFINITION 2.18. The *disjoint union* $\Sigma(\mathbf{F}_i \; : \; i \in I)$ of the set of general frames $\{\mathbf{F}_i = \langle W_i, R_i, T_i \rangle \; : \; i \in I\}$ is the general frame $\mathbf{F} = \langle W, R, T \rangle$ where:

$W = \{\langle w, i \rangle \; : \; w \in W_i\}$,
$R = \{\langle \langle w, i \rangle, \langle i \rangle \rangle \; : \; \langle w, v \rangle \in R_i\}$,
$T = \{X \subseteq W \; : \; \{w \; : \text{for each } i \; \langle w, i \rangle \in X\} \in T_i\}$.

The *disjoint union* $\Sigma(\mathbf{M}_i : i \in I)$ of the set of models $\{\mathbf{M}_i = \langle W_i, R_i, V_i \rangle : i \in I\}$ is the model $\mathbf{M} = \langle W, R, V \rangle$ where W and R are defined as above and, for each p, $V(p) = \{\langle w, i \rangle : w \in V_i(p)\}$.

THEOREM 2.19. *Let* $\mathbf{F} = \Sigma(\mathbf{F}_i : i \in I\}$ *and* $\mathbf{M} = \Sigma(\mathbf{M}_i : i \in I\}$. *Then*

(i)   *For each* $i \in I$ *and each* $w \in W_i$, $Th(w, \mathbf{F}_i) = Th(\langle w, i \rangle, \mathbf{F})$ *and* $Th(w, \mathbf{M}_i) = Th(\langle w, i \rangle, \mathbf{M})$,

(ii) $Th(\mathbf{F}) = \cap(Th(\mathbf{F}_i) : i \in I)$ *and* $Th(\mathbf{M}) = \cap(Th(\mathbf{M}_i) : i \in I)$.

Proof. Clearly, for each i, $\mathbf{F}_i \cong \mathbf{F} \restriction \{\langle w, i \rangle : w \in W_i\} \subseteq_+ \mathbf{F}$. Analogously for the models. Thus (i), and therefore (ii), follows from Theorems 2.7 and 2.11.

COROLLARY 2.20. *For each* L, $KM(L)$, $GF(L)$ *and* $KF(L)$ *are closed under disjoint unions.*

   Proof. As regards $GF(L)$ and $KM(L)$ the result follows from Theorem 2.19 (ii). Moreover, if each $\mathbf{F}_i$ is a frame then $T = \mathcal{P}(W)$ and hence $\mathbf{F}$ is a frame.

3. CANONICAL STRUCTURES AND COMPLETENESS

Canonical Models and Canonical Frames

DEFINITION 3.0. For each $\alpha \leq \omega$ and each consistent logic L, the α-*canonical model* for L is the model

$$\mathbf{M}_L^\alpha = \langle W_L^\alpha, R_L^\alpha, V_L^\alpha \rangle$$

defined as follows:

$W_L^\alpha = \{\Delta : \Delta \text{ is a } (L, \alpha)\text{-maximal consistent set}\}$

(see Definition 0.18; as shown in Theorem 0.19, $W_L^\alpha \neq \emptyset$);

$R_L^\alpha = \{\langle \Delta, \Delta' \rangle: \text{for each } \varphi \in Wff_\alpha, \Box\varphi \in \Delta \text{ implies } \varphi \in \Delta'\}$;

$V_L^\alpha(p) = \{\Delta \in W_L^\alpha : p \in \Delta\}$

(observe that $\beta \geq \alpha$ implies that $V_L^\alpha(p_\beta) = \emptyset$).

If $\alpha = \omega$, then $\mathbf{M}_L^\omega$ is denoted by $\mathbf{M}_L$ and called the *canonical model* for

L. The models $\mathbf{M}_L^\alpha$ for $\alpha < \omega$ are called *weak canonical models*.

LEMMA 3.1. (Fundamental Lemma). *For each* $\alpha \le \omega$, *each* $\varphi \in Wff_\alpha$ *and each* $\Delta \in W_L^\alpha$,

$$\mathbf{M}_L^\alpha \models \varphi[\Delta] \text{ iff } \varphi \in \Delta .$$

Proof. By induction on the complexity of φ. By definition of V_L^α, we have that $\mathbf{M}_L^\alpha \models p[\Delta]$ iff $p \in \Delta$. The inductive steps for $\neg$ and $\wedge$ are obvious. Suppose that $\Box\varphi \in \Delta$ and $\Delta R_L^\alpha \Delta'$. By definition of R_L^α we obtain that $\varphi \in \Delta'$ and, by inductive hypothesis, $\mathbf{M}_L^\alpha \models \varphi[\Delta']$. Thus $\mathbf{M}_L^\alpha \models \Box\varphi[\Delta]$. Suppose now that $\mathbf{M}_L^\alpha \models \Box\varphi[\Delta]$; then, for each Δ' such that $\Delta R_L^\alpha \Delta'$, $\mathbf{M}_L^\alpha \models \varphi[\Delta']$ and, by hypothesis, $\varphi \in \Delta'$. Therefore, by Theorem 0.19 (iii), $\Box\varphi \in \Delta$.

THEOREM 3.2. (Fundamental Theorem). *For each* L, *each* $\alpha \le \omega$ *and each* $\varphi \in Wff_\alpha$, $L \vdash \varphi$ *iff* $\mathbf{M}_L^\alpha \models \varphi$ *(in particular, $Th(\mathbf{M}_L) = L$).*
Proof. By Theorem 0.19 (ii) and Lemma 3.1.

Finally we can show, as announced at the end of Definition 0.14, that

COROLLARY 3.3. *If* $\mathbf{M} \models \varphi$ *for every* $\mathbf{M}$, *then* $\varphi \in \mathbf{K}$.
Proof. $\mathbf{M_K}$ is a model.

This result, together with Theorem 0.16, shows that $\mathbf{K} = \cap\{Th(\mathbf{M})$ for $\mathbf{M}$ a model$\}$ (and hence $\mathbf{K} = \cap\{Th(\mathbf{F})$ for $\mathbf{F}$ a frame$\} = \cap\{Th(\mathbf{F})$ for $\mathbf{F}$ a general frame$\}$).
    $\mathbf{M}_L$ is an example of a model whose theory is a logic. But canonical models are not the only models having this property, as is shown in the following exercise.

Exercise 3.4. Let L be axiomatized by $\mathbf{K} \cup \{\Box\bot\}$. (i) Describe $\mathbf{M}_L$. (ii) Find a proper submodel $\mathbf{M}$ of $\mathbf{M}_L$ such that $Th(\mathbf{M}) = L$.
    Solution. (i) Let Δ be any element of W_L. Since $\Box\bot \in \Delta$ we have that $\{\Delta' : \Delta R_L \Delta'\} = \emptyset$. Thus $\mathbf{M}_L$ is (isomorphic to) $\langle W, R, V \rangle$, where $W = \mathcal{P}(\mathbf{P})$, $R = \emptyset$ and $V(p) = \{w \in W: p \in w\}$. (ii) Consider the model $\mathbf{M}' = \langle W', R', V' \rangle$ where W' is the set of all *finite* subsets of $\mathbf{P}$, $R' = \emptyset$ and $V'(p) = \{w' \in W': p \in w'\}$. Observe that from $|W'| = \chi_0$ and Ex. 0.20 it follows that there is no logic of which $\mathbf{M}'$ is the canonical model. We show that $Th(\mathbf{M}') = Th(\mathbf{M}_L)$. Since $\mathbf{M}' \subseteq_+ \mathbf{M}_L$ we have, by Theorem 2.7, that $Th(\mathbf{M}_L) \subseteq Th(\mathbf{M}')$. Hence suppose that $\varphi \notin Th(\mathbf{M}_L)$ and assume that φ is

in normal form (see Theorem 0.17). Since $R_L = R' = \emptyset$, for each subformula of type $\square\psi$ and $\Diamond\psi$, we have that $\square\psi$ is true in both $\mathbf{M}_L$ and $\mathbf{M}'$, and $\Diamond\psi$ false at each point of $\mathbf{M}_L$ and of $\mathbf{M}'$. Hence we are left to consider only the subformula ψ that is a PC-formula. Suppose that there exists $w \in W$ such that $\mathbf{M}_L \not\models \psi[w]$ and let X be the set of propositional variables occurring in ψ. There exists $w' \in W'$ such that, for each $p \in X$, $w \models p$ iff $w' \models p$. Therefore $\varphi \notin Th(\mathbf{M}')$.

DEFINITION 3.5. For each $\alpha \leq \omega$ we denote by T_L^α the following set:

$$T_L^\alpha = \{X \subseteq W_L^\alpha : X = \{\Delta : \varphi \in \Delta\} \text{ for a } \varphi \in \textit{Wff}_\alpha\}.$$

By the fundamental lemma, $T_L^\alpha = \{V_L^{\alpha+}(\varphi) : \varphi \in \textit{Wff}_\alpha\}$, and so, by Theorem 1.1 (ii), T_L^α is a τ-field of sets over $\langle W_L^\alpha, R_L^\alpha \rangle$. We denote by $\mathbf{F}_L^\alpha$ the general frame $\langle W_L^\alpha, R_L^\alpha, T_L^\alpha \rangle$, which we define the α-canonical general frame. The following result holds.

THEOREM 3.6. $Th_\alpha(\mathbf{F}_L^\alpha) = Th_\alpha(\mathbf{M}_L^\alpha)$.

Proof. As regards the non trivial direction, suppose that $\varphi(p_0, \ldots, p_\beta)$ is an α-formula which is false in $\mathbf{F}_L^\alpha$. Then there exist a valuation V on $\mathbf{F}_L^\alpha$ and a point $\Delta \in W_L^\alpha$ such that $\langle W_L^\alpha, R_L^\alpha, V \rangle \not\models \varphi(p_0, \ldots, p_\beta)[\Delta]$. By definition of T_L^α, for each $i < \alpha$, there exists an α-formula ψ_i such that $V_L^{\alpha+}(\psi_i) = V(p_i)$. Therefore, by Theorem 0.6, $\langle W_L^\alpha, R_L^\alpha, V_L^\alpha \rangle \not\models \varphi(p_0/\psi_0, \ldots, p_\beta/\psi_\beta)[\Delta]$, i.e. $\varphi(p_0/\psi_0, \ldots, p_\beta/\psi_\beta) \notin Th_\alpha(\mathbf{M}_L^\alpha)$. But, by the fundamental theorem, $Th_\alpha(\mathbf{M}_L^\alpha)$ is closed under the rule of substitution among α-formulas, and therefore $\varphi(p_0, \ldots, p_\beta) \notin Th_\alpha(\mathbf{M}_L^\alpha)$.

In other words, the canonical general frame has this particularity: there is a valuation, V_L^α such that, for each other valuation V, $Th_\alpha(\langle W_L^\alpha, R_L^\alpha, V_L^\alpha \rangle) \subseteq Th_\alpha(\langle W_L^\alpha, R_L^\alpha, V \rangle)$. We observe that an immediate consequence of the fundamental theorem and Theorem 3.6 is the following result, which we call

THEOREM 3.7. (Fundamental Corollary). *For each L, each $\alpha \leq \omega$ and each $\varphi \in \textit{Wff}_\alpha$, $L \vdash \varphi$ iff $\mathbf{F}_L^\alpha \models \varphi$.*

The result of the above fundamental corollary relates to the general frame $\langle W_L^\alpha, R_L^\alpha, R_L^\alpha \rangle$ and not the frame $\langle W_L^\alpha, R_L^\alpha \rangle$. It is in fact possible that $\langle W_L^\alpha, R_L^\alpha \rangle \not\models L$.

DEFINITION 3.8. If $\langle W_L, R_L \rangle \models L$ then L is said to be *canonical*.

Exercise 3.9. Show that if L is consistent then $KM(L)$, $GF(L)$ and $KF(L)$ are non-empty.

Solution. By the fundamental theorem and the fundamental corollary, $\mathbf{M}_L \in KM(L)$ and $\mathbf{F}_L \in GF(L)$. As regards $KM(L)$, from Theorem 2.17 it follows that $\mathbf{F}_L$ is g-reducible either to $\mathbf{F}_0$ or to $\mathbf{F}_{00}$; therefore, by Corollary 2.16, either $\mathbf{F}_0 \in GF(L)$ or $\mathbf{F}_{00} \in GF(L)$. But both $\mathbf{F}_0$ and $\mathbf{F}_{00}$ are frames, and hence at least one of them belongs to $KF(L)$.

Exercise 3.10. (i) Show that, for each L and each $\alpha \leq \omega$, $\mathbf{M}_L^\alpha \models L$ and $\mathbf{F}_L^\alpha \models L$.

(ii) Show that if $L \subseteq L'$ then, for each $\alpha \leq \omega$, $\mathbf{M}_{L'}^\alpha \subseteq_+ \mathbf{M}_L^\alpha$ and $\mathbf{F}_{L'}^\alpha \subseteq_+ \mathbf{F}_L^\alpha$.

Solution. (i) Let $\varphi \in L$. If φ is an α-formula then, by Theorem 3.2, $\mathbf{M}_L^\alpha \models \varphi$. Hence suppose that φ is a β-formula for $\alpha < \beta$, and let ψ be obtained from φ by replacing each p_i, for $i \geq \alpha$, with $\bot$. By definition of V_L^α and Theorem 0.6 we obtain that $\mathbf{M}_L^\alpha \models \varphi$ iff $\mathbf{M}_L^\alpha \models \psi$. Since L is closed under the rule of substitution, from $\varphi \in L$ we find that $\psi \in L$. Moreover, ψ is an α formula and hence, by Theorem 3.2, $\mathbf{M}_L^\alpha \models \psi$ and $\mathbf{M}_L^\alpha \models \varphi$. Therefore $\mathbf{M}_L^\alpha \models L$ holds and this, via Theorem 3.6, implies $\mathbf{F}_L^\alpha \models L$.

(ii) From $L \subseteq L'$ it follows that $W_{L'}^\alpha \subseteq W_L^\alpha$ (so condition 2.5.0 is satisfied); by definition of V_L^α and $V_{L'}^\alpha$, condition 2.5.4 follows. Moreover, for each $\Delta, \Delta' \in W_{L'}^\alpha$, $\Delta R_{L'}^\alpha \Delta'$ iff $\Delta R_L^\alpha \Delta'$, thus obtaining 2.5.1. As regards 2.5.3, suppose that $\Delta \in W_{L'}^\alpha$ and $\Delta R_L^\alpha \Delta'$: we show that $\Delta' \in W_{L'}^\alpha$, i.e. $L' \cap Wff_\alpha \subseteq \Delta'$. In fact, if $\varphi \in L' \cap Wff_\alpha$ then, by necessitation rule, $\Box\varphi \in L' \cap Wff_\alpha$ and hence $\Box\varphi \in \Delta$; thus, from $\Delta R_L^\alpha \Delta'$ it follows that $\varphi \in \Delta'$. Therefore we have shown that $\mathbf{M}_{L'}^\alpha \subseteq_+ \mathbf{M}_L^\alpha$. This, together with the Generation Theorem 2.7 (ii) and definitions of T_L^α and $T_{L'}^\alpha$, implies that these τ-fields of sets satisfy condition 2.5.2, thus obtaining the fact that $\mathbf{F}_{L'}^\alpha \subseteq_+ \mathbf{F}_L^\alpha$.

The definitions of $\mathbf{M}_L^\alpha$ and $\mathbf{F}_L^\alpha$ are absolutely non-constructive. In Bellissima (1990, 1984 and 1985a) it is shown that, for the most important logics, submodels of $\mathbf{M}_L^\alpha$ ($\alpha < \omega$) which are equivalent to $\mathbf{M}_L^\alpha$ can in fact be effectively constructed.

Completeness

DEFINITION 3.11. Let L be a logic and let X be a class of structures (models or general frames). L is *complete with respect to* X (or *is determined by* X) if $L = \cap(Th(\mathbf{S}) : \mathbf{S} \in X)$, i.e. $L \vdash \varphi$ iff $\mathbf{S} \models \varphi$ for each $\mathbf{S}$ of X.

THEOREM 3.12. *Let L be a logic. Then:*

(i)   *L is complete with respect to* $KM(L)$.
(ii)   *L is complete with respect to* $GF(L)$.

   Proof. (i) By definition, if $L \vdash \varphi$ then $\mathbf{M} \models \varphi$ for each $\mathbf{M} \in KM(L)$. On the other hand, suppose that $\mathbf{M} \models \varphi$ for each $\mathbf{M} \in KM(L)$; since, by the fundamental theorem, $\mathbf{M}_L \in KM(L)$, then $\mathbf{M}_L \models \varphi$ and, by the same theorem, $L \vdash \varphi$. (ii) By the fundamental corollary, $\mathbf{F}_L \in GF(L)$. Then the proof is as in case (i).

Unfortunately, the classes $KM(L)$ and $GF(L)$ cannot be handled, except for trivial cases. So, given L, it is important to investigate the possible completeness of L with respect to proper subclasses of $KM(L)$ and $GF(L)$.

DEFINITION 3.13. A model $\mathbf{M} = \langle W, R, V \rangle$ is *singly-generated* (or, $\mathbf{M}$ is an *s.g.*-model) if there exists a $w \in W$ such that $\mathbf{M} = \mathbf{M}/\{w\}$ (i.e. cf. Definition 2.5, $W = \{w\} \cup S^*(w)$). Similar definitions are given for general frames.

A first reduction of $KM(L)$ and $GF(L)$, which holds for every logic L, is expressed in the following:

Exercise 3.14. Show that each logic L is complete with respect to the class of all s.g.-L-models and of all s.g.-L-general frames.
   Proof. By Theorem 2.7, for each $\mathbf{M}$, $Th(\mathbf{M}) = \cap \{Th(\mathbf{M}/\{w\}) : w \in W\}$, and similarly for general frames. By the same theorem, if $\mathbf{M} \models L$ then $\mathbf{M}/\{w\} \models L$. Thus the result follows from Theorem 3.12.

   A significant subclass of $GF(L)$ is $KF(L)$. A logic L which is complete with respect to $KF(L)$ is said to be *Kripke-complete*. (Sometimes Kripke-complete logics are simply called *complete*. We prefer to avoid this use, which has caused many misunderstandings). Other significant subclasses of $KM(L)$ and $GF(L)$ are $fKM(L)$, $fGF(L)$ and $fKF(L)$, where f stands for 'finite'. A general result regarding these classes is found in the next exercise.

Exercise 3.15. Show that, for each L, the following are equivalent:

3.15.0. L is complete with respect to $fKM(L)$,
3.15.1. L is complete with respect to $fGF(L)$,

3.15.2. L is complete with respect to $fKF(L)$.

Solution. From Theorem 2.3 and 2.4 it follows that $\cap(Th(\mathbf{M}) : \mathbf{M} \in fKM(L)) = \cap(Th(\mathbf{F}) : \mathbf{F} \in fKF(L)) = \cap(Th(\mathbf{F}) : \mathbf{F} \in fGF(L))$.

This result shows that the concepts of *finite model property, finite general frame property* and *finite frame property* (i.e. conditions 3.15.0–2, respectively), are equivalent. The first one (i.e. finite model property) is, for historical reasons, the most popular, and we shall use it.

Of course, from considerations similar to those of Ex. 3.14 it follows that

COROLLARY 3.16. (i) *If L is Kripke-complete then it is complete with respect to the class of its s.g.-frames.*

(ii) *If L has the finite model property then it is complete with respect to the class of its finite s.g.-models (frames, general frames).*

(iii) *If L has the finite model property then L is Kripke-complete.*

Exercise 3.17. Show that if L is canonical then L is Kripke-complete.

Solution. By hypothesis, $\langle W_L, R_L \rangle \in KF(L)$, that is, $L \subseteq Th(\langle W_L, R_L \rangle)$. By the fundamental theorem, $L = Th(\langle W_L, R_L, V_L \rangle)$, while, obviously, $Th(\langle W_L, R_L \rangle) \subseteq Th(\langle W_L, R_L, V_L \rangle)$. So we obtain $L = Th(\langle W_L, R_L \rangle)$.

Exercise 3.18. Show that if $L \subset L'$ then $KM(L') \subset KM(L)$, $GF(L') \subset GF(L)$ and $KF(L') \subseteq KF(L)$.

Solution. Obviously, $L \subset L'$ implies that $KM(L') \subseteq KM(L)$, $GF(L') \subseteq GF(L)$ and $KF(L') \subseteq KF(L)$. That the first two inclusions are proper follows from Theorem 3.12. If both L and L' are Kripke-complete, we have also $KF(L') \subset KF(L)$.

We now examine some logics from the viewpoint of completeness.

4. THE LOGIC K

Since (see Theorem 0.16) $\mathbf{F} \models \mathbf{K}$ holds for each $\mathbf{F}$, then $\langle W_{\mathbf{K}}, R_{\mathbf{K}} \rangle$ (i.e. the frame of the canonical model for $\mathbf{K}$) is a $\mathbf{K}$-frame. So $\mathbf{K}$ is canonical and hence (see Ex. 3.17) is Kripke-complete.

THEOREM 3.19. $\mathbf{K}$ *has the finite model property.*

Proof. Suppose that $\mathbf{K} \not\vdash \varphi$. We must show that there exists a finite model which falsifies φ. Now, by the fundamental theorem, $\mathbf{M_K} \not\models \varphi$, where $\mathbf{M_K}$ is the canonical model for $\mathbf{K}$. Let Σ be the set of all subformulas of φ (obviously Σ is closed under subformulas) and let $\mathbf{M}$ be the finest filtration of $\mathbf{M_K}$ through Σ (by Theorem 3.16 such filtration exists). Since Σ is finite, also $\mathbf{M'}$ is finite (see 3.15.4), whereas, from Theorem 3.17 and the fact that $\varphi \in \Sigma$, we have that $\mathbf{M'} \not\models \varphi$ (observe that, up to this point, the proof works for each L). Finally, from the fact that $fKM(\mathbf{K})$ is the class of all finite models (see Theorem 0.16) we obtain that $\mathbf{M'} \in fKM(\mathbf{K})$, thus concluding the proof.

In the above proof, the delicate point which prevents the theorem from being valid for each logic is the following: given L, it may happen that $\mathbf{M'} \notin fKM(L)$. In fact $KM(L)$ may also not be closed under finest filtrations. In such a case, one might attempt to find a filtration $\mathbf{M'}$ of $\mathbf{M}$ different from the finest one, in order to obtain $\mathbf{M} \in fKM(L)$ (an example will be given in Lemma 3.29). But this attempt may also fail; in fact (see, for instance, Bellissima (1989)), there exists a continuum of logics without the finite model property.

Exercise 3.20. Find the error: for each L, if L is finitely axiomatizable, then it has the finite model property.

(False) proof. Suppose that $L \not\vdash \varphi$. Let Σ be the set of subformulas of $\Gamma \cup \{\varphi\}$, where Γ is a finite set of axioms for L. Let $\mathbf{M'}$ be the finest filtration of $\mathbf{M}_L$ through Σ. $\mathbf{M'}$ is finite (since Σ is finite), $\mathbf{M'} \not\models \varphi$ and $\mathbf{M'} \models \Gamma$ (by the filtration theorem and the obvious fact that $\mathbf{M}_L \models \Gamma$); hence $\mathbf{M'} \models L$, i.e. $\mathbf{M'} \in fKM(L)$.

Solution. See Remark 1.5.

The Logic **KT**

As shown in Ex. 0.9, if $\mathbf{F}$ is a frame then $\mathbf{F} \models \Box p \to p$ iff R is reflexive. So $KF(\mathbf{KT})$ is the class of all reflexive frames.

THEOREM 3.21. **KT** *is canonical and hence Kripke-complete.*

Proof. We have to show that $R_{\mathbf{KT}}$ is a reflexive relation. Let $\Delta \in W_{\mathbf{KT}}$. Since $\mathbf{KT} \subseteq \Delta$, $\Box\varphi \to \varphi \in \Delta$ for each Δ. So, since Δ is closed under modus ponens, $\Box\varphi \in \Delta$ implies $\varphi \in \Delta$, thus obtaining (see Definition 3.0) $\Delta R_{\mathbf{KT}} \Delta$.

THEOREM 3.22. **KT** *has the finite model property.*

Proof. First we observe that an immediate consequence of condition 1.13.1 is that each filtration of a reflexive model is reflexive. Now, for each φ we obtain, as in Theorem 3.19, a finite model $\mathbf{M}'$ which falsifies φ. Moreover, since $M_{\mathbf{KT}}$ is a reflexive model, $\mathbf{M}'$ is also reflexive and hence $\mathbf{M}' \in fKM(\mathbf{KT})$.

The fact that $KF(\mathbf{KT})$ is the class of all reflexive frames implies neither that $KM(\mathbf{KT})$ is the class of all reflexive models, nor that $GF(\mathbf{KT})$ is the class of all reflexive general frames, as is shown in the following exercise.

Exercise 3.23. Show that **KT** is complete with respect to a class of antireflexive general frames.

Solution. By Theorem 3.21, $\mathbf{KT} = Th(\langle W_{\mathbf{KT}}, R_{\mathbf{KT}} \rangle)$ holds, and $R_{\mathbf{KT}}$ is a reflexive relation. Let $\mathbf{F}_{\mathfrak{H}}$ be obtained from $\langle W_{\mathbf{KT}}, R_{\mathbf{KT}} \rangle$ as in Ex. 2.14. By the same exercise, $Th(\langle W_{\mathbf{KT}}, R_{\mathbf{KT}} \rangle) = Th(\mathbf{F}_{\mathfrak{H}})$, and therefore **KT** is complete with respect to $\{\mathbf{F}_{\mathfrak{H}}\}$.

The Logic **K4**

As we have shown in Ex. 0.10, for each frame $\mathbf{F}$, $\mathbf{F} \models \Box p \to \Box\Box p$ iff R is transitive. Thus $KF(\mathbf{K4})$ is the class of transitive frames.

THEOREM 3.23. **K4** *is canonical and hence Kripke-complete.*

Proof. We have to show that $R_{\mathbf{K4}}$ is transitive. Let Δ, Δ' and Δ'' be elements of $W_{\mathbf{K4}}$ such that $\Delta R_{\mathbf{K4}} \Delta' R_{\mathbf{K4}} \Delta''$, and suppose that $\Box\varphi \in \Delta$. From $\Box\varphi \to \Box\Box\varphi \in \Delta$ we obtain $\Box\Box\varphi \in \Delta$ and hence, from $\Delta R_{\mathbf{K4}} \Delta'$ and Definition 3.0, $\Box\varphi \in \Delta'$. This, together with $\Delta' R_{\mathbf{K4}} \Delta''$, implies that $\varphi \in \Delta''$; therefore $\Delta R_{\mathbf{K4}} \Delta''$.

Exercise 3.24. Find a transitive model $\mathbf{M}$ and a set of formulas Σ such that the finest filtration of $\mathbf{M}$ through Σ is not transitive.

Solution. A very simple example is the following: let $\Sigma = \{p_0, p_1\}$ (Σ is closed under subformulas) and $\mathbf{M} = \langle W, R, V \rangle$ be as follows: $W = \{w, v_0, v_1, z\}$, $R = \{\langle w, v_0 \rangle, \langle v_1, z \rangle\}$ (R is transitive), $V(p_0) = \{w, v_0, v_1\}$ and $V(p_1) = \{v_0, v_1, z\}$. Now, $W_\Sigma = \{[w]_\equiv, [v_0]_\equiv, [z]_\equiv\}$, where $[w]_\equiv = \{w\}$, $[v_0]_\equiv = \{v_0, v_1\}$ and $[z]_\equiv = \{z\}$, and $R' = \{\langle [w]_\equiv, [v_0]_\equiv \rangle, \langle [v_0]_\equiv, [z]_\equiv \rangle\}$. R' is not transitive.

Now we show that **K4** has the finite model property. The proof cannot simply be a repetition of that of Theorem 3.19 because, as shown in Ex. 3.24, finest filtrations do not preserve transitivity.

LEMMA 3.25. *Let* $\mathbf{M} = \langle W, R, V \rangle$ *be a transitive model, let* Σ *be a set of formulas closed under subformulas and let* R' *be the relation on* W_Σ *defined as follows:*

$[w]_\equiv R'[v]_\equiv$ *iff, for each* $\Box\varphi \in \Sigma$, $\mathbf{M} \models \varphi[w]$ *implies that*
$\mathbf{M} \models \Box\varphi \wedge \varphi[v]$.

Then (i) $\mathbf{M}' = \langle W_\Sigma, R', V_\Sigma \rangle$ *is a filtration of* $\mathbf{M}$ *through* Σ, *and* (ii) R' *is transitive.*

Proof. (i) Obviously R' satisfies 1.13.2. Moreover, if wRv and $\Box\varphi \in \Sigma$ and $\mathbf{M} \models \Box\varphi[w]$, then, from $\mathbf{M} \models \Box\varphi \rightarrow \Box\Box\varphi$, it follows that $\mathbf{M} \models \Box\Box\varphi[w]$ and therefore $\mathbf{M} \models \Box\varphi \wedge \varphi[v]$ and $[w]_\equiv R'[v]_\equiv$, thus satisfying 1.13.1 and obtaining (i). As regards (ii), if $[w]_\equiv R'[v]_\equiv R'[z]_\equiv$ then, again using axiom 4, we have that $\mathbf{M} \models \Box\varphi[w]$ implies $\mathbf{M} \models \Box\varphi \wedge \varphi[z]$ and hence $[w]_\equiv R'[z]_\equiv$.

THEOREM 3.26. **K4** *has the finite model property.*

Proof. Similar to those of Theorems 3.19 and 3.22, but using the filtrations of Lemma 3.25.

SYMBOLS

TERMS

Dipartimento di Matematica,
Università di Siena

REFERENCES

Bellissima, F.: 1984, 'Atoms in modal algebras', *Zeitschrift für mathematiske Logik und Grund-lagen der Mathematik* **30**, 303–312.

Bellissima, F.: 1985a, 'An effective representation for finitely generated free interior algebras', *Algebra Universalis* **20**, 302–317.

Bellissima, F.: 1985b, 'A test to determine distinct modalities in the extensions of S4', *Zeitschrift für mathematiske Logik und Grundlagen der Mathematik* **31**, 57–62.

Bellissima, F.: 1988, 'On the lattices of extensions of the modal logics KAlt$_n$', *Archive for Mathematical Logic* **27**, 107–114.

Bellissima, F.: 1989, 'Infinite sets of non-equivalent modalities', *Notre Dame Journal of Formal Logic* **30**, 574–582.

Bellissima, F.: 1990, 'Post complete and 0-axiomatizable modal logics', *Annals of Pure and Applied Logic* **47**, 121–144.

Bellissima, F. and Mirolli, M.: 1989, 'A general treatment of equivalent modalities in normal modal logics', *The Journal of Symbolic Logic* **54**, 1460–1471.

Benthem, J. van: 1983, *Modal logic and Classical logic*, Bibliopolis, Napoli.

Blok, W.J.: 1978, 'On the degree of incompleteness of modal logics', *Bullettin of the Section of Logic* **7**, 167–176.

Blok, W.J.: 1980, 'The lattice of modal algebras is not strongly atomic', *Algebra Universalis* **11**, 285–294.

Chellas, B.F.: 1980, *Modal Logic: An Introduction*, Cambridge University Press, Cambridge.

Fine, K.: 1974, 'Logics containing K4 (Part 1)', *Journal of Symbolic Logic* **39**, 31–42.

Fine, K.: 1985, 'Logics containing K4 (Part 2)', *Journal of Symbolic Logic* **50**, 619–651.

Gabbay, D.M.: 1976, *Investigations in Modal and Tense Logics with Applications to Problems in Philosophy and Linguistics*, D. Reidel, Dordrecht.

Sambin, G. and Vaccaro, V.: 1988, 'Topology and duality in modal logic', *Annals of Pure and Applied Logic* **37**, 249–296.

Segerberg, K.: 1971, *An Essay in Classical Modal Logic*, Filosofiska Studier, Uppsala.

Thomason, S.K.: 1972, 'Semantic analysis of tense logics', *Journal of Symbolic Logic* **37**, 150–157.

EGON BÖRGER

COMPLEXITY OF LOGICAL DECISION
PROBLEMS: AN INTRODUCTION

1. INTRODUCTION*

Logic and Complexity are concerned with two fundamental notions which
Leibniz already recognized as belonging together and whose mathematical development from Frege to Turing has laid the theoretical foundation of computer
science: the concepts of formal language and of algorithms (calculus). In fact
Leibniz seems to have recognised that the creation of a mathematically precise
universal language for the expression of arbitrary statements (*characteristica
universalis*) is related to the development of a sufficiently general concept of
calculus (*calculus ratiocinator*), in which scientific problems can be decided in
a purely formal, algorithmic way. In the development of algorithmic problem-
solving methods on modern computers the main difficulty in the process of
finding a computer program which solves a given problem consists of specifying – in a language called specification language – this problem exactly,
more and more formally and excluding what is not intended to be part of the
problem. The quality of a high level programming language largely depends
on the quality of the specification language as a flexible vehicle of clear and
well documented description and on the reliability of the methods by which
programs are constructed out of the specification. This intrinsic connection
between descriptive and imperative use of language is strikingly explicit in
logic programming languages like PROLOG which rest on a single, flexible
language – first order logic – being simultaneously the specification and the
programming language, one and the same object being able to be a statement
(logical problem description) and a program (algorithm for the solution of the
problem).

Leibniz' (and Boole's) efforts to build such a *characteristica universalis*
were completed by G. Frege who developed a universal language for the
formalization of all (then known) mathematical facts with the specific aim of
separating the 'logical' from the historical, psychological and similar contents
of mathematical thought and to reduce all legitimate mathematical concepts
and deductions to a few clear basic principles. The separation within a logic

71

G. Corsi et al. (eds), Bridging the Gap: Philosophy, Mathematics, and Physics, 71–86.
© 1993 Kluwer Academic Publishers.

language between the aspect concerned with content (semantics) and the formal (syntactical) aspect assumed its still valid form in the work of Tarski (1936) and Gödel (1930) and passed its final fundamental theoretical test in Gödel's completeness theorem for first order logic. Indeed this completeness theorem yields a syntactical (algorithmic, depending only on the external form of the signs occurring, and not on their meaning) characterization of the semantic concept of logical universal validity (truth).

With the realization of Leibniz' idea of a *characteristica universalis* as the language of first order logic, the old dream of an all-embracing problem solving method, which already found expression in the *ars magna* of R. Lullus, took the form of a mathematical task: to look for an algorithm which for each predicate logic expression decides whether or not it is universally valid. Hilbert formulated this problem and called it *Das Entscheidungsproblem*, the decision problem *per se*. The historical background was the Hilbert Programme which was formulated for the purpose of providing all of mathematics with a foundation against the occurrence of paradoxes and to characterize the different branches of mathematics by first order axioms so that in principle each mathematical proof would consist of a logical derivation of the assertion from the axioms.

Intensive work on this *Entscheidungsproblem* led to the impression that such an algorithm may not exist. This contributed to the increasing consciousness of the need to find a mathematically precise and sufficiently general explication of the intuitive concept of algorithms in order to allow rigorous proofs of the impossibility of algorithmic solutions of particular classes of problems. The significant achievement of such an explication, particularly in the form given by Turing (1937) solved Hilbert's *Entscheidungsproblem* in the negative and decisively influenced the development of the first electronic computing machines.

Turing's proof of the unsolvability of Hilbert's *Entscheidungsproblem* consists of an appropriate logical description of undecidable algorithmic phenomena, in his case writing problems of Turing machine programs. We explain in this paper a simple general technique for the loigcal description of programs which directly and naturally transfers many complexity properties from programs to logical formulae. Our descriptions can be viewed as abstract definitions of the meaning of the given programs, as their logical implementation or, better, interpreter. This method reveals deep structural and combinatorial similarities between computations and logical deductions which, in turn, brings out a fundamental and uniform reason for many well known completeness results of both algorithmic and logical decision problems.

The paper is organized as follows: Section 2 deals with a modern proof of the unsolvability of Hilbert's *Entscheidungsproblem*. It is in the spirit of Turing's original proof and establishes a close connection between the *Entscheidungsproblem* and the universality and sharp normal forms for Horn clause (PROLOG) programs. Moreover it gives a method for a simple and natural logical implementation of programs which carries over many interesting complexity properties from program computations to logical deductions and model constructions. Section 3 gives examples of how, by this method for logical interpretation of programs, limited computations are related to recursive analogues of Hilbert's *Entscheidungsproblem* like the NP-completeness of propositional logic satisfiability or the PSPACE-completeness of the satisfiability of certain function free binary Horn clause (PROLOG) programs.

This text is meant as an introduction which does not presuppose more than the knowledge of basic notions of first order predicate logic. We will reference the results which will be borrowed from computation theory in the course of our proofs. Unexplained notions, notations or results, more details, more advanced results and complete references can be found in textbook form in Börger (1989).

2. HILBERT'S ENTSCHEIDUNGSPROBLEM AND UNIVERSAL PROLOG PROGRAMS

We begin with Malcew's observation (1974) that the predicate logic formalizability of word problems of Thue systems gives an especially simple proof of the unsolvability of Hilbert's *Entscheidungsproblem*.

THEOREM (Church (1936) and Turing (1937)). *There is no algorithm to decide for an arbitrary first order formula F whether F is logically valid or not.*

Proof. Let T be an arbitrary Thue system with rules $V_i = W_i$ ($0 \leq i \leq m$) and alphabet $A = \{a_0, \ldots, a_n\}$. Let V, W be arbitrary words over A. Considering $a_0, \ldots, a_n$ as individual constants and concatenation as a binary function symbol, we have a canonical logical description of words over the alphabet A as variable free terms. Thus the following first order formula $F(T, V, W)$ formalizes the assertion that $V = W$ is derivable in T (and therefore in every semigroup with corresponding generators $b_0, \ldots, b_n$ and defining relations $V_i' = W_i'$ ($0 \leq i \leq m$), where U' arises from U by replacing a_j by b_j):

$$F(T, V, W) := \forall x\, \forall y\, \forall z((xy)\, z = x(yz)) \& V_0 =$$
$$= W_0 \& \ldots \& V_m = W_m \to V = W$$

Indeed $V = W$ is derivable in T iff $F(T, V, W)$ is logically valid. Since the former property is algorithmically undecidable and $F(T, V, W)$ can be constructed effectively given T, V, W, the latter property is algorithmically undecidable.

We now give another proof of the theorem of Church and Turing which rests on an equality-free and function symbol-free description of register machine programs developed by Aanderaa (1971) and Börger (1971) continuing the ideas of Turing (1937) (and Büchi (1962)). In order to show the reader the generality of this technique of logical definition of the semantics of programs, following Börger (1988) we give a natural and simple Horn clause implementation of structured programs in a computation universal programming language. Because this Horn simulation of arbitrary computations procedes step-by-step, complexity theoretic properties of programs are preserved by this translation.

Consider a computational universal language L of structured programs. To be specific, let these programs be built up from a finite number of elementary programs e by use of concatenation MN ('first M, then N') and WHILE-iteration WHILE $x_i \neq$ nil DO M with respect to a finite number $x_1, \ldots, x_n$ of 'registers' containing arbitrary data. The following first-order formula U describes the semantics of L; since U is a definite Horn formula, we can also say that U, as a PROLOG program, interprets the programs of L.

DEFINITION. Let Conf be an $n + 1$-ary relational predicate symbol (with $\mathrm{Conf}(M, x_1, \ldots, x_n)$ intended to mean that the configuration consisting of the program M together with the data $x_1, \ldots, x_n$ on which M has to be executed will eventually lead to an acceptable halting configuration). Let M, N be individual variables (to be interpreted by structured L-programs), '*' a binary function symbol (standing for the concatenation of structured programs), and $(\ )_i$ a monadic function symbol (which applied to M is meant to represent the WHILE-program WHILE$x_i \neq$ nil DO M) for $1 \leq i \leq n$. Let e be individual constants (denoting the elementary L- programs). Let $\vec{x} = (\vec{y}, x_i, \vec{z}) = (x_1, \ldots, x_n)$ with individual variables $x_1, \ldots, x_n, u$ (standing for data contained in the registers) and '$\bullet$' a binary function symbol (used to denote concatenation of data). Let nil denote an individual constant (standing for the empty string (of data or of programs)). We define U as the following set (read: conjunction) of clauses (where we use the PROLOG notation $F \leftarrow F_1, \ldots, F_n$

for the logical implication $F_1 \& \ldots F_n \to F$ which the reader should think of as a universally closed first-order formula):

$\mathrm{Conf}(e^* M, \vec{x}) \leftarrow \mathrm{Conf}(M, e(\vec{x}))$ for e elementary where
$\qquad\qquad e(\vec{x})$ is shorthand for (a logical representation
$\qquad\qquad$ of) the result of executing program e on data $\vec{x}$.

Read: To execute program $e^* M$ on data $\vec{x}$ means to execute program M on the data resulting from applying the elementary program e on $\vec{x}$.

$\mathrm{Conf}((M)_i^* N, \vec{y}, u \bullet x_i, \vec{z}) \leftarrow \mathrm{Conf}(M^* (M)_i^* N, \vec{y}, u \bullet x_i, \vec{z})$

$\mathrm{Conf}(\ldots, \mathrm{nil}, \ldots) \leftarrow \mathrm{Conf}(N, \vec{y}, \mathrm{nil}, \vec{z})$ for $1 \le i \le n$

Read: To execute program $(\mathrm{WHILE} x_i \ne \mathrm{nil\ DO}\ M)^* N$ on data $\vec{x}$ means to execute N on x (if $x_i = \mathrm{nil}$) respectively, to execute M on x once and then $(M)_i^* N$ again on the data obtained from computing M on $\vec{x}$ (if the ith register contains compound data, i.e. of the form $u \bullet x_i \ne \mathrm{nil}$).

$\mathrm{Conf}(\mathrm{nil}, \overleftarrow{x}) \leftarrow$ ('Halting clause')

Read: The empty program together with arbitrary data represent by definition an accepting configuration; nothin has to be computed any more. The resolution calculus simulation of computations of L-programs will terminate by producing the empty clause from a halting configuration goal $\leftarrow \mathrm{Conf}(\mathrm{nil}, \vec{t})$ and the halting clause.

THEOREM ON A SIMPLE COMPUTATION UNIVERSAL HORN CLAUSE. *For $n \ge 2$ the above defined Horn clause program U is computation universal, i.e., for every (e.g. Turing machine) program M and every input A for M one has for some L-program m (expressing M) and data a (representing A) the equivalence:*

$\qquad$ *M accepts A iff U accepts $\leftarrow \mathrm{Conf}(m, a)$.*

Here acceptance of the input clause $\leftarrow \mathrm{Conf}(m, a)$ (so-called goal clause) by U means that using the resolution rule, one can derive the empty clause from the union of U with $\leftarrow \mathrm{Conf}(m, a)$.

Proof. By our assumption that L is computation universal we can encode Turing machine programs M and tapes A of M into structured L-programs m and data a on which m simulates the behavior of M on A. (This is a version of Böhm's theorem (1964).) For a short direct proof see §AI1 in Börger (1989). By construction of U, the computation of m on a is simulated by the (resolution calculus computation of) program U started with the goal clause $\leftarrow \mathrm{Conf}(m, a)$. This computation in the resolution calculus will terminate successfully (by producing the empty clause) iff it uses the unique halting clause

Conf(nil,$\vec{x}$) $\leftarrow$ of U at least once. This is the case iff m on input a terminates successfully. The condition $2 \leq n$ is fulfilled by Minsky's theorem (1961) that two registers suffice to compute arbitrary partial recursive functions.

Remark. From the construction of U, it is obvious that in the above theorem the resolution calculus can be restricted to so-called unit resolution, where at each application of the resolution rule at least one of the two clauses involved is a prime formula (i.e., a literal without negation sign). A direct proof for this completeness of (positive) unit resolution for (definite) Horn clauses has been given by Henschen and Wos (1974). It is also obvious from the construction of U that exactly one (positive unit) resolution step in the simulation of m by U corresponds to each step of the given structured program m and vice versa. Thus, by our reduction technique, hierarchy results of computational complexity theory directly find natural logical analogues, namely, interpretations in terms of complexity questions for logical calculi (in particular resolution-based systems).

Remark. The preceding includes a proof of the computation universality of binary deterministic PROLOG programs; weaker forms of this have been rediscovered and proved by other methods in Hill (1974), Andreka and Nemeti (1976), Tärnlund (1977), Itai and Makowsky (1983) and Sebelik and Stepanek (1982). See also the second corollary below.

The following corollary states that the computation universality of U as a program with inputs $\leftarrow$ Conf(m, a) is related to the *universality of the logical decision problem* of the related class of computation formulae $\Lambda(U, \leftarrow \text{Conf}(m, a))$ (remember that the comma here stands for logical conjunction).

COROLLARY ON A STRONG REDUCTION CLASS. *The class* $\bar{U} :=$ $\{\Lambda(U, \leftarrow \text{Conf}(m, a)) \mid m \text{ program}, a \text{ input data}\}$ *of computation formulae defined by the (program) formula U is a reduction class of first-order logic and therefore undecidable. (A class R of first-order formulae is called a reduction class (with respect to satisfiability) iff there is some recursive function associating to each first order formula F a formula $\bar{F}$ which is an element of R and such that: F is satisfiable iff $\bar{F}$ is satisfiable. R is called a conservative reduction class if F has a finite model iff $\bar{F}$ has a finite model. R is called undecidable if there is no algorithm to decide for each element F of R whether it is satisfiable or not. ΛF denotes the universal closure of F.)*

Proof. By Gödel's completeness theorem one can compute for each first-order formula F a (structured Turing machine) program m_F which accepts the empty input nil iff F is not satisfiable. Therefore:

F is satisfiable iff m_F does not accept the input nil
   iff U does not accept $\leftarrow \text{Conf}(m_F, \text{nil})$
   iff $\Lambda(U, \leftarrow \text{Conf}(m_F, \text{nil}))$ is satisfiable.

The last equivalence is due to the completeness of resolution for first-order logic.

The computation universality of U incorporates a *normal form for PROLOG programs* corresponding to the normal form of the *Entscheidungsproblem* represented by U. By more careful choice of the simulated universal programming language one can obtain in a natural way much stronger normal forms for PROLOG programs. To come up with our favorite example (found as conservative reduction class by Aanderaa (1971) and Börger (1971)), it suffices to read $\text{Conf}(i, \vec{x})$ in U as binary relation with variables x_1, x_2 ranging over natural numbers and with i interpreted as 'instruction parameter'. Then our PROLOG program $U(M)$ – which now depends on a given program M (on the Minsky machine (1961) with only two number registers) and defines its interpretation – consists of the following binary clauses containing only number terms $x, y, 0$ (for zero), Sx (for successor $x + 1$ of x):

$\text{Conf}_i(x, y) \leftarrow \text{Conf}_j(Sx, y)$ for M-instructions (i: DO add$_1$ GOTO j)

$$\left. \begin{array}{l} \text{Conf}_i(0, y) \leftarrow \text{Conf}_j(0, y) \\[6pt] \text{Conf}_i(Sx, y) \leftarrow \text{Conf}_k(x, y) \end{array} \right] \quad \begin{array}{l} \text{for } M\text{-instructions} \\ (i: \text{IF } x_1 = 0 \text{ GOTO } j \text{ ELSE DO sub}_1 \text{ GOTO } k) \end{array}$$

Symmetrically for add$_2$ and test$_2$/subtract$_2$ instructions with respect to the second register, interchanging first and second component.

$\text{Conf}_1(x, y) \leftarrow$ for the accepting state (say) 1 of M

COROLLARY ON A PROLOG NORMAL FORM. *(Aanderaa (1971), Börger (1971), Aanderaa and Lewis, (1973)). Every PROLOG program can be brought into the normal form $U(M)$ of the above definition (for some M) containing only number terms $x, y, 0, Sx$, binary relation symbols, and binary clauses. This normal form cannot be improved further by leaving out either one variable or the individual constant or the function symbol or by allowing only monadic* *predicates or only literals.*

If M is universal on the Minsky machine, then U(M) is a computation universal PROLOG program. The class $\bar{U}(M) := \{\Lambda(U, \leftarrow \text{Conf}_0(a, b)) \mid a, b \in N\}$ *of U-computation formulae is a minimal reduction class for each computation universal M.*

Proof. The normal form claim follows from the preceding theorem and Minsky's theorem (1961) that the register machine with only two number registers is computation universal. For the reduction class claim, apply the reasoning of the proof of the preceding corollary to an arbitrary but fixed 2-register machine program M which enumerates first-order logic as follows (where F is an arbitrary formula and $\bar{F}$ an encoding of F as register input to M using 0 and S): F is satisfiable iff M does not accept the input $\bar{F}$, iff $U(M)$ does not accept $\leftarrow \text{Conf}_0(\bar{F})$, and iff $\Lambda(U(M), \leftarrow \text{Conf}_0(\bar{F}))$ is satisfiable. The minimality part follows from the decidability of the decision problem for those classes of formulae which result from leaving out one of the above indicated syntactical components of $U(M)$ (see Aanderaa and Lewis (1973)).

Remark. The step-by-step simulation of M by $U(M)$ transforms (sub-) recursive inseparabilities of M-halting problems into (sub-) recursive inseparability properties of corresponding logical $U(M)$-decision problems. Thus one obtains from the existence of recursively inseparable, recursively enumerable sets immediately the stumbling block for database theory, namely Trakhtenbrot's theorem (1953) that the sets of contradictory, finitely satisfiable and only infinitely satisfiable formulae of first order logic are recursively inseparable. Also one can easily exhibit Horn clauses which are satisfiable but without recursive models, or which as unique non-logical axioms constitute an essentially undecidable (and therefore incomplete) theory. Similar results hold for subrecursive complexities. For more information the reader may consult Börger (1988) and (1989), §FI1.

Remark. The proof method of the above corollary can be refined to a schema for proving the undecidability of many decision problems for PROLOG programs. See Börger (1987).

3. RECURSIVE ANALOGUES OF HILBERT'S ENTSCHEIDUNGSPROBLEM AND LIMITED COMPUTATIONS

In the preceding logical description of programs we placed no restrictions on the length or space requirement of the described computations. Thereby we carried over the undecidability of unrestricted halting problems to the associated classes of formulas. We now want to refine the method to a formalization of time- or space-restricted halting problems in restricted classes of expressions whereby those recursively solvable decision problems turn out to be complete for the underlying time- or space-complexity classes. Thus we obtain in particular the propositional logic analogue of the theorem of Church and Turing – namely Cook and Levin's theorem on the NP completeness of the propositional logic satisfiability problem – by a propositional logic interpretation of polynomial time-bounded computations of non-deterministic Turing machines. This proof naturally extends to a type-theoretical interpretation of n-fold exponential time-bounded computations of non-deterministic TM-programs and thus yields a characterization of type-theoretical spectra (cardinality classes of finite models of a given expression of the n-th order) elementary computation time hierarchy with n-fold exponentiation $2^x, 2^{2^x}, \ldots$ as time bound – a result which is the starting point for a series of important complexity-theoretic investigations into finite model theory (see Börger (1984), (1989), §FIII2 for exact references). Finally we show for the case of polynomial space-bounded TM computations how an appropriate logical description of such bounded halting problems by restricted first order formulae leads to recursive logical decision problems which are complete for the describing complexity classes. 'Natural' complexity classes thereby correspond to 'natural' recursive sub-problems of the Hilbert *Entscheidungsproblem.*

As in Section 2 and following Börger (1989) we will so define program formulas (in prenex normal form with matrix) π_M and initial and end formulas α, ω which depend on the current input and prevailing halting problem, that each computation step of M is reflected in each model of π_M by a logical consequence (e.g. a resolution step) and α and ω determine the underlying initial and final configuration, respectively. Since we want te describe time-bounded computations in particular, we must explicitly assume the time parameter in the formalization. To elucidate again the uniformity of the reductions in the various halting and decision problems, we rely in this section on the Turing machine model (possibly non-deterministic, without loss of generality half-tape) and describe π_M as a propositional logic formula in the 'basic formulas' $A(t, u)$, $B_k(t, u)$, $Z_i(t)$, $S(t, t')$ with parameters t (for time), u (for the numbers of tape cells), k (for letters), i (for M-states) and the following *intended interpretation*:

$A(t, x)$ iff working-cell at time t is (that with number) x

$B_j(t, x)$ iff letter a_j is in cell x at time t

$Z_i(t)$ iff i is the state of M at time t

$S(t, t')$ iff t' is the direct successor (instant) of t.

For an M-configuration C_t at time t let $\underline{C_t}$ be the 'conjunction' of the above basic formulas describing C_t. With respect to these we define the configuration formulas π_M so that their models simulate the M-computations in the sense of:

SIMULATION LEMMA. *Let* $\tilde{A}$ *be a model of* π_M. *For arbitrary M-configurations* C_0 *and arbitrary* t, *if* $\tilde{A}$ *is a model of* $\underline{C_0}$ *then for at least one M-configuration* C_t *which satisfies* $C_0 \overset{t}{\underset{\longmapsto}{M}}$ *C_t, $\tilde{A}$ is also a model of* $\underline{C_t}$.

DEFINITION. To simplify notation (but without loss of generality) we assume that the program M consists of instructions $I_i = (i, t_j, a_k b, \ell)$ with combined printing and movement operations $a_k b$ ('print a_k and carry out the movement b') for $b \in \{left, right\}$. Let π_M be the 'conjunction' of the following expressions for all parameters i (for M-states), j (for M-letters) and all t, t', x, x', y concerned according to time and space-bounds:

$$[Z_i(t) \wedge S(t, t') \wedge S(x, x') \wedge A(t, x) \wedge B_j(t, x)]$$

$$\supset \bigvee_{(i, t_j, a_k \ right, \ell) \ in \ M} [Z_\ell(t') \wedge A(t', x') \wedge B_k(t', x)]$$

$$[Z_i(t) \wedge S(t, t') \wedge S(x, x') \wedge A(t, x') \wedge B_j(t, x')]$$

$$\supset \bigvee_{(i, t_j, a_k \ left, \ell) \ in \ M} [Z_\ell(t') \wedge A(t', x) \wedge B_k(t', x')]$$

$$A(t, y) \wedge y \neq x \wedge B_j(t, x) \wedge S(t, t') \supset B_j(t', x) .$$

{no alteration outside working-cell}

We can assume without loss of generality that M has only one direction of movement for all (i, j). For (i, j) without an m-instruction $(i, j, \ldots)$ we read $Z_i(t') \wedge A(t', x) \wedge B_j(t', x)$ instead of the disjunction.

From this definition the proof of the simulation lemma is a simple induction on t.

Remark. Consider Z_i and A, B_j, S as one-place and two-place predicate symbols respectively, t, x as variables governed by the universal quantifier, t', x' as Skølem successor terms of t, x with zero term 0 and put

$$\alpha :\equiv Z_0(0) \wedge A(0,0) \wedge \forall_x B_0(0,x) , \quad \omega :\equiv A_t \neg Z_1(t) .$$

This expresses that M, started on the empty tape in state 0 will not halt in state 1. (This is a version of Büchi's simplification (1962) of Turing's proof (1937)).

By a propositional logic interpretation of π_M for polynomial time bounded computations of non-deterministic TM-programs there follows

COOK-LEVIN THEOREM. *The satisfiability problem of propositional logic formulas in conjunctive normal form is NP-complete.*

Proof. The membership of this class of formulas to NP can easily be made clear: on input of α in conjunctive normal form 'guess' a possible evaluation of the propositional variables occurring in α and so evaluate α; α is accepted iff it thereby gets the value 1.

For the reduction of computations of non-deterministic TM-programs M which are polynomial time bounded in the length of the inputs it suffices, for arbitrary s (for step number) and $n \leq s$ (for length of input), to give propositional logic formulas π, α, ω in conjunctive normal form (with distinguished input variables $x_1, \ldots, x_n$), which can be computed by a deterministic TM-program in $c \cdot \max \{$length of $M, n, s\}^3$ steps for some constant c, and whose conjunction $\gamma(M, n, s) :\equiv \pi \wedge \alpha \wedge \omega$ simulates M by:

M accepts input word $q_1 \ldots q_n \in \{0, 1\}^n$ in s steps
iff sat$(\gamma(M, N, s) [x_1/q_1, \ldots, x_n/q_n])$.

As π we choose π_M with the time and space-parameters $t, t', x, x', y \leq s$ and the reading of the basic formulas $A(u, v)$, $Z_i(u)$, $B_j(u, v)$ as propositional variables; the conjuncts $S(u, v)$ and $y \neq x$ are read as external conditions on the admissible parameters – they are not part of the formula itself. Put

$$\alpha :\equiv \bigwedge_{1 \leq j \leq n} (B_1(0, j) \leftrightarrow x_j) \wedge (B_0(0, j) \leftrightarrow \neg x_j) \wedge$$

$$\bigwedge_{n \leq j \leq s} B_2(0, j) \wedge Z_0(0) \wedge A(0, 0)$$

(without loss of generality it can be assumed that $a_0 = 0$, $a_1 = 1$, $a_2 = $ empty sign of M).

$$\omega :\equiv \bigwedge_{i \neq 1} \neg Z_i(s)$$

(at the final instant s not state differs from the accepting state 1)

(1) holds by the simulation lemma and the intended interpretation.

Remark. For deterministic programs M, by appropriate rewriting of the condition $y \neq x$ by means of (non-negated) prime formulas, one can change π_M into a *Horn formula*. In the propositional logic case after trivial substitutions in $\alpha, \pi \wedge \alpha \wedge \omega[x_1/q_1, \ldots, x_n/q_n]$ for all $q_1, \ldots, q_n \in \{0,1\}^n$ is also a Horn formula. The satisfiability problem of propositional logic Horn formulas is solvable in polynomial time with an algorithm given by Aanderaa (1976); this algorithm can be converted to a procedure for systematically carrying out all possible unit resolutions from given first order Horn formulas and so yields a simple proof of the theorem of Henschen and Wos (1974) that *Unit resolution is complete for HORN*, that is

$$(\forall \alpha \in \text{HORN}) : \text{con}(\alpha) \text{ iff } \alpha \mapsto_{unit-res} \quad \square$$

Here, resolution steps can even be restricted to those which get resolvants σsub from a prime formula π and a Horn clause $\neg \pi \vee \sigma$. For propositional logic Horn formulas the unit resolution calculus solves the *Entscheidungsproblem* in polynomial time. (For a proof along these lines see Börger (1989), §FKIII1.)

Another interesting phenomenon which has been discovered on the basis of the above proof of Cook's theorem is the polynomial equivalence of two natural complexity measures of Boolean functions, by which the Cook problem P = NP? is shown to be equivalent to the question of short Horn definitions of Boolean functions (see Aanderaa and Börger (1979), (1981)).

Remark. The preceding propositional logic interpretation of our reduction (program, start and stop) formulae can easily be turned into a type-theoretical interpretation which then yields a description of n-fold exponential time-bounded computations of non-deterministic Turing machine programs. This gives a solution to Scholz' famous spectrum problem of (1952) as obtained piecemeal in Bennett (1962), Rödding and Schwichtenberg (1972), Jones and Selman (1974), and Christen (1974). (For the exact history see §1 in Börger (1984)). For a proof along these lines we refer the reader to Börger (1989), §FIII2. There he will also find a discussion of the particularly interesting case with $n = 2$ due to Fagin (1974), namely a representation of NP by generalized first order spectra, which was the starting point of a series of important complexity-theoretic investigations into finite model theory (see Gurevich (1984), (1988)).

We now turn our attention to the use of our reduction method for the description of time or tape-restricted TM-computations by syntactically restricted first order expressions of the pure predicate calculus. The obtained completeness results throw light, from the point of view of complexity theory, on the naturalness of the relevant syntactical characterization of the strength of logical expressions by prefix- and propositional logic structure. They also link these recent complexity theoretic characterizations of decidable logical decision problems to the intensive study of the latter in the early years of mathematical logic (see Suranyi (1959), Ackermann (1954), Börger (1984a)).

As a simple but characteristic example – for more material and references see again Börger (1989), §FIII3 – we select here the following characterization of the complexity class PSPACE:

THEOREM ON PSPACE COMPLETENESS OF THE BERNAYS–SCHÖN-FINKEL CLASS IN KROM FORMULAS. (Plaisted (1984) and Denenberg and Lewis (1984)). *The class of satisfiable Krom formulas from (even the sub-class* HORN $\cap V^2 \Lambda^\infty$ *of) the Schönfinkel–Bernays class is PSPACE-complete.*

(A *Krom* formula is a formula in prenex conjunctive normal form with alter-nations of length of at most two. The Schönfinkel–Bernays class consists of all closed prenex first order formulas without $=$ or function symbols and with prefix of the form $\exists x_1 \ldots \exists x_n \, \forall y_1 \ldots \forall y_m$ for arbitrary natural numbers n, m.)

Proof. We only show the critical part of the proof, i.e. the PSPACE-hardness of the considered decision problem. Here is the simple idea for a reinterpretation of our reduction formulae: the arity of the configuration predicate encodes the computation space bound, and the content of a cell is presented by an individual constant.

It suffices for arbitrary deterministic TM program M which is tape-bounded by a polynomial p and arbitrary input w, to construct a formula γ of length $\leq p(|w|)$ in $V^\infty \Lambda^\infty \cap$ Krom $\cap$ HORN which is contradictory exactly when M accepts the input w with tape-requirement $k := p(|w|)$. We give the propositional logic kernel of the Skølem normal form of such a γ which again has the form $\pi \wedge \alpha \wedge \omega$ with program formula π, initial formula α and final formula ω.

As indicated above, we represent a configuration $\vec{x}$ of length k by a k-

sequence of letters $j \leq s$ of the tape alphabet and of one pair (i,j) of state $i \leq r$ and tape symbol $j \leq s$ of M, the pair indicating the reading head position. These letters i and pairs (i,j) are logically treated as individual constants (read: existentially bounded variables)l. The global representation of configurations as given in Lecture 1 reads accordingly (in the proof oriented form):

Conf$(\vec{x})$ is true iff $\vec{x}$ is reached by M, started with input w,

with space consumption $\leq k$

The program formula π consists of clauses of the following form (with variable sequence xy of length $k-1$):

Conf $x(i,j)\,y \rightarrow$ Conf $x(i',j')\,y$

for print instructions (i,j,j',i') in M

Conf $x(i,j)\,j'y \rightarrow$ Conf $xj(i',j')\,y$

for right-movement instructions $(i,j,+1,i)$ in M

similarly for left-movement instructions in M.

The input formula α is Conf$(0,w_1)\,w_2 \ldots w_n 0 \ldots 0$ for input $w = w_1 \ldots w_n$ and initial state 0. The stop formula ω is $\neg$Conf$(1,0)\,0 \ldots 0$ for accepting state 1 and acceptance with empty tape.

The universal closure of $\pi \wedge \alpha \wedge \omega$ is, by the simulation lemma, contradictory when M accepts w. If w were not accepted by M, then the intended interpretation yields a canonical model for this expression.

For the reduction to the sub-class $V^2 \Lambda^\infty \cap$ Krom it suffices to carry out the above coding over the two-element alphabet $\{0,1\}$. The length of γ is thereby raised by a constant factor.

Remark. A small change in the above proof yields additionally, because M is deterministic, that the Krom and Horn formula arising from $\pi \wedge \alpha \wedge \omega$ by contraposition of its implications represents a deterministic PROLOG program (in the sense of Clocksin and Mellish (1981) and thereby the PSPACE-completeness also holds for such programs. For details see Börger (1989), §FIII3.

Remark. Plaisted (1984) has shown that PROLOG programs as in the theorem but with ternary clauses characterize the class of deterministic exponential time computable sets; the complexity jump is due to a binary encoding of big tape sections and time moments. Lewis (1980) had shown that without the Horn restriction the prefix class $V^*\Lambda^*$ has a satisfiability problem which is complete for nondeterministic exponential time bounded computations.

Remark. The possibility of having predicates of arbitrarily high rank to encode the computation space bounds is crucial in the preceding theorem. A rather complete analysis of the complexity of PROLOG program classes as above but also with fixed arity or fixed number of predicate symbols is given in Börger and Löwen (1987).

ACKNOWLEDGEMENT

I wish to express my thanks to M. Beers for the pleasure he provided with his fast and skilful editing of this manuscript.

Dipartimento di Informatica,
Università degli studi di Pisa

NOTE

* The ideas which appear in this introduction have been elaborated in a more general perspective of intrinsic connections between logic and computer science and were presented to the Colloquium of IBM Germany, Scientific Center Heidelberg on March 3, 1990.

REFERENCES

Aanderaa, S.O.: 1971, 'On the decision problem for formulas in which all disjunctions are binary', in: *Proc. 2nd Scandinavian Logic Symposium*, pp. 1–18, 19.

Aanderaa, S.O. and Börger, E.: 1979, 'The Horn complexity of Boolean functions and Cook's problem', in: *Proc. 5th Scandinavian Logic Symposium* (B. Mayoh and F. Jensen, eds.), Aalborg University Press, pp. 231–256.

Aanderaa, S.O. and Börger, E.: 1981, 'The equivalence of Horn and network complexity for Boolean functions', *Acta Informatica* 15, 303–307.

Aanderaa, S.O. and Lewis, H.R.: 1973, 'Prefix classes of Krom formulas', *JSL* 38, 628–642.

Ackerman, W.: 1954, *Solvable Cases of the Decision Problem*, Amsterdam, North-Holland Publishing Co.

Andreka, H. and Nemeti, I.: 1976, 'The generalized completeness of Horn predicate logic as a programming language', *DAI Res. Rep.* 21, University of Edinburgh.

Böhm, C.: 1964, 'On a family of Turing machines and the related programming language', *ICC*

Bull. **3**, 3–12.

Börger, E.: 1971, *Reduktionstypen in Krom- und Hornformeln; Dissertation,* Universität Münster (s. Beitrag zur Reduktion des Entscheidungsproblems auf Klassen von Hornformeln mit kurzen Alternationen, *AMLG* **16** (1974), 67–84).

Börger, E.: 1984, '*Spektralproblem* and completeness of logical decision problems', *SLNCS* **171**, 333–356.

Börger, E.: 1984a, 'Decision problems in predicate logic', in: *Logic Colloquium '82* (G. Lolli, G. Longo, A. Marcja, eds.), North-Holland, pp. 267–301.

Börger, E.: 1987, 'Unsolvable decision problems for PROLOG programs', *Springer LNCS* **270**, 37–48.

Börger, E.: 1988, 'Logic as machine: complexity relations between programs and formulae', in: *Trends in Theoretical Computer Science* (E. Börger, ed.), Computer Science Press, 1988, pp. 59–94.

Börger, E.: 1989, *Computability, Complexity, Logic,* North Holland Publishing Co., pp. XX, 592.

Börger E. and Löwen, U.: 1987, 'Logical decision problems and complexity of logic programs', *Fundamenta Informaticae* X, 1–34.

Büchi, J.R.: 1962, 'Turing machines and the *Entscheidungsproblem*', *Math. Ann.* **148**, 201–213.

Church, A.: 1936, 'A note on the *Entscheidungsproblem*', *J. Symbolic Logic* **1**, 40–41 101–102.

Clocksin, W.F. and Mellish, C.S.: 1981, *Programming in Prolog,* Springer.

Cook, S.A.: 1971, 'The complexity of theorem-proving procedures' in: *STOC,* pp. 151–158.

Denenberg, L.A. and Lewis, H.R.: 1984, 'The complexity of the satisfiability problem from Krom formulas', *TCS* **30**, 319–341.

Fagin, R.: 1974, 'Generalized first order spectra aned polynomial time recognizable sets', *Complexity of Computation* (R. Karp, ed.), SIAM-AMS Proc. 7, pp. 43–73.

Gurevich, Y.: 1984, 'Toward logic tailored for computational complexity', in: *Computation and Proof Theory, SLNM 1104,* pp. 175–216.

Gurevich, Y.: 1988, 'Logic and the challenge of computer science, in: *Trends in Theoretical Computer Science* (E. Börger, ed.), Computer Science Press, 1988, pp. 1–57.

Gurevich, Y.: 1988a, 'Algorithms in the world of bounded resources', in: *The Universal Turing Machine – A Half-Century Story* (R. Herken, ed.), Oxford University Press, pp. 407–416.

Henschen, L. and Wos, L.: 1974, 'Unit refutations and Horn sets', in: *JACM* **21**, 590–605.

Hill, R.: 1974, 'Lush resolution and its completeness', in: *DLC Memo 78,* University of Edinburgh.

Itai, A. and Makowsky, J.A.: 1983, *Unification as a Complexity Measure for Logic Programming,* Technion-Israel Inst. of Technology, TR 301.

Lewis, H.R.: 1980, 'Complexity results for classes of quanitificational formulas', *JCSS* **21**, 317–353.

Minsky, M.L.: 1961, 'Recursive unsolvability of Post's problem of Tag' and other topics in the theory of Turing machines', *Ann. Math.* **74**, 437–455.

Plaisted, D.A.: 1984, 'Complete problems in the first-order predicate calculus', *JCSS* **29**, 8–35.

Scholz, H.: 1952, 'Ein ungelöstes Problem in der symbolischen Logik', *J. Symbolic Logic* **17**, 160.

Sebelik, J. and Stepanek, P.: 1982, 'Horn clause programs for recursive functions', in: *Logic Programming* (K.L. Clark, S.A. Täarnlund, ed.), Academic Press, pp. 325–341.

Suranyi, J.: 1959, *Reduktionstheorie des Entscheidungsproblems in Prädikatenkalkül der ersten Stufe,* Budapest.

Tärnlund, S.-A.: 1977, 'Horn clause computability', *BIT* **17**, 215–226.

Trakhtenbrot, B.A.: 1953, 'O Recursivno Otdelimosti', *Dokl. Adad. SSSR* **88**, 953–955.

Turing, A.M.: 1937, 'On computable numbers, with an appliation to the *Entscheidungsproblem*', *Proc. London Math. Soc.* **42** (2), 230–265. 'A Correction' *ibid.* **43** (1937), 544–546.

DAG PRAWITZ

REMARKS ON HILBERT'S PROGRAM FOR THE FOUNDATION OF MATHEMATICS*

A foundation of mathematics has two sides, a philosophical one and a mathematical one. One expects from a foundation of mathematics something more than a philosophical theory. As the term 'foundation' suggests, there should be a basis upon which mathematics can be erected systematically. On the other hand, a foundation of mathematics is something more than just an axiomatization of mathematics. It is also concerned with the nature of the basis upon which mathematics is to be built. A foundation takes a stand upon the question what kind of basis is appropriate.

A foundation of mathematics, as I will use the term, has in other words two ingredients: it contains a philosophy of mathematics and a systematic development of mathematics in accordance with this philosophy. Frege, Russell, Hilbert and Brouwer all aimed at a foundation of mathematics in this sense. I shall here be mostly concerned with the philosophical side of foundational work, i.e. the program for the development of mathematics, rather than the execution of the program or the actual development. As a background to this, I shall first briefly comment on some traditions in the philosophy of mathematics as well as in the axiomatization of mathematics.

The reader should recall that the philosophy of mathematics has not been just an isolated branch of philosophy, but has often stood at the center of philosophy and has been a place in which many major questions of philosophy were born.

Already in Greek philosophy there are striking examples of this. With the rise of deductive practice in Greek mathematics, the problem arose of how it can be possible to establish truths just by reflection, by sheer work of the intellect without empirical investigations, and how it can be that the truths obtained in that way are not only absolutely certain but also useful in practice.

Plato's theory of ideas was an attempt to answer these questions. Mathematics is, according to Plato, about an ideal world that is accessible to us through the intellect. In contrast to the changing world known to us by sensations, about which one can only have opinions and where nothing is certain, the ideal world is stable and can be seen clearly by the mind's eye. We then also learn something, although something less exact and certain, about the world

87

G. Corsi et al. (eds), Bridging the Gap: Philosophy, Mathematics, and Physics, 87–98.
© 1993 Kluwer Academic Publishers.

of phenomena, because that world is a pale shadow, an imperfect copy, of the ideal world.

Plato was in this way offering an explanation of how mathematical truths can have simultaneously the three properties of (i) being possible to arrive at by reflection only, (ii) being certain, and (iii) being useful in practice. And from this explanation he developed his whole philosophy.

Kant is a more recent example of a philosopher whose mature philosophy was the result of reflections on the nature of mathematics. His main question, how synthetic judgements *a priori* are possible, is obviously concerned with much the same problem as Plato, how it is possible to combine the three properties (i)–(iii); and mathematical truths are again the prime examples of such judgements. The whole of Kant's later philosophy is based on the explanation of how such judgements are possible, but the explanation is, of course, quite different from Plato's.

In some forms, Plato's and Kant's questions and answers still remain in the philosophy of mathematics and in philosophy in general. What is called *platonism* in the philosophy of mathematics today is not Plato's full theory of the subject, but it shares with Plato the belief that mathematics is about a realm of objects that are given independently of us and of our capacity to reach them – we may leave open whether or not Plato shared the belief that mathematical truths are also independent of us in the sense that they may be hidden to us, not only in practice, but in principle. In any case, in modern discussions such a *realism* under the name of platonism stands against *anti-relaism* or *epistemic idealism*, which along with Kant takes mathematical truths to be conceptually tied to our epistemic capacities. Here again, mathematics often serves as a paradigmatic example in discussions that have a general import far outside the mathematical context.

There is also a long and important tradition behind the other ingredient of a foundation of mathematics, the systematic development of the subject from a small basis. It, too, starts with the Greeks – Aristotle already stated clearly an axiomatic ideal which was fulfilled shortly afterwards in Euclid's work – but the foundational schools of our century should be understood against a background of a growing rigour in the development of mathematics during the 19th century, in particular. If one examined carefully the formulation of mathematical analysis at the time of Leibniz and Newton, one would find, as e.g. George Berkeley did, that the formulation was inconsistent. There is, however, a steady growth in rigour from then on, which leads to the arithmetization of notions such as limit, derivative and integral by work due to Weierstrass and others in the last century, and finally to the reduction of real numbers to sequences or sets of natural numbers by work due to Cauchy and Dedekind at the end of the last century.

The three influential schools in the foundation of mathematics which were born around the turn of the century, Frege's and Russell's *logicism*, Hilbert's *formalism* and Brouwer's *intuitionism*, all try to continue this development but in quite different ways. Whereas Frege and Russell seek a further reduction of mathematics to logic, both Hilbert and Brouwer are opposed to such a development and agree that mathematics is to be based on the most elementary experiences of counting and artithmetical operations, but they then disagree on how mathematics is to be based on such experiences.

Today one finds almost no adherents of logicism, and although a modified form of Hilbert's program is still flourishing in proof theory, one finds few who really defend the formalist position taken by Hilbert. It is different with intuitionism. The followers of Brouwer have always formed a small minority, but both the philosophical debate for and against intuitionism and the mathematical developments within intuitionism are still vigorous. There is now little of the once strong rivalry between formalism and intuitionism. Philosophical discussion around intuitionism has instead partly fused with the debate between realism and anti-realism mentioned above, and in this respect, the philosophy of mathematics is still linked with major problems of philosophy in general. There is, however, another trend in the philosophy of mathematics attached to foundational works that should be mentioned and that deliberately shuns fundamental philosophical issues.

Regardless of one's views about all of this, something can be learned by considering in more detail some aspects of the major programs in the foundations of mathematics, one of which we shall now turn to.

1. GENERAL CHARACTER OF THE PROGRAM

Stated roughly in a few words, Hilbert's program is the program of carrying further the axiomatic development of the 19th century by formalizing mathematics, in particular mathematical analysis, with the aim of eliminating remaining references to the infinite. This is to be achieved in two steps, namely by first codifying mathematical analysis in a formal system, and then showing that all figures of speech that seem to refer to a completed infinite totality are only a harmless manner of speaking, in the sense that such references can be understood as just meaningless formulas, the manipulation of which is consistent with the meaningful, finitistic part of mathematics.

The problem of proving the consistency of mathematical analysis was already stated by Hilbert in 1900, see Hilbert (1901). It occurs as the second problem in the famous list of open problems that Hilbert presented at the Second International Congress of Mathematics in Paris that year. In 1904 Hilbert, see Hilbert (1905), said something more about how he understood this problem, but only in a series of papers from 1918, see Hilbert (1918), and onwards did he explain in detail how he thought the problem was to be solved and in what way such a solution would constitute a foundation of mathematics. Perhaps the most articulate and forceful statement of the program occurs in a lecture that Hilbert gave in 1925 and published in the paper 'Über das Unendliche' (1926). The standard reference for information on early works on Hilbert's program is otherwise the two-volume *Grundlagen der Mathematik* by Hilbert and Bernays published in the thirties (1934, 1939).

2. HILBERT'S KIND OF FORMALISM

The philosophy of mathematics contained in Hilbert's program is usually referred to as formalism, and the first question that I want to take up in more detail is what kind of formalism is involved here.

Strict formalism in the philosophy of mathematics is usually taken to be a view according to which mathematics can be described as a rule-governed activity, or more precisely as an activity in which formal objects defined by certain *formation rules* are dealt with in accordance with certain *transformation rules*. The mathematician may think of the objects as formulas expressing propositions and of the transformations of them as acts of inference, but according to strict formalism this is not an essential ingredient when one wants to understand the real nature of mathematics, which is rather to be likened to a game, whose objects and moves have no deeper meaning in themselves beyond the significance given to them by the rules of the game.

In some popular expositions, Hilbert's position is conceived as that of strict formalism in the sense just explained. The goal of his program is then described as to find out the rules according to which the activity is carried out or the game is played, which just amounts to formulating mathematics as a formal system, and the demonstration of the fact that the rules are consistent. The point of this demonstration is to guarantee the interest of the game, since an inconsistency would imply that all moves in the game were allowed (all formulas were provable), which would amount to an uninteresting game.

Strict formalism is a doubtful doctrine and does not have many adherents. Regardless of what one thinks of its intrinsic consistency, it should be clear that is was not Hilbert's position and that it is really inconsistent with his insistence on the mathematical character of the consistency problem. The point of proving the consistency of some rules is (both for a strict formalist and) for Hilbert to establish a certain truth, not just to find a formal proof in a given metacalculus, and, for Hilbert, to prove the consistency is also a *mathematical* task. Hence, he must recognize that there are at least some meaningful statements in mathematics for which the question arises whether they are true or false. It would certainly be strange if the only meaningful mathematical statements were the ones used in the consistency proof. But if, for instance, assertions *about* a formal system that codifies mathematics are understood as meaningful mathematical assertions, why then could not also formulas *in* the formal systems be understood as expressing meaningful propositions about numbers and other mathematical objects?

Hilbert's program thus requires that one part of mathematics is understood as meaningful and trustworthy, while another part is considered to be more problematic. This was also Hilbert's viewpoint. He distinguished between one sphere of unproblematic *finitary* propositions that are clearly meaningful and about which we have perfect intuition, and another sphere of more problematic *transfinite* propositions. Typical examples of the former are propositions stating the result of a calculation, while a proposition referring to an infinite set understood as a completed totality typically examplifies the latter. The question where exactly the borderline between the two kinds of propositions is to be drawn is something to which we shall soon return.

Finitary propositions are understood by Hilbert as *real* propositions, while transfinite propositions are called *ideal* propositions (in analogy with so-called ideal elements that are sometimes introduced in mathematics). Only towards the latter does Hilbert take a formalist attitude. Sentences expressing such propositions are understood as only a *façon de parler* or as sentences "which do not really mean at all what the words in them purport to mean" to quote Gentzen (1969, p. 247). The only meaning attached to them is the one given by the rules of the formal system in which they occur.

3. STATEMENT OF THE PROGRAM

If the ideal propositions are not considered to have an independent meaning (outside the formal context) that justifies reasoning involving them, there is no guarantee that they do not interfere with real propositions in an inconsistent way. This is the consistency problem. An inconsistency may arise because there are in general several ways in which a proposition can be proved. For instance, to prove that $2^3 \cdot 2^5 = 2^8$ we may either directly calculate the two sides of the equality and verify that they are both equal to 256 or we may use the law that for all a and b, $2^a \cdot 2^b = 2^{a+b}$ (and only then calculate $3 + 5$). In particular, we may prove a real proposition by a proof that stays completely within the real part of mathematics, but we may also prove it by a proof that makes an excursion into the ideal part. In order to have any trust in a system with ideal proposition, we need to know that a proof of a real proposition by excursions into the ideal part can never come into conflict with proofs in the real part, or more generally, with the content of the real propositions.

With these observations in mind, we may now state Hilbert's program a little more precisely by distinguishing the following tasks that it contains:

(1) Codify mathematical analysis (or some other part of mathematics to which the program is applied) as a formal system S.

(2) Identify a set R of formulas in S that express real propositions and for which we hence know what it means that they are true or false.

(3) Show in the real part of mathematics, either in the real part of S or in some extension of it, that for each $A \in R$, if $\vdash_S A$, then A is true.

Establishing (1–3) is thought of as giving a finitistic foundation of mathematical analysis (or some other part of mathematics treated in this way). It establishes by finitistic means the (finitistic) truth of the provable propositions of the theory to the extent that they are interpreted in a finitistic way. The other propositions can then be looked upon as merely an expedient to establish finitistic truths as exemplified above where using the law that for each a and b, $2^a \cdot 2^b = 2^{a+b}$ (which need not, but may, for the sake of illustration, be interpreted as a transfinite proposition) greatly simplifies the verification of a finitistic proposition.

Having shown (3), the consistency of S obviously follows. Conversely, given certain other conditions (see the end of Section 4), consistency implies the truth of the provable real sentences. To say that Hilbert's program aims to show the consistency of S is thus not entirely wrong, but it is (3) that really matters if we are to speak about a finitistic foundation.

4. THE BORDERLINE BETWEEN REAL AND IDEAL PROPOSITIONS

If the atomic sentences of S have a finitistic meaning, which is the case, for instance, when they are decidable, then so have all sentences of S built up by truth-functional connectives and quantifiers restricted to finite domains.

Quantifiers over infinite domains can be looked upon in two ways. One of them may be hinted at as follows. Let x range over the natural numbers, and let $A(x)$ be a formula such that $A(n)$ expresses a finitary proposition for every number n. Then a sentence $\forall x A(x)$ expresses a transfinite proposition, if it is understood as a kind of infinite conjunction which is true when all of the infinitely many sentences $A(n)$, where n is a natural number, hold.

Similarly, a sentence $\exists x A(x)$ expresses a transfinite proposition, if it is understood as a kind of infinite disjunction which is true when, of all the infinitely many sentences $A(n)$, where n is a natural number, there is one that holds. There is a certain ambiguity here, however, depending on what is meant by 'all' and 'there is one'. To indicate the transfinite interpretation one should also add that the sentences are understood in such a way that it is determined, regardless of whether this can be proved or not, whether all of the sentences $A(n)$ hold or there is some one that does not hold.

If instead an assertion of $\forall x A(x)$ is understood as asserting that there is a method which, given a specific natural number n, yields a proof of $A(n)$, then we have to do with a finitary proposition. Similarly, we have a case of a finitary proposition, when to assert $\exists x A(x)$ is the same as to assert that $A(n)$ can be proved for some natural number n.

It is to be noted that the statement in (3) above is a universal sentence. Hence, the possibility of giving a finitary interpretation of the universal quantifier is a prerequisite for Hilbert's program. Does the possibility of interpreting the quantifiers in a finitary way also mean that one may hope for a solution of the problem stated in (3) when all quantified sentences interpreted in that way are included in R?

A little reflection shows that the answer is no, but that R may always be taken as closed under universal quantification. For it can be seen (uniformly in A) that if we have established (3) when R contains all instances of a sentence $\forall x A(x)$, then (3) also holds for $R^+ = R \cup \{\forall x A(x)\}$. To see this let $\forall x A(x)$ be a formula provable in S whose instances belong to R, and let a method be given which applied to any formula in R and a proof of it in S yields a proof of its truth. We want to show that $\forall x A(x)$ is true when interpreted in a finitistic way, i.e. that we have a method which applied to any natural number n yields a proof of $A(n)$. The existence of such a method is obvious, because, from the proof given of $\forall x A(x)$ we get a proof of $A(n)$, for any n, and hence by specialization of the given method, we have a method which yields the required proof of the truth of $A(n)$, for any n.

Having included universal sentences $\forall x A(x)$ in R such that all $A(n)$ are decidable, it is easy to see that one cannot in general also let existentially quantified sentences be included in R, if (3) is still to be possible. For let

S contain classical logic and assume that R contains undecidable sentences $\forall x A(x)$ with $A(n)$ decidable; by Gödel's theorem there are such sentences if S is sufficiently rich. Then one cannot allow R to be closed under existential quantification. In particular, one cannot allow formulas $\exists y(\forall x A(x) \vee \neg A(y))$ to belong to R for all A: the formulas are provable in S but all of them cannot be expected to be true when interpreted in a finitistic way, because then, for any A, we would get a proof of $\forall x A(x) \vee \neg A(n)$ for some n, which would let us decide $\forall x A(x)$.

In accordance with these observations, the line between real and ideal propositions was drawn in Hilbert's program in such a way as to include among the real ones decidable propositions and universal generalizations of them but nothing more; in other words, the set R in (3) is to consist of atomic sentences (assuming that they are decidable), sentences obtained from them by using truth-functional connectives, and finally universal generalizations of such sentences.

Given that R is determined in this way and that the atomic sentences in the language of S are decidable and provable in S if true (and hence that the same holds for truth-functional compounds of atomic sentences in S), which is normally the case, the consistency of S is easily seen to imply the statement in (3) as follows. Assume consistency and let A be a sentence without quantifiers that is provable in S. Then A must be true, because, if it was not, then $\neg A$ would be true and hence provable in S by the assumption made about S, contradicting the consistency. Furthermore, a sentence $\forall x A(x)$ provable in S must also be true, because there is a method such that for any given natural number n, the method yields a proof of $A(n)$. Indeed, just apply the decision method to $A(n)$; by the consistency and the assumption on S, it must yield a proof of $A(n)$ and not of $\neg A(n)$.

5. AN AMBIGUITY IN THE MOTIVATION OF THE PROGRAM

As we have seen, the *raison d'être* of Hilbert's program is that some propositions, the transfinite ones, are seen as more problematic than the others, the finitary ones. However, the motivation of the program and what one expects from its depends very much on what kind of doubts one has about the transfinite sentences. At this point there is a certain ambiguity in Hilbert's writings and there are also different stands by followers of Hilbert. Some of these positions can be ordered almost linearly according to how problematic one judges the transfinite propositions to be. The two extremes may be described as follows.

The transfinite part of mathematics is completely trustworthy. Hilbert often expresses the conviction that all of classical analysis, including the transfinite part, is in perfect order. In a lecture (Hilbert, 1922), he says that we have a complete and justified certainty about the inferences in analysis. A critic of classical analysis like Weyl is accused of seeing ghosts. Hilbert's project is not to *investigate* whether analysis is consistent, it is just to *establish* the consistency. The problem with the transfinite part of mathematics seems to be only social: there are misled critics of this part of mathematics like Brouwer and Weyl, and they need to be set right by a consistency proof.

The transfinite part of mathematics lacks any meaning and is to be understood instrumentalistically. On the other hand, in other contexts, Hilbert said that the infinite in the sense of an infinite totality, as it is still used in deductive methods, is something fictitious. The transfinite mode of speaking, Hilbert said, is just a *façon de parler*; if one is to ascribe a meaning to the quantifiers (and to the logical connectives used outside the quantifiers), it must be their finitary meaning. But since that meaning is not in accordance with classical logic, and since we should not deny ourselves the great usefulness of classical, transfinite reasoning, the right procedure is to treat sentences expressing transfinite propositions as just meaningless formulas. Hilbert's program is then objectively needed to guarantee that the manipulation of these formulas leads to the right result as far as real, finitary propositions are concerned. This position is what is usually called *instrumentalism*, although Hilbert did not use this term.

Hilbert thus vacillated between these two positions. Between them there are various middle positions that can be characterized in general as follows.

The transfinite propositions have a meaning but it is less clear and the reasoning with them is less secure than in the case of finitary propositions. Middle positions of this kind are represented by many logicians who have worked on Hilbert's program such as Gentzen and Takeuti. Gentzen spoke about the "actual sense" of transfinite propositions. He admits in a paper from 1936–37 that the objection to considering such a sense at all "must in any case not be taken too lightly; it is not entirely without merit" (Gentzen 1969, p. 229). His own position at that time was that the transfinite proposition has a "certain 'sense'" but that the whole question of 'sense' is not ready for a final settlement. In a slightly later paper he writes: "Indeed, it seems not entirely unreasonable to me to suppose that *contradictions* might possibly be concealed even in classical *analysis*" (Gentzen 1969, p. 235). In a similar vein, Takeuti describes the intuition behind classical mathematics by introducing what he calls the standpoint of an "infinite mind". He admits that "mathematicians have an extremely good intuition about the world of the natural numbers as conceived by the infinite mind" (Takeuti 1987, p. 298). Nevertheless, since our minds are finite, "we need reassurance of such a world" (Takeuti 1987, p. 100), which is the point of Hilbert's program, particularly urgent when we come to "our imaginary world of sets" (Takeuti 1987, p. 298).

6. MERITS AND WEAKNESSES OF THE PROGRAM

I said above that the possibility of contradictions is quite generally due to the existence of several ways in which a proposition can be established. This was exemplified (Section 3) by two ways in which an equation could be verified, viz. directly by means of a calculation and indirectly by means of a proof using a general law (whether a transfinite proposition or not). Clearly, if there is a conflict between these two ways, then the direct way has priority; the reason being that it is the calculation procedure that determines the meaning of the equation in question. Hence, step (3) in Hilbert's program is something that one certainly wants to carry out.

Hilbert's way of looking at the atomic sentences is of course in sharp contrast with logicism, according to which the meaning of arithmetical equations is explained in terms of much more abstract concepts. But there is an analogy with the usual hypothetical-deductive method of natural sciences, which gives priority to observations over theoretical proofs in case of conflicts concerning observational sentences, because the very meaning of the observational sentences is understood as given in terms of observations.

The phenomenon that there are different ways in which a proposition may be established and the need to ensure that they are not in conflict – which, to its merit, Hilbert's program drew attention to – got very much clarified by Gentzen's contribution to Hilbert's program, which was also a general contribution to proof theory. Gentzen showed that the distinction between different ways in which a proposition can be established, some proofs being direct because of the meaning of the proposition and other proofs being indirect (here, indirect proof does not mean proof by *reductio ad absurdum* but just proof that is not direct), is relevant not only for atomic sentences but for sentences in general. The point of his *Hauptsatz* or of the normalization of proofs is just to show that these different ways stand in harmony with each other, and in doing so, one establishes a much stronger and more general result than Hilbert's program aimed at, but one which it contained in embryo.

The main weakness of Hilbert's program seems to me to be that it does not go far enough. Unlike some of the theories in the philosophy of mathematics mentioned in the introduction, the philosophical part of the program seems to have little impact on philosophy in general and few interesting connections with current, main issues in philosophy. But also from the point of view of the limited task of accounting for the transfinite propositions, which it sets itself, the program seems rather inadequate.

Shortcomings in this respect were already pointed out by Gödel (1931) in a short note. Gödel remarks that the phenomenon of ω-inconsistency means that a system for which Hilbert's program has been carried out can still have the abnormal feature that a sentence $\exists x A(x)$ is provable in it, although for each natural number n, $A(n)$ is a finitistically false sentence.

There are several lessons to be drawn from Gödel's remark, depending on what attitude one has towards the transfinite propositions.

If one takes what I called a middle position and thinks that transfinite propositions have a meaning but a somewhat problematic one, then it is clear that a successful fulfillment of Hilbert's program for a system does not exclude the possibility that a sentence $\exists x A(x)$ is provable in it although it is false, not only according to a finitary interpretation of it, but also according to a transfinite interpretation. Hilbert's program is thus not a way to obtain greater certainty about the transfinite propositions; although one would show that their use cannot give rise to false finitary propositions, provable transfinite propositions may still simply be false by their own standards. The ambition of the program is thus too low.

If one takes the extreme view that transfinite propositions lack all sense, then the precise criticism above is not relevant. But Gödel's example seems to me to show the implausibility of such an instrumentalistic position. No one would seriously entertain a theory in which $\exists x A(x)$ was provable but $A(n)$ false for all natural numbers n, even if the theory was consistent, and this shows that we are not prepared to take a merely instrumentalistic attitude to existential sentences.

More generally, it may be noted that there is a strange tension in the philosophy behind Hilbert's program. On the one hand, it values highly the transfinite part of mathematics, but on the other hand, it is ready to discard transfinite interpretations and to take transfinite propositions as just formal instruments, retaining only their finitary interpretations when possible. If one thinks that there is a valuable transfinite meaning but that it is unclear, then the reasonable thing to do is to try to clarify this meaning. Even if one thinks that the transfinite part is in perfect order and only wants to defend it against unfounded criticism, it seems that the thing to do is to clarify its meaning to the critics. But, as we have seen, Hilbert's program abandons transfinite meanings and cannot clarify them. If, on the other hand, one thinks that only finitary propositions have a real sense, then it is after all very strange to keep the transfinite part as a kind of black box, a machinery whose function we do not understand. In that case it would be more reasonable to follow the procedure of intuitionism and try to develop as far as possible the mathematics that one considers meaningful.

Department of Philosophy,
Stockholm University.

NOTE

* These lectures notes cover the first half of the lectures on the foundation of mathematics that I gave at the International School of Philosophy of Science and which dealt with not only Hilbert's program. This explains the occurrence of a fairly long introduction. It should also be said that the aim of the lectures was to discuss some selected questions with students who already had some knowledge of the topics treated and not to give a complete, introductory account of the area.

Some remarks in the lectures are taken from my paper Prawitz (1981).

REFERENCES

Gentzen, Gerhard: 1969, *The Collected Papers of Gerhard Gentzen* (ed. M.E. Szabo), Amsterdam, North-Holland.

Gödel, Kurt: 1931, 'Contribution to: Diskussion zur Grundlagung der Mathematik', *Erkenntnis* **2**, 135–155.

Hilbert, David: 1901, 'Mathematische Probleme', *Archiv der Mathematik und Physik 3rd Series* **44–63**, 213–237.

Hilbert, David: 1905, 'Über die Grundlagen der Logik und Arithmetik', in *Verhandlungen des Dritten Internationalen Mathematiker-Kongresses in Heidelberg vom 8. bis 17. August 1904*, Leipzig, Teubner, p. 174–185.

Hilbert, David: 1918, 'Axiomatisches Denken', *Mathematische Annalen* **78**, 405–415.

Hilbert, David: 1922, 'Neubegründung der Mathematik. Erste Mitteilung', *Abhandlungen aus dem mathematischen Seminar der Hamburgischen Unvirsität* **1**, 157–177.

Hilbert, David: 1926, 'Über das Unendliche', *Mathematische Annalen* **95**, 161–190.

Hilbert, David and Bernays, Paul: 1934, *Grundlagen der Mathematik*, I, Berlin, Springer.

Hilbert, David and Bernays, Paul: 1939, *Grundlagen der Mathematik*, II, Berlin, Springer.

Prawitz, Dag: 1981, 'Philosophical aspects of proof theory', *Contemporary Philosophy. A New Survey.* Vol. 1, The Hague, Boston, London, Martinus Nijhoff Publishers, pp. 235–277.

Takeuti, Gaisi: 1987, *Proof Theory*, 2nd edition, Amsterdam, North-Holland.

SOLOMON FEFERMAN

WORKING FOUNDATIONS - '91*

1. WHAT'S THE USE OF FOUNDATIONS?

There is currently a general malaise about the logical approach to the foundations of mathematics. One main reason is that foundational thought in this century has been dominated by a few global views about the nature of mathematics – logicism, formalism, platonism, and constructivism – each of which has proved to be defective in substantial ways, while nothing else has come to take their place. In recent years there has been an increasingly steady barrage of criticism directed against these all too familiar positions. Some of that criticism has been very sophisticated, coming from within the field of logic itself. Other criticism has come from mathematicians who would jettison the whole approach via logic and formal systems. Still more specifically on the mathematical side have been movements to replace logical foundations by purely mathematical foundations (e.g., using category theory)[1]. Though these various critical views diverge from each other in many significant respects, they share two common themes: (i) mathematics is more reliable than any of the foundational schemes which have been proffered by the logicians to 'secure' it, and (ii) the logical analysis of mathematics bears little or no relation to actual mathematical practice[2].

I believe that the critical evidence must be taken seriously, but disagree with the conclusions. On the contrary, I think a very strong case can be made for the logical approach to the foundations of mathematics, even when we cease to be preoccupied with the grand schemes. As part of the support for this, the present paper offers a picture of *logical foundations at work*, detailed by a multitude of examples. The main point to be made is that *this work is a direct continuation by more conscious, systematic means of foundational moves which have been carried on within mathematics itself from the very beginning.* Each of the following sections (2–7) is devoted to a particular kind of foundational activity, and begins with familiar examples from mathematics. Each section then goes on to give a number of metamathematical examples, some of which will be familiar to a general audience but most of which will not[3].

The varieties of foundational work to be exemplified here are categorized as

99

G. Corsi et al. (eds), Bridging the Gap: Philosophy, Mathematics, and Physics, 99–124.
© *1993 Kluwer Academic Publishers.*

follows: *conceptual clarification; interpretation, reduction or elimination of problematic concepts and principles; foundations of problematic methods and results; organizational and axiomatic foundations;* and *reflective expansion of concepts and principles.* Except for the last, each of these is a more or less recognized mode of foundational activity for which there are many classical examples[4]. Current research provides new, interesting examples where such moves can be seen in operation. However, the last category – reflective expansion – is a feature of the progress of mathematical thought which has been insufficiently appreciated. Work whose aim is to capture that process in logical terms will be described.

The word 'working' in the title is meant to be ambiguous. On one hand it is meant to suggest (as above) the active pursuit of foundational questions without regard to any fixed philosophical standpoint. Another related intention is to suggest the provisional character of what is accomplished, as part of the evolution of *all* mathematical understanding. Finally, it is meant (more personally) to indicate the work itself which has engaged my interests and attention. In presenting the illustrative examples below, I have drawn liberally on my own efforts and on the work of many others that has enlightened me in this sphere of interests[6].

My original plans for this paper had been somewhat more ambitious. In addition to the material already described, I had wanted to say more in defense of the logical analysis of mathematics and against the argument that it has nothing to do with our everyday experience. My position is that without logic, one will be unable to give a satisfactory answer to the question: *What is it about the concepts and methods of mathematics that makes it such a distinctive body of thought?* My ideas about the logical analysis of everyday mathematical experience have previously been advanced, in part, in Feferman (1979a, 1981), but I had hoped to enlarge upon them here. As the present paper grew in length and my time to prepare it ran out, I decided to postpone doing that to another occasion. Instead, some indication of the further direction of my ideas is given in a postscript to the present paper.

Now, without further ado, let us launch into this survey of types and examples of foundational work.

2. CONCEPTUAL CLARIFICATION

This usually takes place at a reasonably advanced stage in the organization of a subject (cf. Section 6 below). After basic concepts have been settled on and are well understood, one looks for precise definitions of other frequently used informal concepts in terms of them. In certain cases it can be shown that only one such definition is possible meeting specified *adequacy conditions*. In other cases, the criteria for accepting an exact definition of an informal concept are subtler and involve a whole range of experience to be accounted for.

Some classical examples from mathematical analysis are the definitions of *limit, derivative, continuous function*, and *area*. Early in this century the concept of *dimension* occupied much attention in topology. More recently, the ideas of *natural mapping* and *universal construction* in algebra and topology have been defined in terms of category theory.

The most outstanding examples from logic are the definitions of *truth* and *satisfaction* (Tarski) and of *mechanically computable function* (Turing). The former also gives precise meaning to the concepts of *definability* and *logical consequence*, each with reference to a suitably specified *formal language*. The notion itself has at its core the idea of *well-formed formula*, i.e., a syntactically determinate expression of a property, relation or statement.

In the 50's the concept of truth in a language L was used to explain the informal idea of a *transfer principle* from algebra, of which *Lefschetz' principle* in algebraic geometry was the most famous example. Results stated in L can be transferred from one structure $\mathcal{M}$ to another structure $\mathcal{M}'$ if $\mathcal{M} \equiv_L \mathcal{M}'$, i.e., if $\mathcal{M}$ and $\mathcal{M}'$ satisfy exactly the same statements from L. For the strongest present formulations of Lefschetz' principle in these terms, see Feferman (1972) and Eklof (1973). A variety of interesting transfer principles in algebra are given in Cherlin (1976).

Nowadays we think of the notions of limit, continuous function, etc., as being given relative to a topological space. Similarly, we have emphasized that the semantical notions just discussed are relative to a given language (and, implicitly, a given class of subject structures). On the other hand, the concept of mechanically computable function seems to have an absolute character. Such a function is supposed to be computed by a *finite algorithmic procedure* (f.a.p.) operating on *finitely presented data*. However, there is an informal relative concept of *finite effective construction* (e.g., in geometry or algebra) which does not involve any special assumption about the form in which data is presented. One plausible explanation of f.a.p. in general relative terms, i.e., applying to any structure, has been given by Friedman (1971). Less clear are various ideas about *infinitary constructions* which have been explored in a number of generalizations of recursion theory; for one approach and relevant literature see Fenstad (1980).

A ubiquitous mathematical and logical concept is that of *inductively generated set*. The set-theoretic definition of this in terms of *closure operators*, which has now become standard, is reviewed in Aczel (1977); its use also makes clear the ideas of build-up of the set *from below* and definition *from above* ("the least set closed under ..."). In terms of these notions one can go on to distinguish different kinds of inductive definitions: *finitary, effective, deterministic*, etc. There is a *functional* approach to ordinary recursion theory which uses *inductive schemata*; this can be generalized to much wider contexts, as is advanced in Feferman (1977a), Moschovakis (1977 and 1991). An alternative relational approach uses *effectively inductively generated relations*; for generalization of this see (the same references as well as) Fitting (1981). Finally, a notion of *inductively presented formal system* has been advanced in Feferman (1982), improved and extended in Feferman (1989); this is needed particularly for a proper general treatment of Gödel's 2nd incompleteness theorem, where the *canonical* construction of *formal consistency statements* had problematic aspects.

There are several important informal concepts in logical work for which one has not yet arrived at any reasonably convincing precise explanation, though there are technical concepts which serve as partial substitutes. The following are two such examples. First is the idea of *identity of proofs*, which is actually used informally by mathematicians all the time, e.g., when they speak of two proofs being essentially the same or of giving a new proof of a known result. Relevant notions from the general theory of proofs are explained in Prawitz (1971)[6]. The second is the idea of *natural well-ordering*, of which the most familiar example is the ordering of type ε_0 provided by Cantor's normal form. Since Gentzen used this to give a consistency proof of arithmetic, technical proof theory has been dominated by the use of effective transfinite methods and arguments. A precise definition of the concept of natural w.o. is needed to explain what is accomplished by such consistency proofs (since one can always give trivial demonstrations of consistency by induction on suitable 'cooked-up' recursive well-orderings). Approaches to this question through category theory can be found in Feferman (1968) and Girard (1981).

3. INTERPRETATION OF PROBLEMATIC CONCEPTS AND PRINCIPLES

At each stage in the development of mathematics there is a body of ideas and methods which are considered to be well-understood and accepted (though their place in our understanding may subsequently undergo considerable revision), with which mathematics in the main at that stage is confidently carried on. However, concepts and principles often emerge which have some intrinsic plausibility, or have proved themselves to be particularly useful, and yet which are felt to be less certain or secure than what is then commonly accepted. There are different kinds of foundational moves which are taken to deal with such situations. In this section and the next we separate out two of the main kinds of moves.

Basically this section has to do with cases of troublesome notions or propositions whose use is justified or in which one's confidence is increased by means of interpretations in terms of the body of mathematics accepted at the time. Classical examples of such notions from mathematics are *zero, negative numbers, imaginary* and *complex numbers, points at infinity*; examples of such statements are the *parallel postulate* (and its negation) and, more recently, the *well-ordering principle*. Nowadays, we do not view such problematic cases in isolation, but rather as part of an axiomatic system of requirements which must be met as a whole. For example, we say that the use of complex numbers is justified by explicit construction of a *field* C extending the field of real numbers by adjunction of a root i of the equation $x^2 = -1 : C = \mathbb{R}[i]$. But for the original interpretations (by Wessel, Gauss, Argand and others) which secured the use of complex numbers, the conditions to be met were not explicitly formulated. The exception to this observation, of course, is the case of geometry, with its longstanding Euclidean axiomatic setting. The use of axiomatic organization for other parts of mathematics only began to receive conscious attention in the latter part of the 19th century, along with a critical re-examination of axiomatic geometry itself (cf. Section 6 below).

We turn now to examples from logic; the best known cases are the *models* given by Gödel (1939, 1940)[7] for the *axiom of choice*, AC, as well as the *continuum hypothesis*, CH (and its generalization GCH), and by Cohen (1966) for their *negations*. Here the usual axiomatic framework taken as basic is ZF, the system of Zermelo–Fraenkel set theory, and the results are frequently formulated as *consistency* and *independence* results, e.g., that (i) $ZF + AC + GCH$ is consistent, (ii) AC is independent of ZF, and (iii) CH is independent of $ZF + AC$. Further, analysis of the arguments shows that only the principles of ZF need to be assumed in constructing these models, so that one can formulate the conclusions here as *relative consistency results*: if ZF is consistent then: (i)' so also is $ZF + AC + GCH$, and (ii)' so also is $ZF + \neg AC$, etc. Finally, one can look at these models still more formally as giving a *syntactic interpretation* of $ZF + AC + GCH$ in ZF, etc. The methods of Gödel and Cohen have led to a welter of model-constructions and thence relative consistency and independence results for a mass of problematic set-theoretical statements.

Less familiar are the uses to which such models can be put. Here the best way to formulate the matter is in terms of the logical notion of one (axiomatic) theory T being a *conservative extension* of another such theory S for a class of (well-formed) formulas $\mathcal{F}$. This is said to hold if: (i) every formula of $\mathcal{F}$ is in the language L_S of S, and if $L_S \subseteq L_T$; (ii) every theorem of S is a theorem of T; and (iii) every theorem of T which lies in the class $\mathcal{F}$ is already a theorem of S. This notion will be applied frequently in the following.

It was observed by Kreisel that $ZF + AC + GCH$ is a conservative extension of ZF for the class $\mathcal{F}$ of formulas of elementary number theory (as expressed in ZF). The reason is that Gödel's constructible-sets model for $ZF + AC + GCH$ is *standard* for $I\!N$ (but not for sets in general). This conservation result turned out to be applicable to the existence of a decision method for 1st order statements in p-adic fields. That was established by Ax and Kochen (1965) using model-theoretic methods which can be formalized in $ZF + AC + GCH$. But the existence of a decision method for p-adic fields can be expressed in the given class $\mathcal{F}$ of number-theoretic formulas. Hence by Kreisel's observation, it can already be proved in ZF. Later, Cohen (1969) actually produced an explicit decision method for p-adic fields, provably in a weak subtheory of ZF.

We turn now to less familiar examples of interpretations in logic. First are models of the λ-*calculus* produced by Scott (1972) and (following Plotkin) Scott (1975)[8]. The λ-calculus is a formalism in which the universe of discourse is conceived to consist of functions defined on the whole universe. The basic operation is that of *application*, written fx (or $f(x)$). There are several basic principles, but the central one is that any well-formed expression (term) $\tau[x]$ in this symbolism containing a free variable x serves to define a function $f = \lambda x \cdot \tau[x]$ whose values are given by $fx = \tau[x]$. The main feature of this formalism which makes it problematic is the possibility of *self-application*, i.e., formation of ff. This is not possible for any familiar universe of functions, where the elements of the domain of a function are 'prior' to (or of a 'lower type' than) the function itself. A form of functional self-application had been met in recursion theory, by using numerical codes f for mechanical algorithms A_f; then fx is defined to be the value of A_f at input x. However, fx need not always be defined in that case; in particular ff might not be defined for a given f. The difference is expressed by saying that the functions in the λ-calculus are supposed to be *total*, while those in recursion theory may be *partial*. Scott's models for the λ-calculus have been particularly useful in providing (so-called *denotational*) *semantics for programming languages*, which serve to give each element of a program a meaning as a mathematical object, rather than as a procedure; see Stoy (1979) and Mosses (1990)[9].

Different kinds of *self-reference* or *self-membership* appear in ordinary language, in the theory of classes and in algebra. The Liar Paradox and Russell's Paradox show the problems that can arise in the first two cases with *negative* uses of self-application. But is the statement: *This statement is true*, problematic? And are the notions, *the class of all classes*, and *the category of all categories*, problematic? It turns out that one can develop formal theories in which these and other such notions are represented and which are conservative extensions of known consistent theories. For a survey of work in this direction see Feferman (1984). An explicit aim in the latter is to develop theories which are mathematically useful as a result of their greater flexibility of expression. See also Martin (1984) for a representative collection of different approaches to the area of self-referential theories of truth and the problem of the Liar Paradox, as well as the expository survey paper by Visser (1989). The paper by Feferman (1991) deals further with formal theories for such notions of truth, though its main concern falls under Section 7. A workable modification of ordinary set theory allowing self-membership has been developed by Aczel (1987). This has been used by Barwise and Etchemendy (1987) for another approach to the Liar Paradox.

The question of self-membership also arose in connection with an interesting but problematic extension of typed and untyped λ-calculi which emerged in the theory of programming languages in recent years, involving variable types and the notion of *polymorphism*. This borders on having a *type of all types*, and is prima facie *impredicative* (cf. below for the latter idea). In Feferman (1990), I have taken an approach to these via axiomatic theories of operations and classes (considered as variable types)[10]. It was argued there that a predicatively reducible subtheory accounts for all applications of polymorphism met in practice.

Another interesting example of a problematic mathematical notion is Brouwer's concept of *(free) choice sequence*, in his *intuitionistic redevelopment of mathematics*. The idea is that a choice sequence $\alpha = (\alpha_0, \alpha_1, \ldots, \alpha_n, \ldots)$, (say, of natural numbers) may be generated without any known law, e.g., by some random phenomenon or an arbitrary sequence of choices by an individual. The mathematician working with such a sequence will have only a finite amount of information $\bar{\alpha}(n) = (\alpha_0, \ldots, \alpha_{n-1})$ available about it at any given time. The basic principle for choice sequences used by Brouwer is that if $P(\alpha, m)$ is a property of choice sequences α and numbers m such that $\forall \alpha \exists m P(\alpha, m)$[11] then $\forall \alpha \exists n, m \forall \beta [\bar{\beta}(n) = \bar{\alpha}(n) \rightarrow P(\beta, m)]$. From this and some other (more readily acceptable) principles, Brouwer was able to prove that every function of real numbers is continuous. (Real numbers are represented as choice sequences of rationals satisfying Cauchy's convergence condition). Since this contradicts the classical statement of existence of discontinuous functions there must be an escape hatch for Brouwer's principles to be coherent. This is provided by the *elimination* of the *law of excluded middle* as a basic logical principle, so that, e.g., $\forall\alpha[\forall n(\alpha_n = 0) \lor \exists n(\alpha_n \neq 0)]$ is not accepted. See Troelstra and Van Dalen (1988) for an introduction to intuitionistic theories of constructive mathematics (among others), and in particular Vol. II, Ch. 12 (op. cit.) concerning various approaches to the problematic notion of choice sequences.

In particular, Kleene and Vesley (1965) formulated Brouwer's ideas about choice sequences in an axiomatic theory in intuitionistic logic (denoted here) CS, and proved that CS is consistent by giving what is called a *realizability interpretation* for it. This then can be used to justify Brouwer's redevelopment of analysis, which as we have noted is *prima-facie* contradictory with classical analysis. See Troelstra and Van Dalen (1988) for an introduction to a variety of forms and applications of the realizability method. (For another way of handling CS, see the next Section 4.)

Brouwer's and other approaches to constructive mathematics suggest consideration of a number of principles which appear plausible in a constructive framework though they contradict classical statements. For example, what is called *Church's thesis* in this context is some form or other of the statement that every constructive number-theoretic function is recursive (i.e., is mechanically computable). Since for the constructivist, $\forall n\exists m P(n, m)$ means that m can be found as a (constructive) function of n to satisfy $P(n, m)$, the thesis can be expressed as a scheme $CT : \forall n\exists m P(n, m) \rightarrow \exists e\forall n P(n, \{e\}(n))$, where $\{e\}$ denotes the recursive function with (Gödel-) number e. The method of realizability, originally developed by Kleene, can be used to show consistency of CT with practically all intuitionistically accepted systems; see Troelstra (1977) or Troelstra and Van Dalen (1988) for an introduction to this and other such results. Loosely speaking, one significance of the consistency of CT is that theorems proved with its use have in principle an effective mechanizable content.

If one is only willing to accept the constructivist philosophy of mathematics, various principles of *classical* mathematics become problematic in the light of that position. Obviously among such are the *well-ordering principle*; but even some more mundane principles like *extensionality* and the *power-set axiom* (existence of the set of all subsets of any set) must be reexamined. For a survey of work on these and other questions that arise from systematic investigations of the foundations of constructive mathematics, see Beeson (1982).

Even within classical mathematics there are different philosophies which accept parts but not all of what is represented in current systems of set theory. For example the French "semi-intuitionistic" school of Baire, Borel and Lebesgue accepted transfinitely iterated countable constructions, and (implicitly) the countable axiom of choice, but not the full axiom of choice; for a good description of their ideas see Moore (1982) (Section 2.3 and App. 1). Even more restrictive were Poincaré's views, which influenced Weyl to see how far analysis could be developed using only purely arithmetical definitions (Weyl, 1918). (A translation into English of Weyl's 1918 monograph has appeared in Weyl (1987); the paper by Feferman (1988a) provides a critical exegesis of that work, followed by an explanation of modern developments which constitute a realization of Weyl's program.)

Poincaré had formally rejected *impredicative definitions of sets*, which picked out a subset of a given set A by reference to the supposed completed totality of all subsets of A (i.e., the power set of A). The simplest example of an impredicative definition of a set is that as the intersection of all sets X satisfying $\Gamma(X) \subseteq X$, where Γ is a set operator; this is used with monotone Γ in the general approach to inductive definitions from above. Feferman (1970, 1982a) shows how to interpret axiomatic theories for the sets successively generated by monotone inductive operators, into corresponding theories for sets explicitly generated from below. For a survey of work on reductions of theories of inductive definitions 'from above' to theories of sets inductively generated 'from below', cf. Feferman (1988). For work on more strictly predicative principles, see Section 7.

We conclude this section with a famous open problem in logic, namely whether Quine's system NF of *New Foundations* is consistent. This system can be thought of as the image of the simple theory of types obtained by erasing all type distinctions between variables. It permits construction of a universal set V with $V \in V$. Some efforts were made in the past to see whether such features make NF mathematically useful, but it has not been paid much attention in recent years. Jensen (1969) produced an interesting model of $NF- Ext$, i.e., of the system obtained from NF by deleting the axiom of extensionality. However, this has so far given no clue as to how to establish consistency of the full system NF. At present the interest in this problem seems principally to be as a logical challenge.

4. REPLACEMENT OR ELIMINATION OF PROBLEMATIC
CONCEPTS AND PRINCIPLES

The general situation is the same as that brought out at the beginning of
the preceding section. Only the foundational moves described here are of a
different character. Instead of trying to interpret the problematic concepts or
principles (by some kind of model) one tries to find *substitutes* for them which
do the same work, or even tries to *eliminate* them entirely. In the latter case
the aim is to save the baby from being thrown out with the bath water, i.e, to
preserve the useful consequences. In modern logic this is accomplished by
conservation results established by syntactic transformations.

The classical example of a concept which was eliminated while saving
its applications is that of *infinitesimal*, when the associated notions of *limit,
derivative*, etc., which had been explained in terms of it were given substitute
'ε, δ' definitions. (For modern theories restoring infinitesimals as mathematical
objects, see Section 5.) Another example of substitution was for *multiple-
valued functions* in complex analysis, which were replaced by *single-valued
functions on Riemann surfaces.*

More recent examples tying up with logic are the treatments of the notions:
the *class of all classes* and the *category of all categories*, in the context of
the formal theory BG (Bernays–Gödel) of *sets* and *classes*; these are replaced
respectively by: the *class of all sets* and the *category of all small categories*.
MacLane (1961, 1971) used the distinction between set and class in BG to
distinguish between *small* and *large categories*, which led to acceptable substitutes for otherwise problematic general theorems in the subject such as *Yoneda's
lemma*, the *adjoint functor theorem*, etc. Another solution to this problem situation had been given in the framework of ZF by the use of *Grothendieck
universes*. However, that required the existence of many inaccessible cardinals. The *reflection principle* in ZF was applied in Feferman (1969) to avoid
such assumptions; in addition a conservation result was given there which
shows how 'smallness' can be formally introduced in ZF to facilitate the
applications to category theory, and then eliminated.

A *syntactic translation* was used by Kreisel and Troelstra (1970) to *eliminate*
(Brouwer's notion and principles for) *choice sequences*. This gives conservation of a theory CS over a constructive theory without choice sequences: it
also strengthens the results of Kleene and Vesley mentioned in Section 3. The
use of syntactic translations in the metatheory of constructive systems and
related classical systems goes back some years. In 1933 Gödel and Gentzen
independently found the so-called *negative translation* of classical (Peano's)

arithmetic PA into intuitionistic (Heyting's) arithmetic HA[12]. Since then the same translation has been applied to much stronger classical systems; see Troelstra (1977) for references.

Another interesting syntactic transformation was introduced by Gödel (1958) to *eliminate quantified statements* in HA in favor of *quantifier-free* (open) statements about certain *constructive functionals of finite type*. An English version of Gödel (1958) with supplementary notes was worked on by Gödel for publication but never appeared in print. Its final form (as found in his *Nachlass*, and dated 1972) is reproduced in Gödel (1990, pp. 271–280). For a critical examination of this work and survey of further applications of the 'Dialectica' functional interpretation, see the introductory note by A. Troelstra to *1958* and *1972* in Gödel (1990, pp. 217–241). Gödel's functional interpretation was extended to analysis (2nd-order arithmetic) in a series of works by Kreisel, Spector, Howard, Girard and others. In combination with the negative translation, this serves to reduce certain subsystems of classical analysis to constructively accepted quantifier-free systems (see Troelstra, 1977, Section 11 for a survey of this work).

Still another relatively early syntactic method of eliminating quantified statements, which had been considered to be problematic in Hilbert's *finitist program*, was that originating with Herbrand in his 1930 thesis. In its simplest form, Herbrand's theorem tells us that if an existential statement $\exists x A(x)$, with A quantifier-free, is provable (= valid) in classical logic then for some finite sequence of terms $t_1, \ldots, t_n$ we have $A(t_1) \vee \ldots \vee A(t_n)$ provable; moreover, these terms can be found primitive recursively from any given proof of the statement. This has had interesting mathematical applications, e.g., to the extraction of primitive recursive bounds from Artin's (*prima facie* noneffective) solution to Hilbert's 17th problem; for this and other applications of Herbrand's theorem see Kreisel (1958).

A general and very important syntactic method for eliminating statements of complexity higher than that of the conclusion from logical proofs was discovered by Gentzen in 1934 (see his collected works, Gentzen, 1969). This is called *cut-elimination* when applied to one kind of formalism introduced by Gentzen (the sequential systems) and *normalization* when applied to another (natural deduction systems). The method of cut-elimination was subsequently extended to classical arithmetic by Gentzen. Modern applications of this method have extended it much further to give constructive elimination proofs for various subsystems of analysis; see the books Takeuti (1975) and Schütte (1978)[13]. Among other things, that work serves to give *precise ordinal measures* to the *effective content* of classical theories in which much of analysis and descriptive set theory can be formalized. The monograph by Buchholz and Schütte (1988) and the survey by Pohlers (1991) provide systematic expositions of (*prima facie*) uncountably infinitary extensions of Gentzen's cut-elimination procedure, used to establish the ordinal strengths of various systems of analysis and set theory (see also Pohlers (1989) for an introduction).

The results of Jäger and Pohlers (1982) obtained in this direction are noteworthy. Their work serves to reduce the classical theory $\sum_2^1 - AC + BI$ to the constructive theory T_0 of functions and classes introduced in Feferman (1979) as well as determine its proof-theoretic ordinal. An easier reduction had previously been established for $\sum_2^1 - AC$ to a restricted form of T_0 by Feferman and Sieg in Chapter II of the volume Buchholz *et al.* (1981). That volume also presents an arsenal of proof-theoretic methods and results, reducing various classical subsystems of analysis to constructive systems of iterated inductive definitions. (This connects with the work on elimination of impredicative principles described in Section 3 above.) The results obtained by Rathjen (1991) are for still stronger systems than those of Jäger and Pohlers; these make use of effective ordinal notation systems incorporating Mahlo cardinals.

5. FOUNDATIONS OF PROBLEMATIC METHODS AND RESULTS

The foundational questions in Sections 3–4 dealt with troublesome but relatively basic ideas or propositions. In this section the focus of concern is on results and methods of some substantial mathematical content but whose application is uncertain. Though more advanced mathematically, such usually appear early in the development of a subject. The problem, often soon recognized, is to provide them with a rigorous foundation. That takes the form of finding precise *sufficient* conditions under which the methods are applicable or for which the results hold. Since the conditions obtained are rarely *necessary*, there is also the constant possibility of improvement. Thus this type of foundation of a subject usually proceeds over a period of time by successive steps of sharpening, extension and generalization. There is some overlap here with the foundational moves in Section 4, since one frequently needs precise concepts to substitute for informal or intuitive notions. For example, in current foundations of *probability theory* the idea of *random variable* is replaced by *measurable real-valued function*.

Mathematics offers us a wealth of examples of this type of foundational situation and treatment. To mention but a few we have the *foundations of: Fourier series, calculus of variations, Dirichlet's principle, Stokes' theorem, Jordan curve theorem, generalized functions, Descartes–Euler theorem, probability theory* and *algebraic geometry*. When neighboring troublesome subjects have an isolable mathematical part, one sees similar moves as in the *mathematical foundations of quantum mechanics*, etc.

So far, there are many fewer examples of this kind from logic. Two such are quite well known, the first being the *foundations of Cantor's theory of transfinite ordinals and cardinals* as currently given in axiomatic set theory $ZFC(= ZF + AC)$ or in $BG + AC$. The second is the *foundations of infinitesimal analysis* by Robinson (1966). (See also Stroyan (1977) for an introduction and further references.)

Examples of much lesser scope but still requiring attention were: (i) *Gödel's theorem on the unprovability of consistency* (second incompleteness theorem); (ii) *Herbrand's theorem*; and (iii) the *Church–Rosser theorem*. In the case of (i), the first fully worked out proof was given for the case of arithmetic in Hilbert and Bernays (1939). I first gave a precise general form of the theorem in a 1960 paper, and revisited it to give an improved generalization in Feferman (1982) or better (1989). The problem with (ii) is that Herbrand's own proof contained subtle and not-easily correctable errors, though nonconstructive proofs (using the completeness theorem for logic) were easy. A constructive reconstruction of Herbrand's work was finally managed properly in Dreben and Denton (1966). As to the third example, the Church–Rosser theorem is fundamental to the λ-calculus and related combinatory calculi; according to it, normal forms of terms are unique when they exist. Early proofs of this were exceptionally difficult and a number of attempts to simplify them and to extend the result to other calculi were bedeviled by errors. A simple, manageable and generalizable proof was finally given by Tait and Martin-Löf; see Barendregt (1984).

6. ORGANIZATIONAL FOUNDATIONS; AXIOMATIZATION

This is one of the best known aspects of foundational work, which is accordingly much over-emphasized. It comes at a relatively advanced stage in the development of a subject, when that is ready for *consolidation* and *systematic exposition*. Indeed, it is often undertaken as part of the *pedagogical enterprise*. While this takes considerable experience, skill and effort, it is hard to get excited about finished work of this kind, certainly in comparison to the top work described in Sections 2–5. Excitement comes when one is actively engaged in such an enterprise and especially when there are competing schemes in the offing; current examples of this will be given below.

This type of foundational work involves a *choice and ordering of concepts and results*, with basic results being taken as *axioms*. Usually there is also involved a systematization of *methods of proof*, though that is rarely made explicit outside of logical work. In addition there is overlap with the foundational moves described in the preceding Sections 2–5.

The classical example of axiomatic organization is of course provided by *Euclidean geometry*. Though long considered a model for rational thought, the value of axiomatization was not recognized and pursued again until the 19th century. No doubt the onset of this had to do with the growing professionalization of mathematics and the institution of organized higher mathematical education for engineers and scientists. Besides Hilbert's reexamination of the axiomatic foundations of geometry, modern mathematics provides a wealth of examples of axiomatic organization. Just consider: in algebra – *groups, rings, fields, vector spaces*, etc.; in analysis – *topological spaces, metric spaces, Banach spaces, Hilbert spaces, measure spaces*, etc. One can go on in field after field, e.g., in topology, algebraic geometry, etc.

The work in the 19th century on the *axiomatic characterization of the basic number systems* $I\!N$, Q, $I\!R$, C by Méray, Weierstrass, Dedekind, Cantor, Peano, Frege and others form a bridge to 20th century work on the logical approach to the foundations of mathematics. Here one has a shift away from axiomatics of traditional subject areas (such as algebra, geometry, analysis) to the study of principles which underlie *all* areas. The most well-known example of that is the development of *axiomatic set theory* at the hands of Zermelo, Skolem, Fraenkel, van Neumann, Bernays, Gödel and others. Relative to the resulting systems (such as ZFC or BGC) there has been much work on principles whose status and justification is less certain, e.g., those called *large cardinal axioms*; see Kanamori and Magidor (1978) for a recent survey.

Axiomatic set theory is the current expression of the Platonistic philosophy of mathematics, and there is considerable agreement about its core from those who accept that position. *The axiomatic foundation of constructive mathematics* is not nearly so settled, and is still being actively pursued. One reason is that the body of informal constructive practice in mathematics to be accounted for is much smaller than the body of everyday nonconstructive mathematics accounted for in set theory. Another reason is that the constructive approach in this century was long dominated by Brouwer's ideas, which he propounded in startling and in some cases mystifying ways (see Brouwer's *Collected Works*, 1975). Since then other, more understandable approaches within the constructivist frame have been developed, notably those due to Markov and the Russian school on the one hand (see Aberth, 1980) and to Bishop (1967) and his followers on the other hand. However, the various differences in these approaches have not led to a common accepted axiomatization.

As a result of work already referred to in Sections 3–4 by Kleene, Vesley, Kreisel, Troelstra and others, Brouwer's *constructive* redevelopment of mathematics is much better understood now and has to a good extent been tamed axiomatically. The *axiomatic representation of practice in the Russian school of constructivity* is also well in hand; basically, constructive functions are there identified with recursive functions and so *Church's Thesis* is accepted as an axiomatic principle. What makes this approach distinct from so-called 'recursive mathematics' is the restriction of argumentation to that which can be carried out in intuitionistic logic plus a special axiom called *Markov's principle*. For some information on the formal systems associated with this school see Troelstra (1977), Beeson (1981, 1985), and Troelstra and Van Dalen (1988).

The development of constructive mathematics along the lines set out by Bishop was continued for analysis in Bishop and Cheng (1972), and Bridges (1979). The basic work of Bishop in analysis was itself up-dated in the book Bishop and Bridges (1985). The text of Mines, Richman and Ruitenberg (1988) provides an exposition of constructive algebra following Bishop's approach. Two, very different, competing ways have been proposed for the *axiomatic representation of practice in Bishop's approach to constructive mathematics*. One, due to Myhill (1975), provides a subsystems CST ('constructive set theory') of IZF (= intuitionistic ZF). The second, due to Feferman (1975), provides a system T_0 of functions and classes in a new specially designed formalism. These systems are compared in Feferman (1979), which also elaborates and extends my earlier work. The kind of competitive proposals mentioned and the various considerations which go into axiomatization provide an excellent case study in such foundational work.

Independently of the preceding, Martin-Löf has long been developing his own distinctive viewpoint about constructive mathematics. It has been given axiomatic formulation in his *constructive theories of types*; see, Martin-Löf (1982) and (1984). Though not motivated by the question of axiomatizing Bishop's approach, Martin-Löf's systems (here denoted ML) offer still another competing candidate. The book by Beeson (1985) contains a detailed comparative study of the systems CST, T_0, ML and interesting extensions and subsystems (as well as the already mentioned axiomatics of the Russian school).

It is quite common, once axiomatic foundations for a given subject are fairly settled, to undertake a *fine analysis* of the role of different axioms or groups of axioms in different parts of practice. We mention here only a few examples of this type of work. Friedman was engaged for some years in tying down the *exact logical strength* of various results from mathematical analysis, algebra, etc.; see Friedman (1975, 1976). This program has been continued by Friedman *et al.* (1983) and Simpson (1984). A survey of the work by Friedman, Simpson and others in the 'reverse mathematics' program is given in Simpson (1987). What is surprising is how much everyday mathematics can be done in formal systems which are conservative over arithmetic – PA or HA, according to whether one is looking at nonconstructive or constructive practice. A variety of such systems conservative over PA and explanation of what mathematics can be done in them have been given by Takeuti (1978) (written somewhat earlier), Feferman (1977) and Friedman (1980). See the latter part of Feferman (1988a) for further work on this. On the constructive side, systems conservative over HA have been given by Friedman (1977) and Feferman (1979) (by Beeson, 1980). One direction of current research is in the role of very weak subsystems of arithmetic; for such cf. Simpson (1988). In the opposite direction, Paris and Harrington (1977) opened up the way to show the unexpected *strength of various finite combinatorial partition principles*; see among others Friedman *et al.* (1982). Simpson (1987a) gives a good introductory survey of this work. For a critical discussion of the foundational significance of these results, see Feferman (1987).

7. REFLECTIVE EXPANSION OF CONCEPTS AND PRINCIPLES

Mathematical concepts are not fixed once and for all. Every so often a new one appears, and if it establishes itself by its intuitive appeal and great utility, or even necessity, it must be accounted for somehow in mathematics. In Sections 2–4 we spoke about several ways of doing this. One may try to clarify the concept, defining it precisely in previously understood terms, or may try to interpret it (in those terms) in a way that preserves its expected properties. Finally, one may try to replace it or eliminate it altogether. However, after all this, there are a *few* concepts left over which are *genuinely new at each stage*. Examples of such concepts are those of *natural number*, of *point, line* and *plane*, of *ordered pair*, of *infinite sequence*, of *function* and of *set*. The first two arose in the prehistory of mathematics, perhaps with the very emergence of pure mathematics as a separate intellectual discipline. It may appear that a concept is genuinely new at a certain stage and irreducible or uneliminable at that stage, but later reconsideration allows us to shift its place in our conceptual scheme of things. Such was the case with the concepts of *infinitesimal* and of *choice sequence*. Nowadays all the abovementioned examples are reduced to the concept of set in the set-theoretic scheme of things. But this is not the case in (some) constructive conceptual schemes; e.g., the idea of function as given by a *rule* is not reducible to the idea of set as given by a *defining property*, nor is the converse reduction possible. For a discussion of such differences in the two schemes, cf. Feferman (1979). Be that as it may, if a concept appears genuinely new and forces itself on us, the only way to deal with it is to accept it as a kind of *given* and build around it, all the while trying to sharpen our understanding of it as well as possible. Basically, that falls under axiomatic organization which was just discussed in Section 6.

All of the preceding is fairly well-recognized in one way or another. The purpose of this section is to concentrate on a different kind of conceptual innovation which frequently takes place, and which I propose to call *reflective expansion*. Basically, this is a form of *generalization*, but of the following particular character: at a certain point one reflects on what has led one to accept and work with certain concepts, and sees that a much more general concept is *implicit* in accepting that. For example, Euclidean geometry of the plane and space were eventually explained in terms of $\mathbb{R}^2 (= \mathbb{R} \times \mathbb{R})$ and $\mathbb{R}^3 (= \mathbb{R} \times \mathbb{R} \times \mathbb{R})$ by using the real number system $\mathbb{R}$ and the pairing and tripling operations. Reflection on that step led to the concepts of *n-tuple* and *n-dimensional space* $\mathbb{R}^n$. Reflection on the formation of derivatives and integrals as operations on functions led to the idea of *function operators* or *functionals*. Reflection on the idea of the enumeration of the natural numbers in order, where at each stage there is a first element beyond all the previous ones led to the idea of *ordinal number*; here the novelty lay in passing to the first number ω beyond all the natural numbers. Reflection on the concept of limits of sequences of continuous functions (which may be discontinuous) led one to the idea of the iteration of such limits, i.e., to the *Baire hierarchy of functions*; a similar process of reflection led about the same time to the *Borel hierarchy of sets*.

All of the foregoing applies *mutatis mutandis* to *principles* in place of *concepts*. An elementary example is provided by the basic principle for pairs, $(a_1, a_2) = (b_1, b_2) \rightarrow a_1 = b_1 \wedge a_2 = b_2$, which is generalized to the corresponding principle for *n*-tuples. The Cauchy convergence principle for $\mathbb{R}$ is generalized to *completeness* for $\mathbb{R}^n$. Reflecting on the principle of induction for natural numbers leads one to accept the *principle of transfinite induction for ordinals*, and so on.

Note that what is introduced by reflective expansion is rarely genuinely new: the concept of *n*-tuple may be defined in terms of that of pair, the concept of ordinal number in terms of well-ordered set, the principle of completeness of

$I\!R^n$ can be deduced from completeness for $I\!R$, etc.

From a logical point of view, our interest here is in whether we can make theoretical sense of *describing all the concepts and principles that one ought to accept if one has accepted given concepts and principles*, or, put more succinctly, *describing all the concepts and principles implicit in given ones*: the general proposal to pursue such characterizations is due to Kreisel (1970). This followed his own work on characterizing *finitist* notions and principles, and the work of Schütte and myself on characterizing *predicative* notions and principles (the leading ideas in these cases had been suggested informally by Hilbert and Poincaré, respectively).

The study of *predicativity* has been carried out rather fully in a series of papers and thus offers a good case study for the general proposal. Here one wants to say just what is implicit in accepting the *structure of natural numbers* $I\!N$ together with the general *principle of induction* as given, and (in contrast to finitism) with *quantification over* $I\!N$ as a means of defining properties. The first approach to explaining this used *transfinite autonomous progressions of theories* T_α, where α runs through certain effectively given well-orderings. At each stage α, T_α incorporates principles obtained by reflecting on what has been accepted in the earlier stages; e.g., the induction principle may be applied to properties defined by quantifying over previously determined enumerated collections of properties (technically this is a form of *ramified 2nd order number theory*). The *autonomy* or *boot-strap* condition allows one to pass to stage α (and only when) the well-ordering property of the ordering giving α has been established at an earlier stage $\beta < \alpha$. The least nonautonomous ordinal Γ_0 has been characterized by Schütte and myself in terms of hierarchies of normal ordinal functions; cf. Feferman (1968a) or, my appendix to Takeuti (1987) for a survey of this work.

The use of transfinite iteration of reflective processes is a *prima facie* unrealistic element in the modelling of actual reflective expansion. Thus in my work on predicativity since 1964 I developed alternative single formal systems which serve to characterize the same body of mathematical thought. One which (to my mind) most persuasively embodied the idea of the *reflective closure* of the concept of the structure $(I\!N, 0, Sc)$ with the principles of induction and definition by quantification over $I\!N$, is given in Feferman (1978). Three ideas there are central:

(i)   Reflection on what leads one to accept successively the definitions of $+, \cdot, \exp$, etc., as determining total functions on $I\!N$ leads one to the definition of the sequence of primitive recursive functions and thence its enumeration; this can be generalized appropriately to recursively defining sequences of operations *and* of properties.

(ii)   With any property $X(x, x_1, \ldots, x_n)$ is associated another $Y(x_1, \ldots, x_n) := \forall x \in I\!N.\ X(x, x_1, \ldots, x_n)$ obtained by quantifying universally over $I\!N$.

(iii)   The *scheme* of induction rests on the recognition that if one has established a general proposition $A(X)$ of properties $X(x)$, one ought to accept each particular substitution instance B for X by a property $B(x)$ expressed in the system.

It is shown, *op. cit.*, that a system based on these ideas has the same strength as the autonomous (ramified) progression $\cup T_\alpha (\alpha < \Gamma_0)$ described above.

In 1978–79 I developed a general notion of *reflective closure of a schematic theory* S which for $S = PA$ gives the same result (up to mutual interpretation) for predicativity as that just described[14]. This work finally appeared as Feferman (1991). The notion of reflective closure is formulated for and can be applied to much more general S than PA. In particular, it should be of interest to consider it relative to a system S of set theory, where specifically *set-theoretical reflection principles* have led one to consider stronger and stronger extensions of S; such extensions are currently based on so-called *large cardinal axioms* (see particularly the already mentioned survey article of Kanamori and Magidor, 1978). The idea of reflective closure should serve to explain how much of this is implicit in the concepts and principles accepted in given S, and what then requires genuinely new considerations in order to be accepted. That is a research program which at present is completely open.

POSTSCRIPT: FOUNDATIONAL WORK AND PHILOSOPHY OF MATHEMATICS

Most presentations of logical work of a foundational character center on one of the grand philosophical schemes, which are taken as their justification (at least implicitly), e.g., set-theoretic foundations, constructive foundations, formalist (proof-theoretic) foundations, etc. Instead, current foundational activity has been analyzed here along lines independent of any particular philosophical position. In each case we have shown by examples that this is simply a continuation of various types of foundational activity carried on by mathematicians as a matter of course for the progress and improvement of mathematical understanding. Of course, the preoccupations which serve to prod this kind of activity are not, at any given time, universally shared. What one mathe-

matician sees as a need for clarification, for improved organization or to deal
with a problematic concept (resp. principle) is not what may concern another
mathematician. Even so, the results of such work in the past have eventually
been absorbed into generally accepted mathematics.

It is undeniable that in various of the logical examples given above, the
motivating concern *has* been relative to a particular foundational position; but
little was needed in each case to appreciate the nature of that concern and the
steps taken to meet it. To what extent the results of such work will take their
place as part of generally accepted mathematics is a matter for the future to
settle. This paper will have served much of its purpose if the reader has at least
been brought to see the *point* of the work arising out of such specific concerns,
without trying to judge its eventual significance.

With all this stress on appreciating logical work independently of fixed
foundational standpoints I do not mean to reject interest in reaching a basic
philosophical position as to the nature of mathematics. On the contrary that is
required in order to make the full case for the logical approach to foundations
which I advanced in the introduction. As explained there, elaboration of this
will have to wait for another occasion. My only purpose here, in conclusion,
is to give an indication of the direction these ideas have taken.

In answer to the question: *What is it about mathematics that makes it such
a distinctive body of thought?* logic is clearly most successful in analyzing
the underlying characteristics of its *language* and its *verificational methods
(proofs).* In these respects it is closer to an analysis of everyday logical
experience than its critics realize (cf. my papers (Feferman, 1979a, 1981) for
some arguments in support of this). But to give a full answer to the question just
posed one must also deal with the following more difficult question: *What is
the nature of the conceptual content of mathematics?* I agree with the critics of
the traditional positions of logicism, formalism, Platonism and constructivism,
that each of these has failed to give us a satisfactory, convincing answer to that.
In recent years I have been trying to develop and build a coherent case for an
alternative and, to my mind, more satisfactory view. Roughly speaking, this
comes to the following. I am in agreement with the constructivist position as
to the subjective source of basic mathematical conceptions; but for me these
are supposed to be conceptions of certain kinds of ideal worlds, including ones
which are not countenanced constructively (such as 'Platonistic' worlds of sets).
These worlds (or world-pictures of mathematical structures) are presented more
or less directly to the imagination, from which basic principles are derived by
examination. All else (in each picture) is obtained by rational reflection on, and
from, basic concepts and principles. It may be that at the outset only relatively

crude features of a world-picture can be discerned in this way. My main slogan here is that nevertheless, *for mathematics, a little bit goes a long way.*

I have made several extended attempts over the last fifteen years to organize, detail and support the preceding ideas, and had originally thought to use the occasion of the present paper at least to put some of the result on record. Again, that is something which I shall have to put off to the future.

Departments of Mathematics and Philosophy,
Stanford University

NOTES

* My lectures for the 1989 Trieste School in Logic and the Philosophy of Science were based in part on the paper 'Working Foundations' which appeared in *Synthese* **62** (1985), 229–254. (That, in turn, was a greatly modified (and accordingly retitled) form of a paper originally presented to the workshop 'The present state of the problem of foundations of mathematics' arranged by the Florence Center for the History and Philosophy of Science at the Villa di Mondeggi, 15–19 June 1981.) I very much appreciate the willingness of the editors to have that paper reprinted, with some modifications to bring it up to date, in the present volume as a partial representation of my contribution to the Trieste School. Permission to reprint was itself granted by Kluwer Academic Publishers.

I have taken this opportunity to make a few corrections in the original, to add references to the considerable relevant literature which has appeared since its publication, and to expand the text at various points accordingly. Due to these modifications, it was decided to add an indication of the updating to the title. The assistance of Dr. Giovanna Corsi with the integration of the new material with the text is greatly appreciated.

It should be noted that another, 'slimmed down', version of the original paper appeared in Feferman (1984a).

[1] My arguments against this have been presented in the first part of Feferman (1977b).

[2] The following is a grab-bag of references representing a variety of critical approaches: Davis and Hersh, 1981; Goodman, 1979; Hersh, 1979; Kreisel, 1967, 1976, 1977; Lakatos, 1976, 1978; MacLane, 1981; Tymoczko, 1986; Wang, 1974.

[3] Bibliographic references are generally given only for less familiar logical work, but the sources provided are not always the last word on a given subject.

[4] The categorization given here of types of foundational work is my own; at least I have not seen such a side-by-side breakdown elsewhere. The borderlines and application to specific cases are not always clear-cut, and a more refined analysis may be warranted.

[5] I must apologize for the rough-and-ready way in which some of this work is described; to keep the presentation manageable I have had to make some deliberate simplifications.

[6] See my review in *J. Symbolic Logic* **40** (1975), 232–234.

[7] These papers have been reprinted in Gödel (1990) with an introductory note by R.M. Solovay. All of Gödel's published work has been reprinted, with translations and introductory notes in Gödel (1986, 1990). Two further volumes, devoted to his unpublished papers, lectures, scientific notes and correspondence, are in preparation.

[8] See also Barendregt (1977) or his book (1984) on the λ-calculus.

[9] Mosses (1990) provides an up-to-date survey with a comprehensive list of references.

[10] Approaches to the problems raised by polymorphism formed the second part of my lectures at the Trieste School.

[11] In the following we use the logical notation $\forall, \exists, \neg, \wedge, \vee, \rightarrow, \leftrightarrow$ for 'for all', 'there exists', 'not', 'and', 'or', 'implies', 'if and only if', respectively.

[12] A related translation had first been found for the propositional calculus in 1925, by Kolmogoroff.

[13] See also my reviews of these books in *Bull. Amer. Math. Soc.* **83** (1977), 351–361 and *Bull. (New Series) Amer. Math. Soc.* **1** (1979), 224–228, respectively. See Takeuti (1987) for the second edition of his 1975 book, which also contains appendixes by Kreisel, Pohlers, Simpson and myself on further applications and extensions of the method of cut-elimination, as well as other approaches to proof theory. See Sieg (1991) for applications of Gentzen's method combined with Herbrand analyses.

[14] This work was presented in a talk for a symposium on the work of Kurt Gödel at the meeting of the association for Symbolic Logic in San Diego, March 1979.

REFERENCES

Aberth, O.: 1980, *Computable Analysis*, McGraw-Hill.

Aczel, P.: 1977, 'An introduction to inductive definitions', in Barwise (1977), pp. 739–782.

Aczel, P.: 1987, *Lectures on Nonwellfounded Sets, CSLI Lecture Notes*, No. 9, Center for Study of Language and Information, Stanford.

Ax, J. and Kochen, S.: 1965, 'Diophantine problems over local fields I, II', *Amer. J. Math.* **87**, 605–630.

Barendregt, H.: 1977, 'The type free lambda calculus', in Barwise (1977), pp. 1091–1132.

Barendregt, H.P.: 1984, *The Lambda Calculus, its Syntax and Semantics* (2nd ed.), North-Holland.

Barwise, J. (ed.): 1977, *Handbook of Mathematical Logic*, North-Holland.

Barwise, J. and Etchemendy, J.: 1987, *The Liar: An Essay in Truth and Circularity*, Oxford University Press.

Beeson, M.: 1980, 'Extensionality and choice in constructive mathematics', *Pacific J. Math.* **88**, 1–28.

Beeson, M.: 1981, 'Formalizing constructive mathematics: why and how?', in *Lecture Notes in Math.* **873**, 146–190.

Beeson, M.: 1982, 'Problematic principles in constructive mathematics', in *Logic Colloquium 1980*, North-Holland, pp. 11–55.

Beeson, M.: 1985, *Foundations of Constructive Mathematics: Metamathematical Studies*, Springer-Verlag.

Bishop, E.: 1967, *Foundations of Constructive Analysis*, McGraw-Hill.

Bishop, E. and Bridges, O.: 1985, *Constructive Analysis*, Springer-Verlag.

Bishop, E. and Cheng, H.: 1972, 'Constructive measure theory', *A.M.S. Memoirs* No. 116.

Bridges, D.: 1979, *Constructive Functional Analysis*, Pitman.

Brouwer, L.E.J.: 1975, *Collected Works*, I. North-Holland.

Buchholz, W., Feferman, S., Pohlers, W., and Sieg, W.: 1981, *Iterated inductive definitions and subsystems of analysis: recent proof-theoretical studies, Lecture Notes in Math.* **897**.

Buchholz, W. and Schütte, K.: 1988, *Proof Theory of Impredicative Subsystems of Analysis*, Bibliopolis.

Cherlin, G.: 1976, *Model theoretic algebra: selected topics, Lecture Notes in Math.* **521**.

Cohen, P.J.: 1966, *Set Theory and the Continuum Hypothesis*, Benjamin.

Cohen, P.J.: 1969, 'Decision procedures for real and p-adic fields', *Comm. Symp. Pure Appl. Math.* **22**, 131–151.

Davis, P.J. and Hersh, R.: 1981, *The Mathematical Experience*, Birkhäuser.

Dreben, B. and Denton, J.: 1966, 'A supplement to Herbrand', *J. Symbolic Logic* **31**, 393–398.

Eklof, P.: 1973, 'Lefschetz's principle and local functors', *Proc. Amer. Math. Soc.* **37**, 333–339.

Feferman, S.: 1968, 'Systems of predicative analysis II. Representations of ordinals', *J. Symbolic Logic* **33**, 193–220.

Feferman, S.: 1968a, 'Autonomous transfinite progressions and the extent of predicative mathematics', in *Logic, Methodology and Philosophy of Science*. III, North-Holland, pp. 121–135.

Feferman, S.: 1969, 'Set-theoretical foundations of category theory' (with an Appendix by G. Kreisel), in *Lecture Notes in Math.* **106**, 201–247.

Feferman, S.: 1970, 'Formal theories for transfinite iterations of generalized inductive definitions and some subsystems of analysis', in *Intuitionism and Proof Theory*, North-Holland, pp. 303–326.

Feferman, S.: 1972, 'Infinitary properties, local functors, and systems of ordinal functions', in *Lecture Notes in Math.* **255**, 63–97.

Feferman, S.: 1975, 'A language and axioms for explicit mathematics', in *Lecture Notes in Math.* **450**, 87–139.

Feferman, S.: 1977, 'Theories of finite type related to mathematical practice', in Barwise (1977), pp. 913–971.

Feferman, S.: 1977a, 'Inductive schemata and recursively continuous functionals', in *Logic Colloquium, 1976*, North-Holland, pp. 373–392.

Feferman, S.: 1977b, 'Categorical foundations and foundations of category theory', in *Logic, Foundations of Mathematics and Computability Theory*, Reidel, pp. 149–169.

Feferman, S.: 1978, 'A more perspicuous formal system for predicativity', in *Konstruktionen versus Positionen* I, Walter de Gruyter, pp. 87–139.

Feferman, S.: 1979, 'Constructive theories of functions and classes', in *Logic Colloquium, 1978*, North-Holland, pp. 159–224.

Feferman, S.: 1979a, 'What does logic have to tell us about mathematical proofs?', *The Mathematical Intelligencer* **2**, 20–24.

Feferman, S.: 1981, 'The logic of mathematical discovery vs. the logical structure of mathematics', in *PSA 1978* **2**, 309–327. (Phil. Sci. Assoc.).

Feferman, S.: 1982, 'Inductively presented systems and the formalization of metamathematics', in *Logic Colloquium, 1980*, North-Holland, pp. 95–128.

Feferman, S.: 1982a, 'Monotone inductive definitions', in *The L.E.J. Brouwer Centenary Symposium*, North-Holland, pp. 77–89.

Feferman, S.: 1984, 'Toward useful type-free theories, I', *J. Symbolic Logic* **49**, 75–111.

Feferman, S.: 1984a, 'Foundational ways', in *Perspectives in Mathematics*, Birkhäuser, pp. 147–158.

Feferman, S.: 1987, 'Infinity in mathematics: is Cantor necessary?', in *L'infinito nella scienza*, Inst. della Enciclopedia Italiana.

Feferman, S.: 1988, 'Hilbert's program relativized: proof-theoretical and foundational reductions', *J. Symbolic Logic* **53**, 364–384.

Feferman, S.: 1988a, 'Weyl vindicated: *Das Kontinuum* 70 years later', in *Temi e prospettive della logica e della filosofia della scienza contemporanee*, CLUEB, pp. 59–93.

Feferman, S.: 1989, 'Finitary inductively presented logics', in *Logic Colloquium 1988*, Elsevier, pp. 191–220.

Feferman, S.: 1990, 'Polymorphic typed lambda-calculi in a type-free axiomatic framework',

Logic and Computation, Contemporary Mathematics **106**, pp. 101–136.

Feferman, S.: 1991, 'Reflecting on incompleteness', *J. Symbolic Logic* **56**, 1–49.

Fenstad, J.E.: 1980, *General Recursion Theory. An Axiomatic Approach*, Springer-Verlag.

Fitting, M.C.: 1981, *Fundamentals of Generalized Recursion Theory*, North-Holland.

Friedman, H.: 1971, 'Algorithmic procedures, generalized Turing algorithms and elementary recursion theories', in *Logic Colloquium, 1969*, North-Holland, pp. 361–390.

Friedman, H.: 1975, 'Some systems of second order arithmetic and their use', *Proc. Int. Cong. Math., Vancouver, 1974* **1**, 235–242.

Friedman, H.: 1976, 'Systems of second order arithmetic with restricted induction' (abstracts), *J. Symbolic Logic* **41**, 557–559.

Friedman, H.: 1977, 'Set-theoretic foundations for constructive analysis', *Ann. Math.* **105**, 1–28.

Friedman, H.: 1980, 'A strong conservative extension of Peano arithmetic', *The Kleene Symposium*, North-Holland, pp. 113–122.

Friedman, H., McAloon, K., and Simpson, S.G.: 1982, 'A finite combinatorial principle which is equivalent to the 1-consistency of predicative analysis', in *Patras Logic Symposium*, North-Holland, pp. 197–230.

Friedman, H., Simpson, S.G., and Smith, R.: 1983, 'Countable algebra and set existence axioms', *Ann. Pure and Applied Logic* **25**, 141–181.

Gentzen, G.: 1969, *The Collected Papers of Gerhard Gentzen*, North-Holland.

Girard, J.-Y.: 1981, '$\prod_2^1$-logic, Part I: Dilators', *Ann. Math. Logic* **21**, 75–219.

Gödel, K.: 1959, 'Consistency-proof for the generalized continuum-hypothesis', *Proc. Nat. Acad. Sci. U.S.A.* **25**, 220–224.

Gödel, K.: 1940, 'The consistency of the continuum hypothesis', *Ann. Math. Studies* No. 3, The Princeton Univ. Press; rev., notes added 1951, 1966.

Gödel, K.: 1958, 'Über eine bisher noch nicht benützte Erweiterung des finiten Standpunktes', *Dialectica* **12**, 280–287.

Gödel, K.: 1986, *Collected Works, Vol. I. Publications 1929–1936*, ed. by S. Feferman, J.W. Dawson, Jr., S.C. Kleene, G.H. Moore, R.M. Solovay, and J. van Heijenoort, Oxford University Press.

Gödel, K.: 1990, *Collected Works, Vol. II. Publications 1938–1974*, ed. by S. Feferman *et al.*, Oxford University Press.

Goodman, N.D.: 1979, 'Mathematics as an objective science', *Amer. Math. Monthly* **86**, 540–551.

Hersh, R.: 1979, 'Some proposals for reviving the philosophy of mathematics', *Adv. Math.* **31**, 31–50.

Hilbert, D. and Bernays, P.: 1959, 'Grundlagen der Mathematik II', Springer; rev. edn. 1970.

Jäger, G. and Pohlers, W.: 1982, 'Eine beweistheoretische Untersuchung von $(\Delta_2^1 - CA) + (BI)$ und verwandter Systeme', *Sitzungsber. Bayer. Akad. Wiss., Mat.-Nat. Klasse*, 1–28.

Jensen, R.: 1969, 'On the consistency of a slight(?) modification of Quine's *New Foundations*', in *Words and Objections: Essays on the Work of W.V.O. Quine*, Reidel, pp. 278–291.

Kanamori, A. and Magidor, M.: 1978, 'The evolution of large cardinal axioms in set theory', in *Lecture Notes in Math.* **669**, 99–275.

Kleene, S.C. and R.E. Vesley: 1965, *The Foundations of Intuitionistic Mathematics, Especially in Relation to Recursive Functions*, North-Holland.

Kreisel, G.: 1958, 'Mathematical significance of consistency proofs', *J. Symbolic Logic* **23**, 155–182.

Kreisel, G.: 1967, 'Mathematical logic: what has it done for the philosophy of mathematics', in *Bertrand Russell, Philosopher of the Century*, Allen and Unwin, pp. 201–272.

Kreisel, G.: 1970, 'Principles of proof and ordinals implicit in given concepts', in *Intuitionism* and *Proof Theory*, North-Holland, pp. 489–516.

Kreisel, G.: 1976, 'What have we learned from Hilbert's second problem?', *Proc. Sympos. Pure Math.* **XXVIII**, 93–130 (A.M.S.).

Kreisel, G.: 1977, 'Review of *Brouwer 1975*', *Bull. A.M.S.* **83**, 86–93.

Kreisel, G. and Troelstra, A.S.: 1970, 'Formal systems for some branches of intuitionistic analysis', *Annals. Math. Logic* **1**, 229–387.

Lakatos, I.: 1976, *Proofs and Refutations: the Logic of Mathematical Discovery*, Cambridge University Press.

Lakatos, I.: 1978, *Mathematics, Science and Epistemology: Philosophical Papers Vol. 2*, Cambridge University Press.

MacLane, S.: 1961, 'Locally small categories and the foundations of mathematics', in *Infinitistic Methods*, Pergamon, pp. 25–43.

MacLane, S.: 1971, *Categories for the Working Mathematician*, Springer-Verlag.

MacLane, S.: 1981, 'Mathematical models: a sketch for the philosophy of mathematics', *Amer. Math. Monthly* **88**, 462–72.

Martin, R.L. (ed.): 1984, *Recent Essays on Truth and the Liar Paradox*, Oxford University Press.

Martin-Löf, P.: 1982, 'Constructive mathematics and computer programming', *Logic, Methodology and Philosophy of Science* VI, North-Holland.

Martin-Löf, P.: 1984, *Intuitionistic Type Theory*, Bibliopolis.

Mines, R., Richman, F., and Ruitenberg, W.: 1988, *A Course in Constructive Algebra*, Springer-Verlag.

Moore, G.: 1982, *Zermelo's Axiom of Choice*, Springer.

Moschovakis, Y.N.: 1977, 'On the basic notions in the theory of induction', in *Logic, Foundations of Mathematics, and Computability Theory*, Reidel, pp. 207–236.

Moschovakis, Y.: 1991, 'The formal language of recursion', *J. Symbolic Logic* **54**, 1216–1252.

Mosses, P.D.: 1990, 'Denotational semantics', in *Handbook of Theoretical Computer Science, Vol. B. Formal Models and Semantics*, Elsevier, pp. 595–631.

Myhill, J.: 1975, 'Constructive set theory', *J. Symbolic Logic* **40**, 347–383.

Paris, J. and Harrington, L.: 1977, 'A mathematical incompleteness in Peano Arithmetic', in Barwise (1977), pp. 1133–1142.

Pohlers, W.: 1989, *Proof Theory: an introduction, Lecture Notes in Math.* **1407**.

Pohlers, W.: 1991, 'Proof theory and ordinal analysis', *Archive for Mathematical Logic* **30**, 311–376.

Prawitz, D.: 1971, 'Ideas and results in proof theory', in *Proc. Second Scandinavian Logic Symposium*, North-Holland, pp. 235–307.

Rathjen, J.: 1991, 'Proof-theoretic analysis of KPM', *Archive for Mathematical Logic* **30**, 377–403.

Robinson, A.: 1966, *Non-Standard Analysis*, North-Holland; rev. ed. 1974.

Schütte, K.: 1978, *Proof Theory*, Springer.

Scott, D.: 1972, 'Continuous lattices', in *Lecture Notes in Math.* **274**, 97–136.

Scott, D.: 1976, 'Data Types as lattices', *SIAM J. Comput.* **5**, 522–587.

Sieg, W.: 1991, 'Herbrand analyses', *Archive for Mathematical Logic* **30**, 409–441.

Simpson, S.G.: 1984, 'Which set existence axioms are needed to prove the Cauchy/Peano theorem for ordinary differential equations?', *J. Symbolic Logic* **49**, 783–802.

Simpson, S.G.: 1987, 'Subsystems of Z_2 and reverse mathematics', Appendix to Takeuti (1987), pp. 432–446.

Simpson, S.G.: 1987a, 'Unprovable theorems and fast growing functions', in *Logic and Combinatorics, Contemporary Mathematics* **65**, 359–394.

Simpson, S.G.: 1988, 'Partial realizations of Hilbert's program', *J. Symbolic Logic* **53**, 349–363.

Stoy, J.E.: 1979, *Denotational Semantics: The Scott–Strachey Approach to Programming Lan-

guage Theory, M.I.T. Press.

Stroyan, K.D.: 1977, 'Infinitesimal analysis of curves and surfaces', in Barwise (1977), pp. 197–231.

Takeuti, G.: 1975, *Proof Theory*, North-Holland.

Takeuti, G.: 1978, 'A conservative extension of Peano Arithmetic', Part II of *Two Applications of Logic to Mathematics*, Princeton Univ. Press.

Takeuti, G: 1987, *Proof Theory* (2nd. ed. with Appendixes by G. Kreisel, W. Pohlers, S. Simpson and S. Feferman), North-Holland.

Troelstra, A.S.: 1977, 'Aspects of constructive mathematics', in Barwise (1977), pp. 973–1052.

Troelstra, A.S. and van Dalen, D.: 1988, *Constructivism in Mathematics* (in two volumes), North-Holland.

Tymoczko, T. (ed.): 1986, *New Directions in the Philosophy of Mathematics*, Birkhäuser, Boston.

Visser, A.: 1989, 'Semantics and the liar paradox', in *Handbook of Philosophical Logic*, Vol. IV, Reidel, pp. 617–706.

Wang, H.: 1974, *From Mathematics to Philosophy*, Routledge & Kegan Paul.

Weyl, H.: 1918, *Das Kontinuum*, Veit.

Weyl, H.: 1987, *The Continuum: A Critical Examination of the Foundation of Analysis* (translation of Weyl (1918) by S. Pollard and T. Bole), Thomas Jefferson University Press.

K. TAHIR SHAH

MINDS AND BRAINS, ALGORITHMS AND MACHINES

1. INTRODUCTION

The well known philosophical problem of the mind–body relationship meta-morphed into a mind–machine debate when Turing asked the question: 'Can a machine think?' This question led him to define an operational definition of intelligence in terms of a test now known as the Turing test (Turing 1950). J. Searle (1980) used a variant of this test, a *gedankenwelt* experiment, to argue against 'Strong AI' claim. It led to a debate now known as the Chinese Room debate that is still continuing (Searle 1990, Churchland 1990, Denning 1990, Torrance 1984). My theme of discussion is this experiment and some related issues. The arguments I shall present are to support a symbolic paradigm[1] claim – machines are as powerful as minds. However, I shall pursue the opposite path and argue that minds can be explained in terms of a purely symbol-processing paradigm and thus try to establish the equivalence of minds and machines.

I have chosen the above title deliberately to emphasize the following points.

1. The human brain is the most complex of brains that has evolved. It is based on the same electrochemical mechanism as any other brain of lower species. Consequently, the intelligence, the 'Umwelt' of a given species, and the tasks it can perform, are all directly related *only* to their brain's complexity and specific circuitry.

2. A brain has many subsystems, each dedicated to the processing of a specific task. Although they may be extremely complex neuronal sets, these subsystems are not general-purpose computing devices (Gazzaniga 1989 and Knudsen 1987)[2].

3. Recently, it has become clear that many of the tasks performed in real-time by a nervous system or its subsystems are known to be NP-complete (i.e., cannot be solved in polynomial computation time) because of their inherent complexity. For instance, the general task of visual search is inherently in-tractable in the formal sense (Tsotsos 1990). But nature somehow solves them in a reasonable amount of time. In fact, human vision is effortless and precise. No doubt: survival depends on it. This suggests that computational complex-

G. Corsi et al. (eds), Bridging the Gap: Philosophy, Mathematics, and Physics, 125–139.
© *1993 Kluwer Academic Publishers.*

ity leads to constraints on the hardware architecture. One should be careful to distinguish a material implementation of an abstract computing model from the model itself. Although the inherent complexity is defined with respect to a machine model, a Turing machine, say, somehow the efficiency can be varied. In parallel systems this is clear, where some interconnection topologies are more efficient than others for the same class of problems (Jamieson 1987, Uhr 1984). Many intractable problems of *average* data size are solvable by parallel machines in a reasonable amount of time (Bokhari 1981). This does not mean that an intractable problem becomes tractable. It simply means that a tradeoff between the data size and computation time is feasible. Fortunately, there are no 'Chinese' rooms (Searle 1980) in nature: we would not have survived. In raising the issue of hardware I am neither contradicting nor approving the dogma of cognitive science – the theory dualism. Cognitive science neither looks into complexity issues nor does it deal with descriptions at the lowest level of a system. For a recent debate on parallel hardware and mind–body problem, see comments by Ramsey (1989) on Thagard (1986).

In my opinion, these empirical aspects are essential to any serious discussion of intentionality. Searle claims it is due to the *causal power* of the brain. I shall argue that intentionality is nothing but a property of a symbol-processing system, given there are supporting systems present, such as an appropriately represented *knowledge-base*. Moreover, I shall base my arguments on neurobiological evidence since it is clear that philosophical notions based on everyday language are not robust enough to discuss the mind. Specifically, I shall look for similarities between the mechanisms in parallel machines and those in the brain.

The Turing and similar tests cannot be considered as a serious contender for a human versus machine behavioral test, given the complexity of a natural language and the issues involved. The importance of the knowledge-base, its storage, retrieval, activation for language processing, and representational issues were not taken into account properly in Searle's thought experiment. An active knowledge base is essential for what has been called 'understanding', whether by a symbol processing or some other kind of system (if its exists). Human subjects with lesions in some critical areas (i.e., Broca's and Wernicke's areas, memory and other subsystems, such as parts of cortex) of their brain where linguistic tasks are processed do not have 'understanding' in the usual sense (Gazzaniga 1989, O'Grady and Dobrovsky 1987, Boden 1988, Deacon 1988) or as it is intended by Searle. They have only partial understanding. The point I want to make is that there are many level of subsystems in the brain. The intentional attributes are at the top level while neurons are at the bottom level. The number of distinct levels are possibly more than three. One should be careful not to mix issues at different levels. In a parallel distributed system like the brain, a single subsystem, or a sub-subsystem, does not 'understand' higher level tasks. Damage to any subsystem of either a biological brain or an artificial system leads to malfunction or loss of the behavior associated with it.

Therefore, intentionality and other higher cognitive functions are first of all, derived from cooperation between various specialized subsystems. Second, there is no evidence for any mysterious causal power in the brain that is not possible in machines. Both are physical systems and are equally subject to the laws of physics and information.

2. THE PROBLEM OF OTHER MINDS

What is a mind? Obviously this is the first question I should ask myself and must be able to answer before I try to compare it with anything like (the mind of a) machine. Let me assume that I call all my activities like thinking, understanding natural language, etc., mental activities. Terms like 'thinking', 'understanding', and so on are taken to be as they are understood within ordinary language. Despite all the complications of how to define a mind, I shall assume I have a mind – whatever possible attributes it may have. The difficult philosophical problem, however, is *how do I know other minds exist*. This is the problem of the knowledge of other minds that has been discussed by such noted people as J.S. Mill (Malcolm 1981). Mill asks himself the question "By what evidence do I know, or by what consideration am I led to believe, that there exist other sentient creatures. . . possessing Minds?" Our knowledge of the other minds actually follows from an argument by analogy. His answer is the following:

I conclude that other human beings have feelings like me, because, first, they have bodies like me, which I know, in my own case, to be the antecedent condition of feelings; and because, secondly, they exhibit the acts and other outward signs, which in my own case I know by experience to be caused by feelings. . ..

He concludes the paragraph,

I bring other human beings, as phenomena, under the same generalization which I know by experience to be the true theory of my own existence.

Malcolm objected to this, saying that this would be very *weak* inductive reasoning. According to him, Mill states that he has *no criterion* for determining whether another "walking and speaking figure" does or does not have feelings for the only plausible criterion would lie in behavior and circumstances that are open to view. I agree that if I do not know how to establish that he has the *same* as I have when I have pain, it is not possible to improve my understanding of other's feeling without having any criterion (or a set of criteria) of 'sameness'. Consequently, I have difficulty in accepting that others have minds as I have. H.H. Price (see Malcolm 1981) suggests that "one's evidence for the existence of other minds is derived primarily from the understanding of language". Since I am going to discuss whether a symbol processing machine can have a mind, I must therefore either have a criterion of sameness (perhaps in the computational paradigm), or it will be futile to discuss such an issue. Let me assume understanding of a language as a criterion since we need it for *sameness* – for men and machine both. It seems that Searle ignored this problem completely. In fact, my first reaction to Searle's paper was: *How do I know that he understands his native language English and that he does not understand Chinese!* There is no way to determine the validity of his assertion except through what he tells me. Without the solution to this problem it is futile to discuss mind–machine equivalence. If one accepts behaviorism, Searle understands Chinese. Without behaviorism, it not possible to establish that he understands his native tongue, English.

The problem is now shifted to what it means 'to understand'. By narrowing down my argument to Searle's *gedankenexperiment* I shall try to convince you what Minsky asserted; "traditional, everyday, precomputational concepts like believing and understanding are neither powerful nor robust enough..." (comments on Searle 1980). Furthermore, I shall use the case of 'Mr. Chance, the gardner' (Fass 1990) to show that human judgement of who is 'intelligent', and who is not, is indeed faulty. We do use some element of behaviorism to discuss another's mind. This is obvious in Searle's experiment, where it is accepted that a native speaker 'understands' English. However, the same behaviorist criterion is discarded for computer programs. This is a remarkable contradiction.

The other option is, if one accepts that psychology can be reduced to neurophysiology, then it is enough to compare neural circuits and their properties.

My point of view, like the naturalism of Quine and Feyeraband, is that correct philosophical theories are essentially extensions of empirical/scientific theories. Unless there is an adequate empirical description of the brain that clearly shows the existence of 'causal power' the mechanist approach is the only one that can be substantiated by empirical neurobiological data.

One question which has been discussed quite often is the relationship between information processing and physical laws. Or, how does a physical system compute? What is the relationship between information processing and the physical system (material implementation of a machine model) that does information processing? Shall we find the answer among the laws of physics or elsewhere? I think that the laws governing the embodiment of mind should be sought among the laws governing information rather than energy or matter.

Historically, though dualism originated earlier than the information and computation theory, perhaps there is an explanation why the material implementation of a computing machine and the algorithm that runs on that machine are considered as independent, although they are related to each other. One should carefully distinguish information and information processing from the medium that carries this information and processes it by changing its states. Thus, the key is that one must *distinguish* between the two interrelated aspects: information and the medium that holds that information.

The bridge between the two aspects is the notion of the 'state' of a (classical or quantum) physical system. The 'state' of a physical system contains some information. For instance, the position of an object, a spin state of an electron, or the 'ON' state of a flip-flop circuit, all contain some information, but with reference to their external world. The state of a system represents a relationship between itself and its external world. Without its external world, the state is meaningless and does not give any information. The information it contains depends on the interpretation given to it by an external system. It could be interpreted to be anything one likes; perhaps a building, a sheep, or number 7. Thus not only the notion of the 'state' is meaningful in an overall context of the physical world, but it also requires some interpretation in order for it to be useful. It is clear I have semantics in mind. Biological information processing systems, i.e. brains, get their semantics by mapping the external world onto its symbol processing device in the form of neural net structures. The so-called causal powers of the brain are its sensory mechanisms and its neural network's architecture, which contains all relevant knowledge about the external world. It is semantically self-sufficient, i.e., semantics is not due to anything outside its symbol processing mechanism.

Information is thus simply a relationship of an information-carrying medium with its external world. And in most cases it is a matter of convention. Since the relationship between two physical objects or systems is non-physical (e.g., probabilistic or otherwise) and since computation is a sequence of states, mind or a program can be considered non-physical only in this sense. In other words, the mind is a sequence of states of a system – no different from an algorithm. Whether you hard-wire such a sequence of states sequentially or in parallel with an appropriate time dependence, you get a brain – a physical system. The mind is thus influenced by physical laws only indirectly, either as a constraint on its speed or limitations on what it can compute. I shall thus consider that the property of being in a given mental state is identical with the property of being in a given neurophysiological state. Mind is a set of interacting algorithms that are processed by the brain.

3. GÖDELIZATION ARGUMENTS

In the fifties and sixties there were attempts to show the mechanist doctrine to be incorrect and that machines do have limitations, in constrast to brains (Nagel and Newman 1959), for an in-depth discussion, see Hofstadter (1979). These were exercises to show that a program cannot function at the meta-level and thus mind and machine are not equivalent. Lucas invoked Gödel's *incompleteness theorem* (that if you start with consistent axioms and apply sound rules of inference, then the collection of theorems that can be deduced is incomplete) to claim, ". . . to prove that mechanism is false, that is, that minds cannot be explained as machines". Hofstadter (1979) appealed to the Church–Kleene theorem to refute Lucas's conclusion. Most recently, Penrose (1989), using a variety of arguments, centered on this theorem, seeking to show that the thought process is non-algorithmic and cannot be simulated on a computer.

My argument is that artificial intelligence does more than just theorem proving. Moreover, formal systems can be more complex than classical systems, where logic is only two-valued (assuming the validity of the principle of *tertium non datur*), and where the rules of inference are *sound*. For logical systems such as non-monotonic systems (Genesereth and Nilsson 1986), logic of default (Reither 1980), or where knowledge can be added to the knowledge-base (Reiter's closed world assumption), a straightforward application of Gödel's theorem is dangerous. In fact, *there is no provision for knowledge acquisition in Gödel's theorem*. Without the learning facility a human mind will be just as limited as a machine. There is no doubt we can program non-monotonic or multivalued logic on universal symbol processing machines. They are not handicapped due to incompleteness of consistent formal systems. Recently, Shoham (1990) attempted to show that causality can be introduced using non-monotonic reasoning. On the other hand, a generalization of Turing machines has been proposed (Smale 1989) where other than bit-operations, e.g. rational operations, are taken to be the primitive operations of the machine. These types of model can possibly correlate connectionism with the symbol processing paradigm in a rigorous manner.

Besides the question of incompleteness, both Lucas and Searle are not careful about the level of description. As Arbib (1987) puts it very elegantly, "Human thought is informal at the level of English, formal at the biochemical level",

I think there is a large gap between psychology and neurophysiology. One should not mix up the issues at different levels. I agree with Hofstadter, "Lucas' argument applies merely to their bottom level, on which their intelligence – however great or small it may be – does not lie." This applies to Searle's assertion too that machines do not understand.

4. LOSS OR PARTIAL LOSS OF UNDERSTANDING

Now I shall summarize some relevant biological facts to contrast whether there is any need for 'causal powers', or whether intentionality is simply a result of many knowledge-base systems cooperating with each other. I am not pretending that we know how knowledge is represented in the human brain. But this is not essential to my arguments.

Studies of patients with brain lesions suggest strongly that the brain handles information processing in a manner similar to a knowledge-base system at higher level. On the other hand, it is known, from experiments on the nervous systems of simple animals, that there are specialized neuronal circuits for specific, simple tasks. Taken alone each task, at an appropriate granularity, is comparable to what is possible on today's machines. It is known that some of these tasks are performed by artificial neural networks (Shah 1990). These networks are capable of learning by looking at examples, very much like humans.

Language is one of the tasks that the human brain has specialized in recent times in evolutionary history. The cortex, a grey tissue area, contains many of the cognitive abilities that distinguish humans from other mammals. These abilities are localized, such as the localization of cognitive and perceptual functions in a particular hemisphere of the brain, called 'lateralization'. It is known that the left side is specialized in language, analytic reasoning, temporal ordering, reading and writing, and arithmetic, while the right side deals with the perception of non-linguistics sounds, music, visuo-spatial skills, holistic reasoning, and pattern recognition. However, some language-related tasks are carried out by the right brain rather than the left brain, e.g., in the interpretation of the voice tone and intonation cues that signal emotions such as anger and fear. Thus a patient suffering from damage to the right hemisphere may be able to understand the literal meaning of a sentence but fail to recognize whether it is spoken in an angry tone or in a fearful way. A parallel phenomenon manifests itself in the area of semantics. Here, it seems that damage to the right hemisphere can interfere with the ability to understand and appreciate metaphorical use of language. Patients with this type of brain damage are able to provide only a literal or concrete interpretation of figurative sentences. (e.g. He was wearing a loud tie). Remember, in the Turing test, as well as in the Chinese room, there is no mechanism to communicate other than written words. These tests are inadequate since there is no provision for non-linguistic communication, such as gesture, tone of voice, etc.

A similar problem can occur in split-brain patients (Gazzaniga 1987). They may be able to perceive objects presented to the left visual field but unable to describe them, since the information is not passed to language processing sections located in the left hemisphere. Some of the important language centres known are:

Broca's Area. This is responsible for organizing the articulatory patterns of speech. It plays a crucial role in the formation of words and sentences.

Wernicke's Area. This area plays many roles. It is responsible for the reception of auditory inputs. Its most important role is in the representation of meaning and it is involved both in the interpretation of words and in the selection of lexical items for the purpose of sentence production.

The Broca and Wernicke areas are connected through communication fibres known as Arcuate Fasciculus.

Annular Gyrus. This area lies behind Wernicke's area and is responsible for converting a visual stimulus into an auditory form, and vice versa. This area is crucial for the matching of a spoken form with a perceived object, the naming of objects, and the comprehension of written language, all of which require connections between the visual and the speech regions.

For example uttering a word is a sequence of processing by many areas, as given in the diagram below (Dobrovolsky and O'Grady 1987).

Visual area $\longrightarrow$ Angular gyrus (associates visual
form to auditory form) $\longrightarrow$ Wernicke's area $\longrightarrow$ Broca's
area (articulation) $\longrightarrow$ Motor Cortex

It is known that electrical stimulation of various sites on the human cerebral cortex can disturb different language functions. Such stimulation can cause, for instance, slurring, repetition, confused counting, misnaming with and without perseveration, hesitation, problem with phoneme identification, naming, reading, grammar, and short-term verbal memory. Each problem is associated with a specific area (Deacon 1988).

What is understanding, even *grosso modo*? When we try to understand a spoken rather than a written name of an object, the stimulus from the auditory cortex is transmitted to Wernicke's area, where it is then interpreted. In cases where the object can be associated with an image, a message can be sent to the angular gyrus, where it is converted into a visual stimulus, arousing the appropriate pattern in the visual area. The man inside the Chinese room 'understands' partially, like a brain-damaged patient. He can recognize symbols, read instructions and associate them with hand movements so as to sort various symbols and manipulate them and so on, but no access to the knowledge base is given to him where semantics, i.e. real world experience to symbol correlates, are stored. Consequently, he does not associate these symbols with the real world objects.

If the language centres or the connections between them are damaged, the ability to use language deteriorates. Language disorders are called aphasia. Poor articulation is due to the damage to Broca's area. It also leads to deletion of sounds – phonemic paraphasia. A severe disability is the inability to form morphological and syntactical patterns (mainly nouns are uttered). Broca's aphasia is also accompanied by deficits in syntactical knowledge. While patients appear to comprehend well as long as they can rely on their knowledge of word meaning and pragmatics, they are unable to make use of syntactic knowledge to interpret sentences. Thus Broca's area appears to be largely concerned with organizational and structural aspects of language and is, therefore, responsible for the articulatory rules that create sound patterns, as well as for the morphological and syntactical rules that form words and phrases.

Wernicke Aphasia. Since Wernicke's area is responsible for the representation of meaning, as well as the interpretation of words during comprehension and the selection of words during speech production, the most striking feature of this disorder is an inability to comprehend spoken language and to construct meaningful utterances. Although Wernicke patients may sound almost normal, their speech does not make sense. Responses may be well-formed structures, but inappropriate as answers to the questions by the experimenter.

Recently, there was an interesting study by D. Fass (Fass 1990) in which he compares a character from the novel *Being There*, a certain Mr. Chance the gardener, with some of the early AI systems, ELIZA, PARRY, and SAM. Mr. Chance is a brain-damaged patient whose only source of knowledge is TV and he knows only about gardens. This case is a good strike at the assumption that the 'Other's mind' exists. Context forces us sometimes to conclude that another human being understands although, in fact, they simply do not.

The literary commentators remarked "How a dead soul like Chance can be mistakenly seen by others as highly intelligent and can manage to achieve eminence." We normally assume that any human is at least averagely intelligent. We then rely on further clues: their appearance, remarks made about a person by others and so on. Having judged a person to be intelligent, we may well regard any subsequent behavior produced by that person as intelligent. Chance's major achievement is being (mis)taken as a good user of metaphor, even though he does not intentionally produce metaphors nor does he understand them. Here is a case of a damaged brain (incomplete system for a given task) and it is no different than a machine with incomplete knowledge. Where is the 'causal power' of the brain which Searle says is responsible for intentionality? Or can it be it is lost because only a part of the brain's system is non-operational?

Damage to the angular gyrus does not affect vision. Patients with alexia (reading ability impairment) and agraphia (writing ability impairment) can still see normaly. They can even copy letters and words. However, since the angular gyrus contains the information specific to graphic representation of sound patterns, alexic patients perceive the words and letters as meaningless patterns rather than symbols representing linguistic structures.

The role of the angular gyrus in reading and writing is closely tied to the type of writing system a particular language uses (O'Grady and Dobrovolsky 1987). Because the alphabetic writing system of English represents a sound structure, the angular gyrus plays a crucial role in its use. In some societies the writing system may represent *concepts rather than sounds* (e.g. Chinese). A damage to the angular gyrus does not impair the use of such a writing system. In fact, there are recorded cases of individuals who know both a sound-based writing system and a concept-based one losing the ability to use *only* the former after suffering damage to the angular gyrus.

In this connection, I should mention the case of psychedelic experience due to drugs. These drugs are either similar to neurotransmitters or their antagonists. One of the most incredible changes wrought by a psychedelic drug is *synesthesia* (Snyder 1986). This is a phenomenon in which the senses become transmuted, so that touch may be experienced as sound, sound as vision, and so forth. Furthermore, in cases of potent drugs like LSD, there is loss of ego and self-awareness. What this means is that since neurotransmitters are used in the neural communication system, a mixing up of communication lines causes mixing of data channels going to specialized processing units or modules. The behavior is no different in a machine if the data from the speech interface are sent to a vision system (by mistake or due to some fault in the communication lines) rather than to their proper destination. There is a distortion of perception in human subjects due to noisy data – noise due to externally introduced neurotransmitter-like chemicals, e.g. LSD.

To summarize, I would like to state two propositions which are supported by the evidence I have described above.

1. Studies on lower animals suggest that nervous systems do not have any unusual and mysterious causal powers that machines cannot have. In many cases, a precise circuit-to-behavior relationship is known (Selverston 1985, Getting 1989).
2. Brain-lesion studies in humans suggest that our brain works like a system, no different than a man-made system. If some part is damaged then that aspect of behavior is lost.

Some philosophers insists that intentionality is a property of the human mind and it is due to some attributes, but they do not tell us which part of the brain or what mechanisms are exclusive to the human (and animal) brain for the production of intentionality.

5. CONCLUSION

The idea I want to convey is that the brain is a complex system consisting of many subsystem, sub-subsystems and so on.

Ultimately, one gets down to the level of a neuron, the basic unit of information processing. If we start from the bottom and work up, studies on the nervous systems of simple species suggest a precise mechanistic description of how information is processed. The behavior of these species is also simple and mechanistic, and in some cases, once it is started it continues, even under unfavorable circumstances. The brain of higher species works on principles no different than for the lower species. Furthermore, these neuronal nets are dedicated to a specific task and are not general purpose computing devices. As more and more nets are added, more and more functions (behavior) become possible. Add to this a memory mechanism and you get knowledge-based processing. For higher cognitive functions a large number of such nets cooperate, as is illustrated for the natural language case.

At the lowest level, the information processing is no different than symbol processing by a formal system. It does not matter whether processing is done by a parallel or serial system since that will affect only the speed of execution. Thus, at the lowest level, even a complex brain is without doubt only a syntactic system. The so-called computational maps are a good example of specialized computing nets[3]. But as we move upward, the knowledge of the external world (retrieved from the memory) brings in the semantics. The overall system does 'understand' what is being processed, but at the bottom level, neurons, being a single unit, do not have any attribute we call understanding or other intentional properties. We do not know which animal species have intentional attributes like us and which ones do not. A more difficult question is at what level of complexity such attributes begin to show up.

Thus, given what we know about human and animal nervous systems, higher cognitive behavior, such as understanding a natural language or having intentionality, have nothing to do with what happens at the lowest level. No doubt, if the system at the lower levels does not function properly, it can lead to abnormal behavior at the higher levels. Or, perhaps a command from the higher level may stop processing at the lower level. But these levels must be fairly independent in order for them to be distinct. What makes the difference at higher levels is the cooperative behavior of many lower units. A recent computer simulation of rhythmic neural population oscillation confirms that such cooperative behavior is possible not only in biological systems but also in machines (Traub *et al.* 1989). Similarly, the recently discovered 40 H oscillations in the cat visual cortex suggest a visual awareness mechanism (Gray *et al.* 1990). A detailed explanation is provided by Crick and Koch (1990) of how such collective behavior of neurons may be responsible for intentionality, awareness etc. Such oscillations were in fact predicted by van der Malsburg's neural network model.

The other point is that human understanding is strongly context dependent (e.g. the meaning of metaphors depends on the cultural context). The bottom line is that background knowledge and a mechanism to store and retrieve (in a reasonable time) are essential ingredients for understanding in *any* system.

A human would, like a simple computing machine, fail to understand a concept, a story, or whatever unless the above conditions are met. A brain-damaged person is unable to comprehend if the damage is in a critical area. A failure of memory, for instance in the case of senile dementia, causes similar problems.

By this token the inference drawn from the *gedankenexperiment* does not tell the whole story. Not only are levels mixed up, but also many other factors, such as memory and other mechanism, are ignored in the discussion. Above all, common sense notions from everyday language are not appropriate to a serious discussion of the mind.

International Centre for Theoretical Physics,
Trieste, Italy

NOTES

[1] I shall not separate symbolic processing and connectionism despite the fact that connectionism lacks a clear description in terms of recursive computability. Nevertheless, a correlation is possible in the following terms. An algorithm graph can be mapped onto a set of processors, as is well known in systolic (parallel) machines. There is strong evidence that such a mapping occurs in biological systems as well (Shah 1991). The only difference is that the graph is a weighted graph. This mapping is the graph isomorphism problem (Bokhari 1981) which is known to be NP-complete. The learning process in a neural network, which is also a graph mapping problem, has recently been shown to be NP-complete.

[2] *The Chinese Room*. The point of view expressed in Searle (1980) is that:

(i) Intentionality in human beings (and animals) is a product of *causal features* of the brain;

(ii) Instantiating a computer program is never by itself a sufficient condition of intentionality. Searle states, "It is a characteristic of human being's story-understanding capacity that they can answer questions about the story even though the information that they were given was never explicitly stated in the story." Since correct answers to questions about his 'Hamburger story' can also be given by a simple theorem prover, given there is an appropriate knowledge base, perhaps he means more than the simple application of the inference rule.

The crux of the paper is his *refutation* of the following propositions: (i) That the machine can literally be said to understand the story and provide the answers to the questions; (ii) That what the machine and its program do *explain* is the human ability to understand the story and answer questions about it.

This is reminiscent of J.R. Lucas's arguments, based on Gödel's incompleteness theorem, which point out the inadequacy of theorem proving. Obviously, Schank's work is not all that AI does. It is only an attempt to test a theory of computational linguistics, which is subject to improvements.

Searle asserted three other consequences of propositions (i) and (ii). The essence is that the brain produces intentionality because it has causal power. Thus, any mechanism capable of producing intentionality must have *causal powers* equal to those of the brain. Since machines do have this causal power, they could not succeed just by designing programs but would have to duplicate the causal powers of the *human* brain.

There are many problems with this experiment and his assertions (1)–(5). Some problems are due to the vagueness of the terms like 'understand', 'causal power' and 'good' etc. Other problems are due to mixing up levels of description of the brain. Still more problems are due to either the inadequacy of the Turing-like tests, or the fragility of the everyday linguistic concepts. For instance, when he says 'I become so good' and 'programs are so smart', it is not clear why he needs to become 'so good' and 'programs so smart'. Is he learning? Are programs changing? Or what?

[3] *Computational Maps*. The maps, a key building block of the information processing infrastructure in the nervous system, are examples of a problem driven choice of innate neural connectivity. Their existence supports the algorithmic nature of neural systems. They are highly efficient and specialized information processing neural networks and have definite topological and topographical structure. In a computational map, there is a systematic variation in the value of the computed parameter across at least one dimension of the neural structure. The neurons that make up such a map represent an array of preset processors or filters, each tuned slightly differently, that operate in parallel on the afferent signal. *Filters (or preset processors) have specific neural network topologies for their specific tasks they are to perform:*

Computational maps in the visual system include those of line orientation preference and ocular dominance in the primary visual cortex and a movement of the direction in cortical middle temporal visual area MT. Computational auditory maps include maps of interaural delay, interaural intensity difference, and sound source location in the brainstem. These maps are most precisely elaborated and occupy the greatest portions of the brain in species that have highly developed sensory capacities, such as monkeys (vision) and owls and bats (audition). Efficiency in one specific domain requires such structures, that are fast and provide the low-level computation necessary for survival. These are comparable to the fast, real-time algorithms we know in computer science. Computational maps are also involved in motor programming.

REFERENCES

Arbib, M.A.: 1987, *Brains, Machines, and Mathematics*, Springer-Verlag, New York.

Boden, M.: 1988, 'Artificial intelligence and biological intelligence', in *Intelligence and Evolutionary Biology* (ed. by H.J. Jerison and I. Jerison), Springer-Verlag, New York, pp. 45–71.

Bokhari, S.H.: 1981, 'On the mapping problem', *IEEE Trans. on Computers* C–30, 207–214.

Churchland, P.S. and Churchland, P.M.: 1990, 'Could a machine think?', *Scientific American* 262, 26–31.

Crick, F. and Koch, C.: 1990, 'Some reflections on visual awareness', *Cold Spring Harbor Symp.* 55, 953–962.

Deacon, T.E.: 1988, 'Human brain evolution', in *Intelligence and evolutionary biology* (ed. by H.J. Jerison and I. Jerison), Springer-Verlag, New York.

Denning, J.: 1990, 'Is thinking computable', *American Scientist* 78, 100–102.

Dobrovolsky, M. and O'Grady, W.: 1987, *Contemporary Linguistics Analysis*, Copps Clark Pitman Ltd, Toronto.

Fass, D.: 1990, 'Dehumanized people and humanized programs', *SIGART Bull.* 1, 3–7.

Gray, C.M., Engel, A.K., Konig, P., and Singer, W.: 1990, 'Stimulus dependent neuronal oscillations in cat visual cortex: receptive field properties and feature dependence', *Eur. J. Neurosci.* 2, 607. See also, *Nature* 338, 334.

Hofstadter, D.: 1979, *Gödel, Escher, Bach*, Basic Books, New York.

Gazzaniga, M.: 1989, 'Organization of the human brain', *Science* 245, 947–952.

Genesereth, M.R. and Nilsson, N.J.: 1986, *Logical Foundations of Artificial Intelligence*, Morgan Kaufman Publ. Los Altos, CA.

Jamieson, L.H.: 1987, 'Characterizing parallel algorithm', in *The Characteristics of parallel algorithms* (ed. by L.H. Jamieson, D. Gannon and R.J. Douglas), The MIT Press, Cambridge, Mass., pp. 65–100.

Malcolm, N.: 1981, 'Knowledge of other minds', in *The Philosophy of Mind* (ed. by V.C. Chappell), Dover Publications, New York.

Knudsen, E.: 1987, 'Computational maps', *Ann. Rev. of Neuroscience* 10, 41–45.

Nagel, E. and Newman, J.: 1959, *Gödel's Proof*, New York University Press.

Penrose, R.: 1989, *The Emperor's New Mind*, Oxford University Press.

Ramsey, W.M.: 1989, 'Parallelism and functionalism', *Cognitive Science* 13, 139–144.

Reiter, R.: 1980, 'A logic of default reasoning', *Artificial Intelligence* 13, 81–132.

Searle, J.S.: 1980, 'Mind, brains, and programs', *The Behavioral and Brain Sciences* 3, 417–457.

Searle, J.S.: 1990, 'Is the brain's mind a computer program', *Scientific American* 262, 20–25.

Shah, K. Tahir: 1990, 'Lectures on neural computing', in *Proceedings of the College on Recent Developments in Computers and Mathematics* (ed. by R.F. Churchhouse, K.T. Shah and P. Zanella), World Scientific, Singapore, 1991.

Shoham, Y.: 1990, 'Non-monotonic reasoning and causation', *Cognitive Science* **14**, 213–252.

Smale, S.: 1989, 'On a theory of computation and complexity over the real numbers', *Bull. AMS* **21**, 1–47.

Snyder, S.H.: 1986, *Drugs and the brain*, Scientific American Library, New York.

Thagard, P.: 1986, 'Parallel computation and the mind–body problem', *Cognitive Science* **10**, 301–318.

Thistlewaite, P.B.: 1988, *Automated Theorem-Proving in Non-Classical Logics*, John Wiley & Sons.

Torrance, S.: 1984, *The Mind and the Machine*, Ellis Horwood Ltd.

Traub, R.D., Miles, R., and Wong, R.K.S.: 1989, 'Model of the origin of rhythmic population oscillations in the hippocampal slice', *Science* **243**, 1319–1325.

Tsotsos, J.K.: 1990, 'Analyzing vision at the complexity level', *The Behavioral and Brain Sciences* **13**, 423–458.

Turing, A.M.: 1950, 'Computing Machinery and Intelligence', *Mind* **LIX**, 433–460.

Uhr, L.: 1984, *Algorithm-Structured Computer Arrays and Networks*, Academic Press, New York.

GIUSEPPE LONGO

REMARKS ON INFORMATION AND MIND

1. CREATURA AND PLEROMA

It was not until the forties that the existence of the world of information was explicitly recognized, i.e. it was recognized that along with the world of physics, the world of forces, masses and impacts, there is the realm of communication, difference, organization and meaning, where the laws which hold sway are quite different from those of physics, and are sometimes surprising. A major contribution to the recognition of the world of information came from the work of Claude E. Shannon, an engineer and mathematician then working at Bell Laboratories, who in 1948 wrote a seminal paper on the mathematical theory of communication. That paper was important not only because it contained beautiful theorems concerning the coding of information sources and transmission channels, but also because it attracted people's attention to such key concepts as entropy, redundancy, capacity etc.

But Shannon's theory, so important as it is from a theoretical as well as from a practical point of view, does not exhaust all aspects of information. There are many concepts which are not stated, or even used, in Shannon's theory. These limitations stem mainly from the fact that the mathematical theory of communication is only concerned with the syntactic aspects of information, and deliberately neglects its semantic and pragmatic aspects. Being so clearly bounded, Shannon's theory yields a variety of important and deep theorems. At the same time, those theorems have no validity outside the scope of the theory. This is why many attempts to apply such concepts as entropy or capacity in fields different from communication engineering have led to meaningless results.

In what follows I shall try to give some hints about the concept of information assuming that the reader is familiar with the fundamentals of Shannon's theory.

Following Gregory Bateson, Carl Gustav Jung and the Gnostics, I shall call *Pleroma* the world of matter and forces, and *Creatura* the world of information and structure. In the Pleroma each thing always stands for itself, whereas in the Creatura each thing can stand for another thing, thus becoming a symbol: every thing can bear a *meaning*, which is not inherent in the thing itself but is

141

G. Corsi et al. (eds), Bridging the Gap: Philosophy, Mathematics, and Physics, 141–146.
© 1993 Kluwer Academic Publishers.

given to it by man, i.e. by a meaning-giving and meaning-using being.

Meaning, in turn, is strictly related to *redundancy*, i.e. to repetitive information. Actually it is when we perceive a redundant pattern (e.g. a circle) that we can grasp or understand its form even before we can see it all: as happens with associative memories, we can reconstruct the whole form from a part of it. The same applies to a numerical sequence: if the sequence is redundant, i.e. is not completely random, we can guess at the next term with good success and even encompass the whole (infinite) sequence in a finite recursive formula. In other words, we again know or 'understand' what the redundant sequence 'means' before we see it all.

In the Creatura there is no conservation law for information (information does not divide, rather it multiplies or replicates among several users) and the absence of information can itself be information. In the world of communication and organization the letter which you did not write can precipitate an angry reaction because zero is different from one and zero can therefore be a cause, which is obviously not true in physics. So in the Creatura the *differences* are of paramount importance, but for a difference to be effective, i.e. to make a sequence of differences along the communication and causation pathways, we also need a *context* and an *observer* capable of appreciating these differences. Even sameness can be a cause, because sameness differs from difference! (We now understand why Pleroma is a good word for the world of physics: the Pleroma is 'full', i.e. there is no room for difference.)

To a large extent, Creatura is the world of living organisms. We organisms, like some of the machines we make, are able to store energy, and this slowly accumulated energy (or 'collateral' energy) we can spend abruptly in response to an external stimulus, irrespective, to a large extent, of the size of the stimulus. The same amount of sound energy associated with a word can evoke an affectionate reply, a puzzled look or a furious reaction according to the word we pronounce, to whom, and in which language. If you kick a stone, it moves according to the laws of dynamics with the energy it got from your kick. If you kick a dog, it moves (and can attack you) with a collateral energy, an energy it gets from its own metabolism and it can direct towards several different goals.

The explanatory principles of Creatura are different from those of Pleroma. These principles are essentially ideas. *Ideas are differences*, and *differences are information*. Information can be communicated only by means of a material support, but cannot be reduced to that support. Actually, the same information can be encoded in many different ways using as many different material supports. In Pleroma there is no information, although Pleroma is the matrix of all (potential) information and the source of all (potential) differences. The whole energy and matter structure of Pleroma is unable to explain the situation in which you and me are right now, with me having written (or better typed on my computer keys) this text for you and you reading (a complex but reversible transform of) it.

Actually most situations in which men find themselves can be described, explained and understood in a significant way only in terms of information, meaning, and communication. Take a very common situation: a person delivering a lecture in front of an audience. In principle a quantum mechanical description is possible, as is a description at the molecular level. However, such physical descriptions, which belong to the Creatura but are close to the Pleroma, badly miss the point: a lecture is not (only) a material phenomenon, it is basically an informational one. If you consider this example, you will easily realize that for almost any phenomenon there is an 'appropriate' level of description, at which the phenomenon becomes 'simpler' and easy to 'understand'. The choice of the appropriate description, however, assumes a background, a past experience of similar phenomena, in which most of the descriptional complexities are confined. Without such previous experience (i.e. redundancy) even very simple events might be described in an utterly 'inappropriate' way. Actually, the events themselves, outside their (repetitive, historical, experiential) context are very 'poor' and are compatible with a variety of descriptions.

As an example, consider how an Alien visiting the planet Earth (Strangeland) might give a (coherent) description of a lecture:

I entered a large room dimly lighted, where many Strangelandians were sitting in comfortable armchairs, all of which faced the same direction. As I looked more closely into the vast hall, I noticed that the floor was gently sloping and converging, as it were, to a central point, brightly lighted, where a stately, handsome Strangelandian was standing. I surmised he might be the City Mayor, or even the King, for he was so noticeably distinguished and imposing. When the confusion that had seized me upon entering the room abandoned me and I could perceive the details out of the whole, I was bitterly surprised to see and hear that the majestic Strangelandian, whom I thought of as the King, was uttering strange sounds, shrilling and buzzing and shrieking, from his masticatory apparatus, as if he were in great physical discomfort. But, to my outmost bewilderment, nobody helped or assisted him in the least: quite to the contrary. Actually, looking at the faces of the Strangelandians in the audience, I could see some grinning, some evidently enjoying the scene, some lightheartedly smiling as if the great pain of their Monarch was a source of joy to them, which I took as a sign of disgusting cruelty and I was thus confirmed of the many tales I had heard about the horrible population of Strangeland.

2. THE MENTAL WORLD

Let me now state that *Creatura is the mental world*. If we adopt this viewpoint, many obscurities an puzzling intricacies concerning the so-called *Mind–Body*

Problem (MBP) are likely to be somewhat clarified.

Mind and pattern as explanatory principles were excluded long ago from scientific biology, but this exclusion probably contributed to the maintenance of a radical and dangerous dichotomy in epistemology, the age-old Cartesian dualism between mind and body. Many of you are well aware of the difficulties that this dualism raises, but the attempts to eliminate the dualism in one way or another have also raised intricate problems.

The MBP can be stated in several different ways, each emphasizing a particular side of it. The first, utterly inadequate impression that the MBP evokes is that there is something called 'body' to which something called 'mind' is attached. The point is that there are two possible ways of conceiving a man: either the *machine-man*, immersed in the Pleroma and subject to its laws, and the *person-man*, active in the Creatura where he performs symbolic and informational activities. These two views seem difficult to conciliate.

The MBP can be formulated, for example, in the following ways:

(i)   Does there 'exist' something which is non-physical?
(ii)   Can physical sciences describe and explain all 'existing' things or are there things which need a different kind of science?
(iii)   Does the success of biology and the neurosciences oblige us to conclude that man is nothing but a pleromatic body?
(iv)   Is the mental dimension of man something real, and to some extent autonomous, or is it a metaphor for something else (a convenient way of saying, for instance)?
(v)   Are the mental aspects a shorthand for 'the totality of the human person' in its complicated relationships with other persons, culture, society, symbols, mathematics, existence, inheritance and so on?
(vi)   If (iv) is true, the mental aspects should not and could not be reduced to the physical, pleromatic aspects of the body, but should rather be enlarged to include the rich perspective of subjectivity, personhood, culture, history and individuation.

Although what we call mind rests on physical grounds, one has the impression that it exceeds a purely physical description. The question: What is mind? can be given and has been given various answers. Certainly it is not a material thing, since it does not obey the pleromatic laws. One possible definition is: *Mind is a function.* Functionalism separates the hardware aspects of the physical support, which are not crucial, from the vital software aspects. Mind can be embodied in various, even bizarre substrates, so far as its characteristics are preserved. This viewpoint opens up a wide perspective for artificial intelligence (AI) and is reminiscent of the Turing test for mechanical intelligence. Of course it suffers from the objection that it is a purely behavioral hypothesis, since it falls into the domain of the 'other minds problem': how can I ever be sure that the person or artifact in front of me really has a mind and feels what I subjectively feel (joy, love, pain and the like)? No amount of successful Turing test can ever go beyond the perhaps unsatisfactory behavioral level. (In the original version of the Turing test for conscious intelligence in a symbol-manipulating machine, the inputs to the machine are questions and remarks typed on sheets of paper, and the outputs are the typewritten answers from the machine. The test is passed by the machine if its responses cannot be discriminated (by a human) from the responses of a real person.)

Another possible answer is: *Mind is a word*. Wittgenstein was the first to adopt this linguistic approach, as he pointed out that many words do not refer to things or objects which we can, at least in principle, see and touch. If we approach and investigate the word 'mind' with a careful and delicate attitude, it can reveal much of the non-pleromatic nature of mind, it can reveal that mind refers to Creatura, to meaning, to language, to information and difference rather than to material objects and deterministic physical laws.

Actually, if we use information, communication and pattern as explanatory (no metaphysical) principles, many of the difficulties of the MBP would seem less severe. The MBP concerns essentially men: each person is a peculiar focal point where Pleroma and Creatura meet, and each such interface is an ideal source of problems, pathologies, illusions and mistakes. But man can neither be reduced to one component (all mind or all body) nor be separated into components: each man is a *mind-in-a-body* performing many physical and mental activities in many different contexts at the same time, and these activities lend themselves to different kinds of description. Each activity has its own right *level of description*, which is adequate because it meaningfully explains that activity. So what is needed to grasp the wide territory of mind is not a single all-encompassing and all-embracing approach, but rather *a plurality of descriptions*, a multitude of approaches and interpretations.

One can go further, and say that mental phenomena do not concern a 'mind', considered as some component of man, they concern a *man* in its full and rich complexity. There are no mental phenomena without a body. On the other hand, to understand such mental phenomena as joy or pain or love, I cannot study only their physiological and neural counterparts, which are certainly there, can be investigated and give a lot of information to those who are interested in that kind of information. The really understand love and pain, I have to study *love-in-a-man*, or *pain-in-a-man*. A human, a particular man or woman, expresses love or pain through many and different aspects of his or her person: physical, chemical, physiological, behavioural, communicational, cultural and social. A person is a *system*, and its mental activities are *systemic activities*, concerning his or her whole existence.

When we consider the mental activity of a man, we always have to consider this man in a larger context, i.e. in the family, in the society, even in the whole of Pleroma and Creatura. Perhaps this complexity is reducible, but so far the attempts to reduce it have been largely unsuccessful, oscillating between an inadequate dualism and a crude reductionism. This is why the study of mind is so difficult and fascinating: psychological facts are not only linked with a physiological substrate, they are also linked to values, beliefs, laws, ends, all of which belong to the Creatura, although they could not exist and work without the Pleroma. A mind, every mind, is a sort of irreducible microcosm where the 'subject' itself is reflected in its whole complexity.

In the last few decades mind has gradually become less and less an object of idealistic or spiritualistic mysticism and more and more the focus of systematic studies which have begun to confer explanatory power to it. Information theory, system theory and artificial intelligence have contributed in various degrees to this transformation.

Dipartimento di Elettrotecnica Elettronica Informatica
Università degli Studi di Trieste

PART II

PHYSICS AND PROBABILITY

PAOLO BUDINICH

AXIOMS AND PARADOXES IN SPECIAL RELATIVITY

1. FOREWORD

The physical sciences of this century have been characterized by two revolutions: quantum mechanics and relativity.

The arrival of a revolution in physics is generally announced by the appearance of unexpected or paradoxical results of experiments or of theoretical computations in a certain, well-established branch of physics.

For quantum mechanics it was the infinity which classical electrodynamics computation predicted for the black body radiation (which is obviously finite in reality). For relativity it was the breaking of Galilei covariance in the Maxwell equations, confirmed by the result of Michelson–Morley experiment, which showed that, contrary to the rules of Galilean kinematics (and of common sens), no velocity could be added to the velocity of light in vacuum, which is then the maximum conceivable velocity for any physical action compatible with the fundamental principle of causality.

These unexpected results, which appeared as paradoxical in the frame of the at that times well established disciplines of classical mechanics and electromagnetism, forced the abandonment of some of the axioms on which those disciplines were based; and in this consisted the revolutionary change of those scientific disciplines. In fact the axioms and principles are generally defined and accepted in force of their mere evidence and to abandon one of them means to recognize as wrong what was previously considered as evident; or, in other words, to correct an erroneous prejudice due to our limited capability to perceive and conceive the world around us. Therefore a revolution in science may bring about a drastic change in our natural, intuitive conceptions of the world. In the case of quantum mechanics it was the concept of the massive geometrical point of Lagrangian mechanics that had to be abandoned, while, in the case of relativity, it was the Newtonian concept of absolute time which had to be substituted by the new relativistic conception of time.

The purpose of this note is to analyze this manner of acquisition of knowledge which may derive from science, with particular attention to the changes brought

149

G. Corsi et al. (eds), Bridging the Gap: Philosophy, Mathematics, and Physics, 149–173.
© *1993 Kluwer Academic Publishers.*

by relativity in the conception of time.

2. INTRODUCTION

Physics is the branch of natural sciences which deals with the study of matter and forces in space and time.

This study needs the definition of measurable quantities referring to the state of matter, its position, its movements, its structure and to forces or other entities acting on matter and generating its changes of position or structure in the course of time.

On the measure of those quantities, expressed by numbers, mathematical and geometrical operations may be performed which may give origin to equalities or equations. These are assumed to characterize and represent the natural laws ruling the physical phenomena under study. An example is the Newtonian law of motion (second principle of dynamics) of a massive body under the action of a force represented by the equation where the mass of the body m, multiplied by its acceleration $\vec{a}$ (change of velocity), is equated to the force $\vec{F}$ acting on it:

$$(1) \qquad \vec{F} = m\vec{a}$$

(the arrow over a and F indicates the direction of the acceleration and of the force, respectively). Another example is the famous law represented by the equation (derived from the theory of special relativity) giving the energy E of a body of mass m as a product of m times c^2:

$$E = mc^2$$

where c indicates the velocity of light (300 000 km s^{-1}).

The validity of a physical law expressed by an equation may be generally verified by performing experiments (where the quantities are measured) to test its validity. And just one experiment which do not agree with the equation representing the law is enough to disprove the validity of the equation and of the corresponding law. As an example the law of composition of velocities, in Galilean relativity, may be expressed in the simple form

$$\vec{v}_{observed} = \vec{v}_{observer} + \vec{v}_{body} ,$$

establishing that the *observed* velocity of a body equals the velocity of the body summed (vectorially) to the velocity of the observer. This law is easily verified in all ordinary motions[1], however when it was experimentally tested (Michelson and Morley (1887) with the velocity of light it gave a wrong result: no velocity may be added to the velocity c of light to give a velocity either larger or smaller than c: therefore the laws of Galilean relativity are not true for optical phenomena, that is for Maxwell's theory of electromagnetic waves, as we will see, and the result will be Eindstein's theory of relativity (special).

As in every branch of science, so also in physics; in order to formulate the equations and the corresponding physical laws some axioms and principles are a necessary prerequisite, assumed as true in force of their mere evidence, out of which all propositions may be derived and proved. Some of them are explicit, like the causality principle which is valid for all scientific disciplines, and, in particular branches of physics, the Galilean principle of inertia, valid in classical mechanics, or the Heisenberg principle of uncertainty, valid in quantum mechanics.[2] Some of them are implicit assumptions. In particular, before the advent of relativity, these included the following:

(a) Physical phenomena may be described in a three-dimensional space.
(b) The space where physical phenomena occur may be considered homogeneous and its geometry is the Euclidean one.
(c) Physical phenomena may be labelled by a measurable variable called time. This time variable is characterized by an arrow going from the past to the future, is absolute – that is not dependent either on space or the observer – and may be mapped onto an oriented straight line.

3. SYMMETRY

The concept of symmetry is an old one. It is easy to link the common meaning of the word 'symmetry' with the concept of invariance with respect to certain transformations. As an example, if a solid has a symmetry axis its geometrical configuration remains the same if the body is rotated, by a given angle, around that axis. A sphere is invariant for all transformations consisting of rotations around its centre: it has spherical symmetry.

On can affirm that a good part of the progress in physics since its birth, but especially, in an explicit and conscious manner, during this century, after the advent of relativity, is based on research, discovery and on the consequences of new symmetries; which can be defined as follows: 'invariance of laws with respect to transformations (changes) of physical quantities which intervene in them'.

To describe the property of symmetry of a law or of equations which represents it, physics uses the mathematical instrument of group theory.[3]

It is therefore affirmed that the law, or the corresponding equations, present a given symmetry when the law is invariant or, better, the equations are covariant[4] with respect to a given group of transformations; for example the coordinates in space and time, or other quantities which are represented in the law and in the equations.

It is easy to understand why the symmetry properties of a law are important for physics; in fact the greater the symmetry of a law, in other words the more a law is invariant with respect to groups of transformations, the more it is 'true' in the sense of 'objective' since it is not dependent on the particular values of the quantities which appear in it, which may depend on a particular point of observation and therefore on the particular observer; this is therefore why 'more symmetric' may mean 'more objective' and therefore 'more true' in a strictly scientific sense; since more suitable to explain and to predict to me as well as to other observers, the particularity of natural phenomena as they appear to us.

But there is also another reason why the symmetries of the laws in physics are important. It is expressed by the fundamental theorem formulated by the German mathematician Emmy Nöther (1918), which affirms that to every covariance of the equations of motion of a physical system there corresponds a law of conservation for a corresponding quantity of that system, which means a quantity, which remains constant in time (and therefore in a sense is easily observable). As an example: from the invariance of physical laws with respect to translations (which simply means that the laws and the equations which represent them are the same here in Trieste and in Rome), for Nöther's theorem, the conservation of momentum of an isolated system, that is Galilei's principle of inertia, follows. Similarly the law of conservation of angular momentum follows from the invariance with respect to rotations (in virtue of which the earth rotates, maintaining the axis through the North Pole always directed towards the North Star, and the orbits of the planets lie on planes). Translations and rotations constitute the Euclidean group of transformations and the corresponding symmetry is called Euclidean. This symmetry expresses the isotropy of our empty space; and its consequences, when discovered, constituted important principles: of inertia, of conservation of the angular momenta (second law of Kepler). Thanks to Nöther we now know that those principles are simple consequences of her fundamental theorem, a pillar of modern physics, which connects symmetries to invariances in time (and to the so-called good quantum numbers in quantum mechanics).

4. GALILEAN RELATIVITY

One of the main merits of Galileo is his having established that the equations of mechanics are covariant with respect to the group of transformation which transforms from a fixed system of reference[5] to a moving one with constant velocity; such a system is called inertial. Galilean relativity therefore states that the laws of mechanics are the same in two inertial systems of reference, for example in my laboratory or on a train which travels on a straight railway at constant velocity. It therefore follows that experiments of mechanics, such as the falling of a heavy body, the oscillations of a spring, etc., give the same results if performed in my laboratory or on that train. The equations representing the Galilei transformations and expressing the position or the coordinates of a point in a moving system with respect to those in the fixed one build up the Galilei group.[6] The transformations expressing the velocities of a point in the two systems are given by Equation (2).

Equation (1) represents Newton's secon principle of dynamics (restricted to the motion of a massive body under the action of a force $\vec{F}$) and may easily be generalized to the motions of all mechanical systems. For the second principle of Newtonian mechanics Galilean relativity holds. That is Equation (1) (and its generalizations) are covariant with respect to Galilean transformations.

The reason for this covariance may be easily understood. In fact in Equation (1) the motion of the massive body is represented by its acceleration $\vec{a}$ which is the measure of the change of velocity $\vec{v}$ of the body.[7] Therefore if a constant velocity is added to the massive body its acceleration $\vec{a}$ will not change and (provided the force F is velocity independent) Equation (1) will remain the same in two inertial systems, that is in one moving with respect to the other with constant velocity.[8]

In words: Galilean relativity of mechanics is due to the fact that in the fundamental law of mechanics represented by the second Newtonian principle only accelerations (that is second derivative of the positions), appear.

5. SPECIAL RELATIVITY

For some time the mechanical nature of light was assumed (Newton), in which case the Galilean relativity would have been valid for it. However, when, at the end of last century electrodynamics and, in particular, the electromagnetic nature of light was discovered, it clearly appeared that Galilean relativity could hardly apply to the new theory.

In fact in Maxwell's equations homogeneous (1873), representing the propagation of electromagnetic waves (light) not accelerations (change of velocity) but velocities themselves[9] appeared in an essential way; in particular the velocity of light c appeared in the equation of light propagation. Therefore the addition of a velocity, even if constant (of the reference system), would have

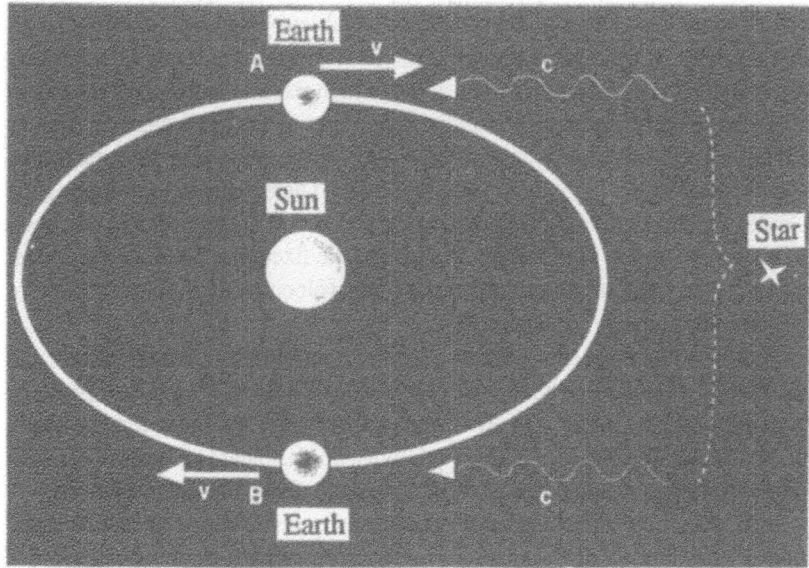

Fig. 1. Symbolic representation of Michelson–Morley experiment. The velocity of light arriving at the earth from the star is always $c = 300.00$ km/sec and not 300,030 in A and 299,970 in B, as Galilean mechanics (and common sense) would have predicted.

altered the Maxwell equations of motion. This was an intriguing problem since it was difficult to accept that the beautiful Galilean principles of invariance of the laws referring to mechanics, a part of physics, could not be extended to electromagnetism, especially because the two parts were connected; in fact electric and magnetic forces may accelerate massive bodies.[10]

A crucial experiment was performed in 1885 by Michelson and Morley (1887) using the earth itself in its motion around the sun as inertial system. The result was surprising: the law of composition of velocity expressed by Equation (2) did not hold for light. In fact the measured velocity of the light coming from a given star S arriving on the earth in A (Figure 1) and the one in B (6 months after) resulted in the same c while, according to Equation (2), it should have been $c + v$ in A and $c - v$ in B where v is the velocity of motion of the earth on its orbit (about 30 km s^{-1}).

This experimental result generated profound disconcertion in the world of physicists not only because it contradicted the validity of the Galilean composition law represented by Equation (2), but also because it paradoxically contradicted common sense. In fact it is naturally intuitive that if I run against the wind this appears to me stronger than if I run in the same direction as the wind, and it was unconceivable that this should not also apply to the velocity of light.

After many complicated attempts, proposed by several distinguished physicists, to explain this unexpected result, it was finally Poincaré and Lorentz who found the way out, and Einstein drew the final physical consequences, formulating the theory of special relativity in 1905–1920 (Einstein 1924). The main observation was that one had to abandon one of the main axioms of physics, precisely (c) of Section 1 referring to time. Time in physics is not an independent, absolute variable, as assumed by Newton. Instead, time, like space coordinates, is a variable relative to the observer and each observer of different inertial systems has its own space *and time* variables.

Galilean transformations from one inertial reference system to another were then substituted by Lorentz transformations which involved not only space but also time variables.[11] As a consequence Maxwell's equations of motions, as well as the equations of mechanical motion, appropriately modified, were the same in two inertial systems.

In this way the symmetry of Galilean relativity was enlarged to embrace all known physical laws, including both those of mechanics and those of electrodynamics and of light; they are both the same in all inertial reference systems. In a way, aesthetically, too, it was a more satisfactory theory since it would have been difficult to swallow that on a moving system (say on a jet plane) one had to employ different Maxwell equations than on a fixed one in order to represent the propagation of light.

Einsteinian relativity does not contradict the Galilean one; on the contrary it extends it to all physical laws. The latter relativity is a particular case of the former, and is identical with it when the velocities concerned are small compared to the velocity of light, which is the case for most known motions of the ordinary mechanics of macroscopic bodies.

The paradoxical aspect of the result of the Michelson–Morley experiment may be easily understood from the fact that we are accustomed to observe slow motions compared to the velocity of light (think that in a second light may tour the earth almost eight times) therefore Galilean relativity is intuitive while Einstein's is not. In other words our limitation in being able to perceive only slow motions introduces prejudices in our mind and in our intuition, which induces error in us. Paradoxes then result from our confrontation with these

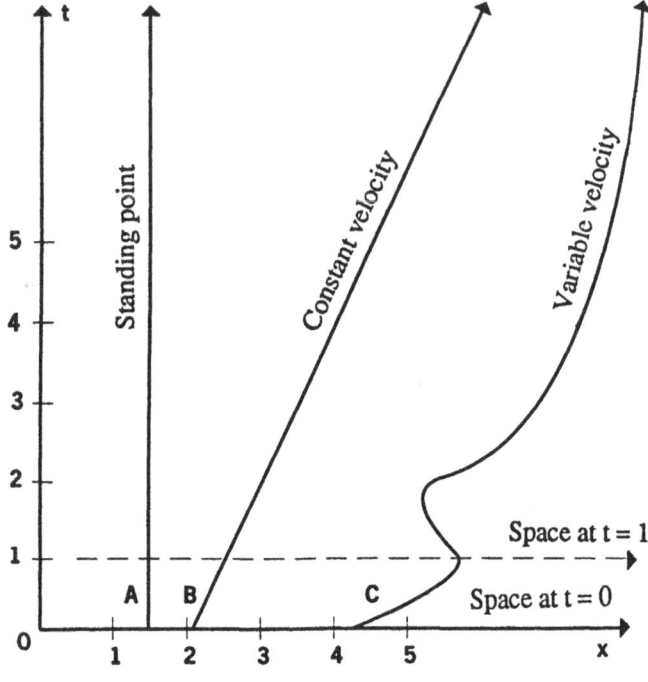

Fig. 2. World lines.

erroneous prejudices.

There are more apparent paradoxes resulting from special relativity; they may be derived by applying Lorentz transformations to physical quantities appearing in equations representing physical laws through rigorous mathematical procedures. They may, however, also be easily derived through graphical methods, which are more accessible to laymen, as I will try to show.

6. CONSEQUENCES OF SPECIAL RELATIVITY: RELATIVITY OF SIMULTANEITY

Let us first represent space-time graphically as in Figure 2 where the horizontal axis x represents space (one-dimensional for case of design) and the vertical axis t time, and the numbers 1,2... arbitrary units of space and time, respectively.

Then a standing point A is represented by a straight line parallel to t, a point B with constant velocity by a straight oblique line, and a generally moving

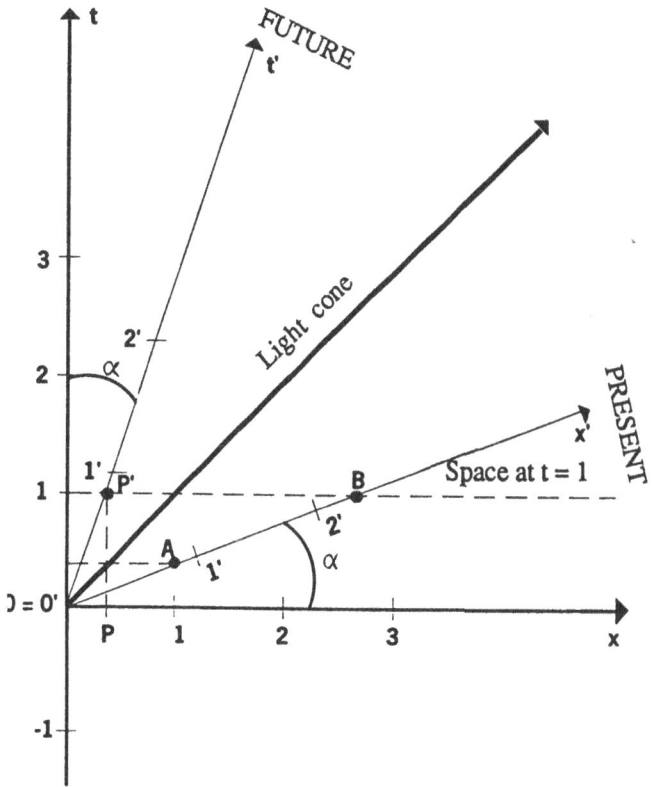

Fig. 3. Lorentz transformations

point C by any (non-straight) line. These lines are termed 'World lines'. A straight line orthogonal to the time axis (or parallel to x) represents the whole space at a certain time: all points of this line represent simultaneous events.

Let us now propose the following units: seconds for time axis t and 300 000 km for space x, as in Figure 3. Then we will represent by a straight line at 45 degrees to both the x and t axes the world line of a point moving with light velocity c (300 000 km in one second); this is called the light world line. All light world lines passing through 0 build up the so-called light cone (see also Figure 5).

Let us now suppose that another observer O' is in straight uniform motion with respect to O but such that at time zero ($t = 0$) the coincides with O, then this world line will be the line t' in Figure 3 (in the same way as the oriented straight line t represents the world line of O).

In order to find the oriented straight line x' representing the space of the moving observer O' let us remember that, according to the result of the Michelson–Morley experiment, the velocity of light for him is the same c as that for the standing observer. Therefore the two observers must have the light world line in common. This happens if we choose to represent the space of the moving observer by an oriented straight line x' symmetric to the time axis t' with respect to the light cone which then bisects both the angle $t\,O\,x$ and the angle $t'\,O\,x'$, as can be seen in Figure 3. Now all lines parallel to t' represent, in the moving reference system, world lines of standing points (moving with the same velocity as O'), while lines parallel to x' represent simultaneous events in the moving system.

From the figure we can immediately see an important fact: let us consider the two point events A and B of the observer O'. They both happen at time $t' = 0$ and are therefore simultaneous events. But for the observer O, B happens after A. From this the relativity of contemporaneity of the two events is clearly seen for observers in motion, one respect to the other. This relativity of simultaneity has had consequences in other fields than physics (contributed to the origin of existentialism).

The relativity of simultaneity is more clearly shown in Figure 4 where the axes t', x' represent the space-time of an observer who moves with a velocity represented by the world line t'. Then the point events A and B are such that while B is posterior to A for the standing observer: $t_B > t_A$ (space-time x, t), B is anterior to A for the observer in motion: $t'_B < t'_A$ (space-time x', t'). This is why the world line OAB, which represents a point which moves with a hypothetical velocity greater than the velocity of light, cannot represent any physical object because if a physical object is in B it can be there only if it was previously in A. But as we have seen, B happens before A for the observer with space-time (x', t'). Therefore the world line OAB, which represents the movement of an object with velocity greater than the velocity of light, cannot represent any physical object. Neither can it represent any physical action as, for example, the action of a gravitation or any other force, because this action would not respect the fundamental principle of causality which imposes that cause preceeds, in time, the effect, for all observers, whether they are at rest or in motion.

This is the reason why all point events in Figure 4 outside of the light cone can be contemporary to O for an observer in motion with appropriate velocity and therefore represent the present (potential).

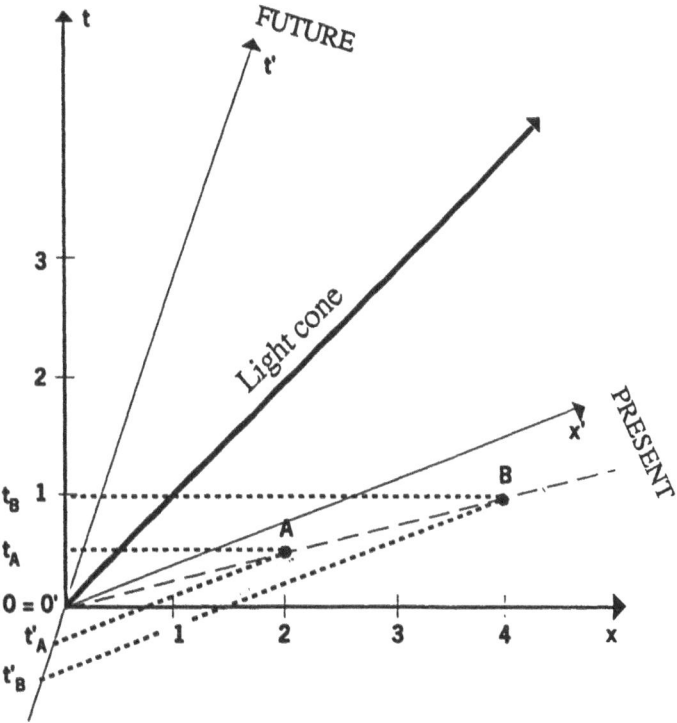

Fig. 4. Relativity of simultaneity.

The space-time of special relativity may be better analyzed if we represent ordinary three-dimensional space at the time $t = O$ with a plane (x, y) as in Figure 5.

All light world lines through the origin O build up the so called 'light cone' and the whole graph is appropriate to represent symbolically so-called Minkowski space-time where relativistic physica phenomena occur.

In this graph the straight lines, generators of the cone, represent the world lines of the points of a light sphere generated by a spark which at time $t = 0$ was created in O. In fact after 1 second the light sphere is 300 000 km away from O; after 2 seconds 600 000 km away, etc. In reality, since the real space is three-dimensional (and not 2-D like in the figure) the surface of the light cone is three-dimensional (and not two). If I intersect the light cone of Figure 5 with a plane parallel to plane xy and intersecting the time axis at $t = -2$ I,

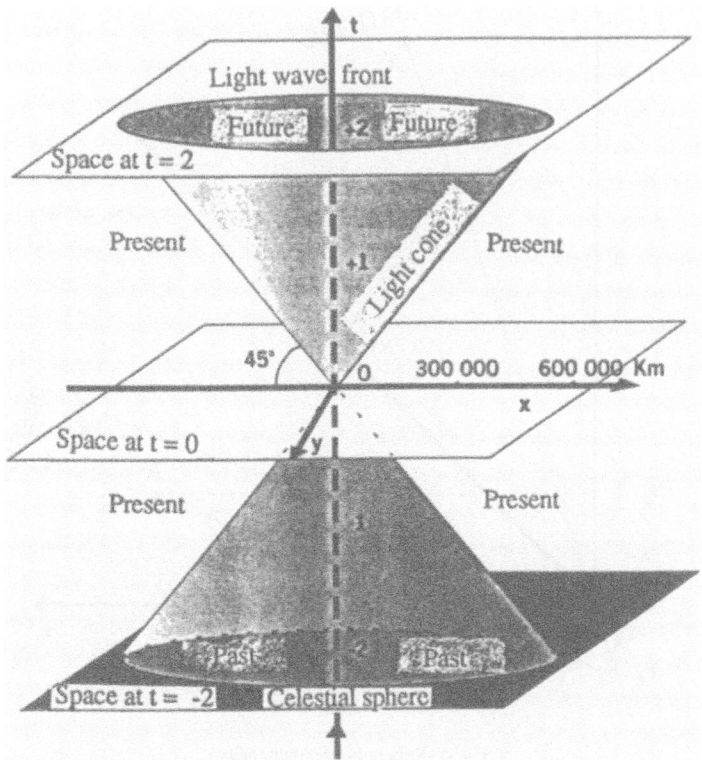

Fig. 5. Minkowski space.

obtain a circle indicated by "celestial sphere" in Figure 5. In reality it is a two-dimensional spherical surface and it represents light which arrives to me in O as it were a couple of seconds before arriving. It therefore represents the light of the stars on arrival to me; it is called the celestial sphere.

In Poincaré-Einstein relativity the world line of any physical agent which is in O at time $t = 0$ must remain inside the light cone (no physical agent can have higher velocity than light), which represents the future for t positive and the past for t negative for the observer standing in O, as indicated in Figure 5.

The relativity of simultaneity may sound paradoxical insofar as our common sense, based on our everyday experience, would suggest to us that if two events are contemporary to me they are also contemporary to any other observer, moving or not. As already stated, this apparent paradox derives from the erroneous prejudice of the absolute nature of time implied by the axiom (c) of Section 1.

7. FURTHER CONSEQUENCE OF SPECIAL RELATIVITY: DILATION OF TIME, CONTRACTION OF LENGTH, EQUIVALENCE OF MASS AND ENERGY, EXISTENCE OF ANTIMATTER

We will now derive furthere 'paradoxical' consequences of special relativity, trying to employ the maximum of graphical and the minimum of mathematical arguments.

Let us first observe that the result of the Michelson–Morely experiment, stating that for all observers, at rest or in motion, the velocity of light is always c (300 00 km s^{-1}), may be expressed by affirming that for all of them the front of a light wave starting at a certain point O will be a sphere with centre in O and radius r increasing with the velocity of light c:[12]

(3) $r = ct$

where r and t are the space and time coordinates of each observer, moving or not.

Therefore Equation (3) must be covariant with respect to the simultaneous change of r and t from one to the other reference syste; that is, with respect to Lorentz transformations. More generally:

$$r^2 - c^2t^2 = \text{constant}$$

or, in orthonormal coordinates, $x^2 + y^2 + z^2 - c^2t^2 = \text{constant}$, must be invariant with respect to Lorentz transformation, which simply affirms that the sphere of light waves generated by a luminous point is the same for all observers.

If we reduce to one space and one time variable, like the case represented in Figure 4, then the invariant quantity reduces to $x^2 - c^2t^2 = \text{constant}$. If we make the constant 1 we obtain the invariant equation

(4) $x^2 - c^2t^2 = 1$.

Notice that for

(4') $t = 0$ we have $x = 1$ (or $= -1$).

In the same way the following equation is invariant

(5) $x^2 - c^2t^2 = -1$

and for

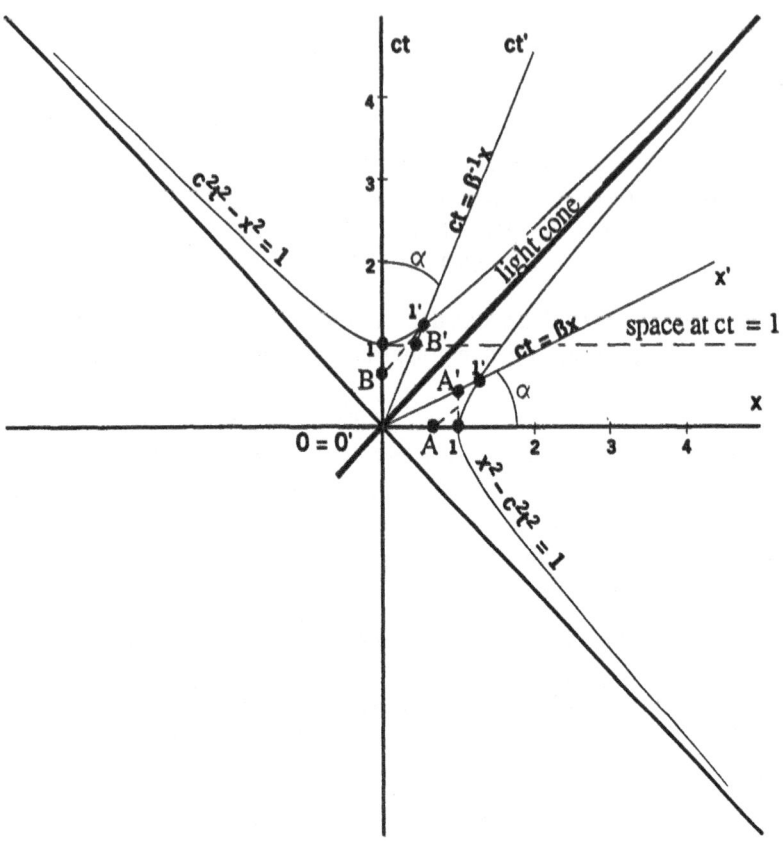

Fig. 6. Lorentz contraction of length and dilation of time.

(5') $x = 0$ we have $ct = 1$ (or $= -1$).

Equations (4) and (5) are represented by two hyperbolas as in Figure 6 and, because of Equations (4' and 5'), they intersect on the x, t, and x', t' axes the unit of length and of time (more precisely of ct) indicated by 1 and 1', respectively, on Figure 6.

It is easy to see that the segments $1B'$ and $1A'$ represent the measure β of the velocity v of the moving system with respect to the standing one in unit of the velocity of light: that is

$$\beta = \frac{v}{c}$$

(β also equals the tangent of the angle α).

Let us now consider the unit of length in the moving system represented by the segment $O1'$ on the axis x'. What is its length, measured in the standing system? Since the world line of $1'$ is $A1'$ its length in the system at rest will be OA shorter than the length at rest $O1$ on the x axis. Similarly the unit length $O1$ at rest will be measured as OA' shorten than $O1'$ in the moving system.

This is the famous relativistic effect of so-called Lorentz contraction of length in moving systems. An easy geometrical construction for this shortening gives the amount:[13]

(6) $\qquad \ell' = \sqrt{1 - \beta^2} \; \ell$

In an analogous way if we consider two clocks in the two systems they will measure a unit of time, say a minute, in the points 1 and $1'$ of the standing world lines t and the moving one t', respectively. Now suppose you are standing in O. After one minute your are in 1. Simultaneous to you will be B' on the t' obtained by intersecting t' with a parallel from 1 to the x axis. B' happens before $1'$ where the moving observer measures one minute of time. Therefore the moving clock runs slower than the standing one. Naturally the effect is reciprocal: when the moving observer measures one minute he is in $1'$ and for him B is simultaneous to $1'$ and B is before 1 in the system at rest. This is the famous relativistic Lorentz dilation of time intervals for moving systems; an easy geometrical construction gives:

(7) $\qquad \Delta\tau' = \dfrac{\Delta\tau}{\sqrt{1 - \beta^2}}$

where $\Delta\tau$ is the time interval measured at rest (also named 'proper time') and $\Delta\tau'$ is measured on the system in motion.

These effects are again paradoxical to our common sense and gave rise to various paradoxical examples (as in the case of two twin brothers, one travelling for say 20 years at high velocity, and after they meet one is 20 years old and the other one, who travelled, could be only 5 years old, say, if the velocity was enough near to c such that the factor $1/\sqrt{1 - \beta^2}$ equals 4). These paradoxical effects are true, however. Consider as an example, particles like the μ mesons created by cosmic rays arriving from outer space when they hit the atmosphere, generally at an altitude of about 20 kilometers. These mesons have a lifetime of $1/100\,000$ of a second, after which they decay into electrons and neutrinos. At the speed of light they could then run only for about 300 meters. Nevertheless they are observed at sea level and below. This is simply due to the fact that at their creation they have often a velocity very near the velocity of light such that in Equation (7) the factor $1/\sqrt{1 - \beta^2}$ is larger than 1 000 and frequently

10 000 in such a way that they live a longer time (1/1000 of a second), enough to travel 20 km and reach the surface of the earth, where they are observed.[14]

In special relativity space x and time ct we build up a 4-vector of 4-dimensional Minkowski space-time with which relativistically invariant equations like (4) and (5) may be formed. In a similar way momentum pc (mass times velocity square) and energy E of a massive body with mass m form a 4-vector obeying an invariant equation, like (4). In this case the invariant constant is determined and is the mass^2 times c^4; precisely

$$E^2 - p^2c^2 = m^2c^4 \ .$$

For a standing body ($p = 0$) the equation reduces to $E^2 = m^2c^4$ or, taking the square root of both sides, the following equation is obtained:

(8) $E = \pm mc^2$

Everybody is familiar with the equation $E = mc^2$. The equation $E = -mc^2$, however, is paradoxical and gave a lot of trouble to theoretical physicists at the beginning of the century. In fact it predicted the existence of a new form of matter (antimatter) which could not have been dreamed of before the discovery of relativity. It represented one of its most paradoxical predictions and consequences.

Even if paradoxical,[15] it was, however, true and in fact antimatter was discovered by Occhialiniand Blackett (1933) in cosmic rays in the form of antielectrons or positrons some years after its prediction. Nowadays antimatter is created in most elementary particle laboratories.

Now we may understand that antimatter is a necessary consequence of the new conception of time.

8. ANTIMATTER CONCEIVED AS MATTER TRAVELLING BACKWARDS IN TIME

In fact let us go back to Figure 1 and suppose that axis x represents the distance from a point in the high atmosphere where a high-energy electron was created. The world line of this electron, which in N collides with an atom without losing velocity, generally it will emit a photon (*Bremsstrahlung*), is for example the one shown in Figure 7.

After the collision the electron has simply deviated without practically losing any velocity (we recall that (x, y) represents three-dimensions). But, since space and time are equivalent coordinates, nothing forbids that in the collision in N the electron goes back in time, as shown in Figure 8, up to the collision with another nucleus in M where it returns back to its world lines towards

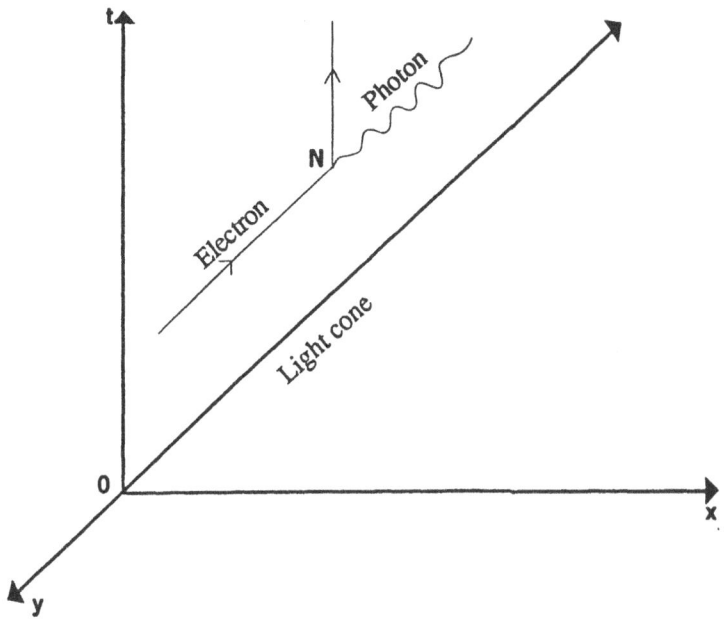

Fig. 7. World lines of an electron colliding with a nucleon.

positive time. How will the phenomenon appear to the observer in O?

At time 2 he will see an electron which arrives from the high atmosphere. At time 3 he will see the same plus two electrons, one normal and one which travels back in time. It is easy to show how this will appear to the observer as a positron or anti-electron which in N annihilates the primitive electron coming from the high atmosphere. After annihilation which occurs in N the observer will see, for example at time 4, just one electron which travels normally ahead in time.

In reality, what happened is that in M there was a photon indicated with an undulating line which created an electron and anti-electron or positron pair and that the latter annihilated in N an electron (of cosmic rays or also of an atom of the air) giving rise to the creation of two photons.

This picturesque description, due to Feynman (1949), perfectly represents the rigorous calculations of quantum mechanics. It shows that antimatter can also be thought of as matter which travels back in time and that this description is legitimate in the framework of relativity which considers time as the fourth

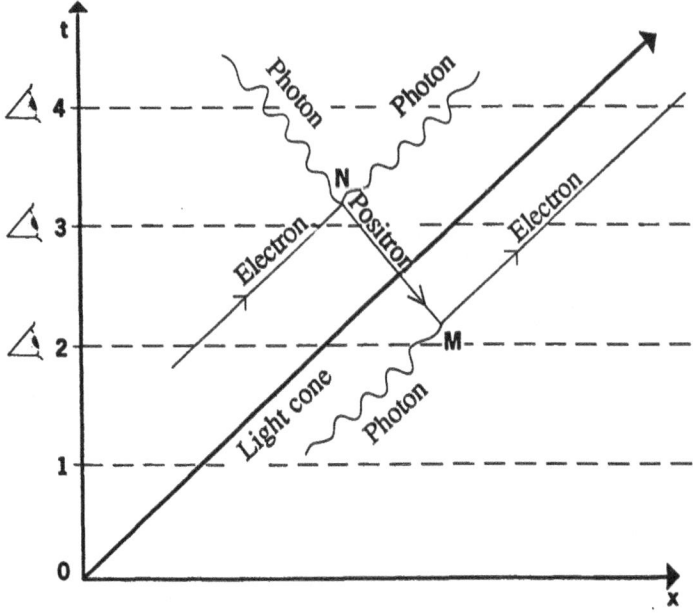

Fig. 8. A positron (antimatter) represented as an electron (matter) travelling towards past time.

dimension of a four-dimensional space. Feynman's graphs have become very popular among physicists and constantly used as a synthetic representation of complex theoretical calculations of quantum mechanics.

9. PARADOXES AND PREJUDICES

Special relativity is a good example of how through science we may require new knowledge about the world around us which would otherwise not have been accessible due to the limitation of our sensory perception. That this may happen in general is obvious since we can see with our eyes only a very limited part of the spectrum of electromagnetic radiation arriving from the world and we can, through telescopes, radio telescopes, microscopes and particle accelerators, discover otherwise invisible aspects of the world.

However the limitation of our capability of perception may also induce errors for other reasons;[16] and precisely may induce us to take for granted, in the form of evident propositions like the ones listed in (a), (b), (c) of Section 1, concepts, which, after a deeper analysis, turn out instead to be wrong. This was the case of special relativity which revised the concept of time which was taken as absolute, as in the postulate (c) of Section 1.

When this happens then paradoxes may appear in the new theory, in which one of the fundamental postulates of the old one is abandoned, which simply express the consequence of the rejection of the previously accepted postulates which were suggested as obvious by our intuition based on our erroneous common sense. The fact is that our common sense is grounded in our everyday experience which, however, is based on our limited capability of perception (in the case considered our everyday experience with the slow velocity of macroscopic bodies) and then what seemed to be obvious and evident is actually an erroneous prejudice.[17] Therefore the paradoxes are an indication of the fact that some of the evident postulates were in fact just wrong prejudices. And if the paradoxes finally turn out to be true, then we have to correct our intuition of reality and consider what was before a self evident postulate just as a wrong prejudice, to be abandoned.

It may be that the first symptoms of a coming change in a branch of science are paradoxes. In a way the result of the Michelson and Morley experiment was paradoxical: that is contradicting the intuitive Equation (2). And it took some difficult and cumbersome attempts before the revolutionary revision of postulates was discovered by Poincaré–Lorentz–Einstein.

In our days physics is still confronted with paradoxes. One may be considered the appearance of infinities in quantum field theory, especially in gravitation. Other ones are concerned with the separability of quantum systems. Attempts to overcome them gave gone on for many years. However no indication of basic axiomatic changes has emerged up to now.

It is interesting to note that special relativity has generated the correction of another wrong prejudice implicit in Galilean relativity. In fact, consider the law of composition of velocities expressed by Equation (2). Suppose that both the observer and the body move on the same straight line. Then Equation (2) simply results in the algebraic sum of velocities or segments representing them. There is no limit in summing segments and therefore there is no finite limit to the velocity that you may reach with this procedure. The limit is infinity. Therefore in Galilean mechanics there is no limit to the progression of possible velocities; the limit, like in the length of a straight line, is infinity. The same happens with the velocity of propagation of an action, like the one of gravity. There is no reason not to consider this action propagation at infinite velocity. In other words that it is instantaneous. And this was effectively assumed by both Galileo and Newton. This is in fact one of the reasons why Galileo refused to accept the hypothesis that the sea tides were due to the moon, as proposed instead by his contemporary Marco de Dominis (Supek 1974).

In fact the instantaneous nature of the gravitational action of the moon on the sea would have rendered ambiguous, if not contradictory, the fundamental principle of causality which imposes that the cause precede the effect in time.

Newton, too, was confronted with the same difficulty of the instantaneous action of gravity and at the end of his great work *Principia* he affirms that his theory presents a fundamental difficulty due to the instantaneous action of gravity, for which he had no remedy to propose.

It was only Poincaré–Einstein, in theory of relativity, who settled this difficulty, and it was done by imposing an upper limit on possible velocity, the velocity of light, and extending this upper limit to the possibility of propagation of any physical agent, including gravity; only after this was the causality principle also respected for the theory of gravitation. In this way, too, an infinity was eliminated from physics – that of possible velocities – which is not a minor result if one considers how damaging the concept of infinity has been and still is in physics.

10. FURTHER SYMMETRIES AND PROSPECTS

The transformations of Euclidean geometry representing rotations and translations of a rigid body build up a group which is known as IO(3) (I stands for inhomogeneous, O for orthogonal and 3 for the dimension of ordinary space). Euclidean symmetry or covariance with respect to the group IO(3) simply represents the homogeneity of ordinary three-dimensional space.

The transformations of special relativity may be simply thought of as an extension of Euclidean transformations to a four-dimensional space which includes both the ordinary three-dimensional one and time (multiplied by the velocity of light c and an imaginary unit i), all of them dealt with on the same footing. They build a group indicated with IO(3,1) called the Poincaré group which includes the Euclidean group IO(3) as a subgroup (Galileo's group may be obtained from IO(3,1) in the limit of $v/c \rightarrow 0$).

There is another group of symmetry in space-time which in turn includes the Poincaré group IO(3,1) as a subgroup. This is the conformal group symmetries represented by IO(4,2), which is very fascinating for many reasons. First, the Maxwell equations, which are naturally Poincaré or relativistic covariant, are also covariant for the conformal group of symmetry, which was discovered as early as 1909 (Bateman 1909). The conformal group includes dilations, that is, changes of units of space and time, in such a way that, in a conformally covariant system, 'large' or 'small' have no meaning, and therefore the concept of infinity may hardly find a place (space-time in this theory would appear compacted, precisely represented by a circle times a three-dimensional spherical surface). There are strong indications that the fundamental laws regulating the behaviour of the elementary forms of matter (elementary particles) may be conformally covariant, although this has not yet been proved.

Another higher symmetry presented by natural phenomena induced by gravitational phenomena is that of general relativity, discovered by Einstein. This symmetry also induces a revision of our intuitive concept of space and time as exposed in propositions (a), (b), and (c) of Section 1. The space-time adopted by general relativity is curved in the presence of masses and Euclidean geometry holds only in the limit of absolutely empty space.

Many more symmetries have been discovered and proposed in recent decades in the study of subatomic particles. Some of them are called internal symmetries, others are supersymmetries. There is a widespread suspicion that these symmetries have also to do with the intimate structure of space-time which is less naive than appears in (a), (b) and (c) of Section 1 even after the already established amendments forced upon it by special and general relativity. Space-time would emerge in these theories from higher dimensional spaces (of dimension 10 or 26, say) and even the concept of a geometrical point should be substituted by an extended one-dimensional geometrical entity called a relativistic string.

There is a variety of such theories (strings, superstring – supergravity theories) which emerged, and still emerge in recent times, in the attempt to obtain a unified theory of the four main forces discovered up to now in natural phenomena: gravitational, electric, weak (responsible for radioactivity), and strong (responsible for neutron and proton binding in the nuclei).

These attempts, started by Einstein himself as early as 1925, with several attempts at unifying gravitational and electromagnetic forces, have failed up to now in general, despite the tremendous effort on the part of the community of theoretical physicists (apart from a successful model: the so-called 'standard' model, which indicates the common origin of the weak and electromagnetic forces). The main difficulties stem from the appearance of infinities in most theoretical computations implied by the various theoretical models proposed at different times; an unambiguous sign of the fundamental unadequacy[18] of the proposed theoretical models.

It is almost universally agreed that the present status of physics is one of profound crisis and that the way out will come only after a rather drastic revision of the postulates (a), (b) and (c) of Section 1, going beyond the corrections already forced upon us by special and general relativity; a revision which may herald a new revolution in physics comparable to those of relativity and quantum mechanics of this century, and which may once again induce a new conception of the world. It is difficult to predict from which direction the new theory will emerge and which of the postulates implicit in (a), (b) and (c) will turn out to be an erroneous prejudice.

Among the many tendencies in thought there is one which indicates Euclidean geometry as the weak point of the axioms (even after its Riemannian generalization to deal with curved space-time). In fact, in the early part of this century (1913), a new geometry, more elementary than the Euclidean one, was discovered by the great French mathematician Elié Cartan (1913); it was succesfully applied to physics by Pauli and Dirac to give reasons for the existence of spin (intrinsic momentum of the electron) and of fermions.[19] Because of this such geometry was afterwards called spinorial or spinor geometry.

Spinor geometry is more elementary than Euclidean in the sense that, with spinors you may build up the vectors and tensors of Euclidean geometry, but not vice versa. It is more elementary, even if it is less accessible to our intuition. In fact a straight line, a point as intersection of two lines, a triangle, a circle of Euclidean geometry may easily be 'seen' (triangle, circle) or imagined (straight line, point) by extrapolating to infinity what we actually can see (segment, small disc) and this possibility of immediate, direct perception and knowledge of Euclidean geometry is simply due to the fact that the main sources of our direct knowledge of the world are our eyes and the main intermediate for this direct knowledge is light. Spinor geometry, instead, cannot be intuitively even imagined (it deals with vectors of zero length, with so-called 'totally null' planes in which every two vectors are orthogonal to each other). And maybe this is the reason why the concept of the fundamental nature of Euclidean geometry is a psychological prejudice. It is not inconceivable that is might be removed and substituted by the acceptance of the fundamental role of spinor geometry which is otherwise extremely elementary and elegant.

In any case revolutionary changes in physics are to be expected and if and when they arrive they may imply important changes of our conception of space and time and of the world in which we live.

Laboratorio Interdisciplinare per le Scienzi Naturali ed Umanistiche, Trieste, Italy

NOTES

[1] If I am on a bridge of a ship the velocity of the wind I feel (and may measure) equals the velocity of the wind with respect to the sea plus the velocity of the ship on the sea.

[2] It may happen that a principle in a particular branch of physics may be proved as a theorem when that particular branch is considered in a wider context. So the Galilean inertial principle is a particular case of the Galilean relativity-principle and may be proved true as a consequence of E. Nöther's (1918) theorem (see below). Heisenberg's uncertainty principle may be obtained as a consequence of Fourier theorem in wave mechanics.

[3] A set of elements build up a group if: (1) an associative composition law (or product) is defined for any two elements a and b, such that if $ab = c$, c is also an element of the set. (2) A unit element e exists in the set such that $ea = ae = a$. (3) For each element a the inverse a^{-1} exists in the set such that $a^{-1}a = aa^{-1} = e$.

[4] Literally the two members of the equality $A = B$ vary in the same way such that if, as a consequence of a transformation, $A \to A'$ and $B \to B'$ we have $A' = B'$.

[5] A system of reference is usually represented by three orthogonal oriented axis: x, y, z, with a common point 0 called, the origin. Such system is called a Cartesian orthonormal reference system. To every point of space there corresponds a 3-number (x, y, z) equal to the measure of the distance from the planes of the reference system; say x indicates the distance of the point from the plane (y, z). And *vice versa*, to every triplet of numbers (x, y, z) there corresponds a point.

[6] In an orthonormal reference system the transformations of the Galilean group are:

$$x'_j = a_j + \sum_{k=1}^{3} \alpha_{jk} x_k + v^0_j t \quad j = 1, 2, 3$$

where $x_1 = x$, $x_2 = y$, $x_3 = z$, a_j represent a translation, α_{jk} the entries of an orthogonal matrix: $\|\alpha_{jk}\| = \pm 1$, representing rotations; v^0_j the components of the constant relative velocity. For $v^0_j = 0$ the transformations of the Euclidean group are obtained.

[7] In mathematical language the velocity $\vec{v}$ of a moving point P is represented by the derivative (of the coordinates) of P with respect to time symbolically:

$$\vec{v} = \frac{dP}{dt}$$

while its acceleration is the derivative of its velocity

$$\vec{a} = \frac{d\vec{v}}{dt}$$

therefore it is the second derivative of its position:

$$\vec{a} = \frac{d^2 P}{dt^2} .$$

[8] In mathematical language: the derivative of a constant is zero therefore

$$\vec{a} = \frac{d\vec{v}}{dt} = \frac{d(\vec{v} + \vec{v}_0)}{dt}$$

if $\vec{v}_0$ is constant.

[9] Not the second but the first derivatives of the electric and magnetic fields with respect to time and position appear in Maxwell's equations.

[10] Historically, the necessity to overcome this inconsistency was the motivation which brought Einstein to the discovery of special relativity. In fact he knew of the Lorentz covariance of Maxwell's equations and therefore, for consistency, mechanics had to present the same covariance and he thus arrived at relativistic mechanics. He hardly knew of the Michelson experiment which,

instead, is useful for a didactical introduction of relativity.

[11] Lorentz transformations express the coordinates (x', y', z', t') of space (in an orthonormal reference system) and time of a moving inertial system in terms of the ones (x, y, z, t) of a fixed one. They are:

$$x'_\mu = \sum_{v=1}^{4} \alpha_{\mu v} x_v$$

where $x_4 = ict$, i is the imaginary unit and $\alpha_{\mu v}$ the entries of an orthogonal matrix: $\|\alpha_{\mu v}\| = \pm 1$ representing rotations in a four-dimensional space-time. They build up the Lorentz group. If a 4-translation is added the Poincaré group, a 4-dimensional extension of the Euclidean group is obtained. See also Section 8.

[12] In a three-dimensional orthonormal reference system where a point P is represented by 3 coordinates xyz, the equation of the sphere is: $x^2 + y^2 + z^2 = c^2 t^2$, or $x^2 + y^2 + c^2 - c^2 t^2 = 0$. In the latter form it may be considered as representing a 4-vector of zero length. Such vectors are the subject of study in spinor geometry (see Section 8).

[13] In the fixed reference system (x, ct) the equation of x' is $ct = \beta x$, therefore the coordinates of $1'$ on x', intersection of x' with the hyperbola $x^2 - c^2 t^2 = 1$ are $x = 1/\sqrt{1 - \beta^2}$, $ct = \beta/\sqrt{1 - \beta^2}$, the coordinates of A' are $1, \beta$. From this (6) is easily found.

[14] With $1/\sqrt{1 - \beta^2}$ reaching 10 000 the μ mesons spend only 1/100th of their lifetime in travelling, the rest is spent in the matter of the earth where they build up mesonic atoms. It is amusing to translate this prolongment of lifetime in human language. Should men have the capability to reach a velocity so near that of light (a conceptual, but not practical, possibility), they could choose to reach in less than a month, say, of travel, a near star (ten days for α Centauri) and spend the rest of their life there.

[15] It is amusing, today, to read what the, otherwise outstanding, theoretical physicist Wolfgang Pauli (1933) wrote in his article in the 'Handbuch der Physik' about the at that time newly discovered relativistic Dirac equation (Dirac 1928) for the electron which in fact predicted the existence of negative energy-states as foreseen in Equation (8) that is of the afterwards discovered positron. Just because of this he affirms that *the equation may not be right* while it afterwards turned out to be one of the most fundamental equations of quantum mechanics.

[16] There may also be other sources of prejudices in science of which there are well known historical examples, like the ones which determined the delayed acceptance of the heliocentric theory and those which gave origin to the Lysenko episode in biology.

[17] A world where paradoxical and obvious are interchanged is not inconceivable. It could be the case of intelligent life (of not too massive living being) in the vicinity of a neutron star or black hole. Then velocity near that of light c ($\beta \approx 1$) would be normal and normal to profit of time-dilatation (Equation (7)) to travel to the next star to see friends while paradoxical would be considered to add velocities like we do according to Equation (2).

[18] The birth of quantum mechanics may also be traced back to the presence of an infinity in classical Maxwell electrodynamics, the one which appeared in the computation of a black body radiation and was eliminated by Planck with the introduction of the quantum of action.

[19] Electron, proton, neutron, neutrinos are fermions, the photon is a boson. Fermions and bosons obey different statistics. With fermions you can build bosons but not vice versa.

REFERENCES

Bateman, H.: 1909, *Proc. London Math. Soc.* **8**, 223.

Blackett, P.M. and Occhialini, G.P.: 1933, *Nature* **32**, 917.

Cartan, E.: 1913, *Bull. Soc. Math. France* **41**, 53.

Dirac, P.A.M.: 1928, *Proc. Roy. Soc. A* **117**, 610.

Einstein, A.: 1924, *The Theory of Relativity*, London.

Feynman, R.: 1949, 'The theory of positrons', *Physical Review* **76**, 749.

Maxwell, J.C.: 1873, *Treatise on Electricity and Magnetism*, London.

Michelson, A.A. and Morley, E.W.: 1887, *Sill. Journ.* **34**, 333.

Nother, E.: 1918, 'Invariante Variationsprobleme', *Nachr. Ges. Wiss., Göttingen*, 235.

Pauli, W.: 1933, *Handbuch der Physik*, Vol. 24, 2nd ed.

Supek, I.: 1974, 'Marcus Antonius de Dominis 1560–1624', *Encyclopedia Moderna* **28/9**, 69

GIAN CARLO GHIRARDI

THE QUANTUM WORLDVIEW: ITS DIFFICULTIES AND AN ATTEMPT TO OVERCOME THEM

This paper consists of two parts. The first one is devoted to a general discussion of the problems which one meets in trying to build up a coherent worldview which takes into account the quantum nature of physical phenomena. This part requires only a general knowledge of the formal structure of the theory. In the second part a recently proposed modification of the formalism aimed at overcoming the difficulties and the limitations of the standard approach is considered. For a better understanding of this part we suggest a prior careful reading of Ghirardi and Rimini (1990) and of the detailed contribution to these proceedings by Professor T. Weber.

PART I

1. ORTHODOXY VERSUS HERESY

In this first part we will expound in a simple, concise and elementary way the consequences in the philosophy, methodology and foundations of physics which derive from the adoption of the quantum axiomatic scheme and of its related, so-called 'orthodox interpretation'. It could be remarked that the expression 'orthodox interpretation of quantum mechanics' (OIQM) is not a very precise one, and that it would therefore have been preferable to avoid its use. In fact, first of all, different authors have designated by such an expression a whole set of positions about the formalism and its conceptual implications, which in some cases may differ on fundamental points. Secondly, and more important, as we will try to show, almost all the conceptual difficulties that quantum mechanics encounters are not due to its interpretation but arise from the quite peculiar features of microscopic phenomena which nature has compelled us to accept. It is the occurrence of such peculiarities which seems not to allow us to base on the formalism a consistent and unified description of natural phenomena at all levels. A tentative conclusion of this paper will

175

G. Corsi et al. (eds), Bridging the Gap: Philosophy, Mathematics, and Physics, 175–197.
© 1993 Kluwer Academic Publishers.

in fact be that it is the theory itself, and not its interpretation, which requires changing.

Why, then, do I insist on using the expression OIQM? There are various reasons for this. First, even though, as remarked above, there are many different specific views which are referred to as the OIQM, they correspond to the taking of similar positions with respect to some basic questions which are central to this paper, such as:

– What are the physical theories about: physical systems or laboratory experiments?
– What is the role of the apparatus and of the observer in physics?

Secondly, it is useful to stress that it is just the kind of answers that have been given to such questions by eminent exponents of the OIQM which have played a fundamental role in leading the majority of the scientific community to accept, for many years, the point of view that the conceptual problems raised by the theory *"were of no great interest and could, in any case, be sorted out with few hours' careful thought"*, see Squires (1986). Fortunately, some great scientists, like Einstein, Schrödinger and Bell, have not given up the firm belief that these problems had to be faced and have expressed the hope that they could be solved, provided one dared to go on asking radical questions. It seems to us that, nowadays, there is an increasing consensus about the fact that the OIQM (whichever of the positions going under this name one considers) has failed in preseting a rational, coherent scheme for the understanding of natural phenomena and that the whole subject requires critical reconsideration.

The last reason for adopting the expression OIQM is quite specific. Recently, an interesting preprint by K. Gottfried (1989) has been circulated, in which the attempts to overcome the abovementioned difficulties by accepting a modification of quantum mechanics, and in particular the one we will discuss in the second part of the paper, have been referred to as HERESIES. We have therefore decided to retain the term orthodox to stress, by contraposition, that we consider it perfectly legitimate to investigate the possibility that heresy, i.e. changing the theory, could help in understanding better *"the mystery of the quantum world"* (Squires 1986).

2. THE QUANTUM FORMALISM

To avoid possible misunderstandings of our position, we start by pointing out that we are perfectly aware of the innumerable bodies of fact in favour of quantum mechanics. The elaboration of the quantum scheme for the description of physical phenomena has required revolutionary and deep conceptual changes which represent important steps in our understanding of nature. In particular the discovery that the superposition principle rules all physical processes has required the acceptance of indeterminism, the recognition that systems which have interacted lose their individuality, and that nonlocal features characterize microscopic processes. The theoretical scheme has been extremely successful. Created to account for atomic phenomena it has revealed itself as a powerful, practical tool to account for physical processes from the microcosmos to the cosmological scale. However, the same superposition principle which, as stated above, represents the most innovative and relevant element of the theory, lies at the root of its conceptual difficulties and has led to very peculiar epistemological positions.

We consider it appropriate to start by sketching briefly the axiomatic structure of the theory. We have:

a. The states of the systems are described by elements of a linear vector space, the physical observables by self-adjoint operators on this space.

b. The only possible results of a measurement of an observable are the eigenvalues of the corresponding operator.

c. A preparation procedure consists in the measurement, for instance, of an observable with a nondegenerate, purely discrete spectrum. The state $|\Psi, 0>$, immediately after the measurement, is the eigenvector corresponding to the eigenvalue which has been found.

d. The evolution of the state vector is governed by the *deterministic* and *linear* Schrödinger equation, whose solution can be written as:

$$(1) \qquad |\Psi, t> = \exp[-iHt] \, | \, \Psi, 0>$$

H being the Hamiltonian operator of the system.

e. Knowledge of the state vector yields all the information we can have about our system. This information is probabilistic: for any considered observable A, the probability $P(A = a_k \,|\, \Psi, t)$ of obtaining, in a measurement of A (at time t), one of its eigenvalues, say a_k, is given by the square of the modulus of the scalar product of the state vector $|\Psi, t>$ with the eigenvector $|a_k>$ of A belonging to the eigenvalue considered:

$$(2) \qquad P(A = a_k \,|\, \Psi, t) = |<a_k \,|\, \Psi, t>|^2$$

f. When a measurement is performed and a specific result is obtained, then wave packet reduction takes place: the state of the system immediately after the measurement is the eigenvector corresponding to the eigenvalue which has been found.

With reference to the above axiomatic formulation, some remarks are appropriate;

1. Rule (e) has important implications about the predictions that the theory is able to make concerning properties of physical systems. If it happens that the state vector $|\Psi, t >$ describing the system is an eigenstate belonging to a specific eigenvalue, say b_k, of an observable B, then the theory allows us to predict with certainty the outcome of a prospective measurement of B. In such a case, one is allowed to say, following Einstein, Podolsky and Rosen (1935) that the system 'possesses' the property b_k. However, since, in general, operators associated with different observables do not commute, the eigenstate $|b_k >= |\Psi, t >$ will not be an eigenstate of another observable, say C. The state $|\Psi, t >$, will then have a non-zero scalar product, with eigenstates of C belonging to different eigenvalues. This is quantum indeterminism: the fact that a system can be asserted to possess a property makes it illegitimate to assert, or even to think, that it possesses definite properties referring to other specific observables.

2. Due to the above situation one can say that, in a sense, quantum mechanics is a theory of potentialities. This is a quite interesting conceptual achievement: the study of microscopic systems has revealed unsuspected features of natural phenomena. Physical systems may possess some properties, but there are always other properties which cannot legitimately be attributed to them at the same time. To describe the full complexity of the actual situation we have to list all probabilities referring to all potentialities of the system. It is a nice fact that such extremely rich information can be summarized in the knowledge of the state vector.

3. The now mentioned facts put into evidence that the theory is going to have to face the problem of the "objectification of the potentialities" (Busch 1990). Specific potentialities will become actual only when a specific measurement is performed, and measurement procedures aimed at actualizing different potentialities are, in general, mutually incompatible.

   This gives a peculiar conceptual structure to the theory: quantum mechanics is only able to make conditional probabilistic predictions about the outcomes of prospective and, in general, incompatible measurements processes.

4. With reference to the previous remark and to assumption (f) (which in a sense is directed at making precise the conditions for obtaining the actualization of potentialities), one can immediately see the emergence of a serious difficulty for the theory: is assumption (f) compatible with the requirement that the

measurement process itself be governed by the laws of evolution of the theory? This problem has been discussed in great detail in Ghirardi and Rimini (1990) to which we refer the reader, but we can grasp its essence: since, as already remarked, the theory makes only conditional predictions about the outcomes of prospective and, in general, incompatible measurements, any attempts to apply quantum mechanics itself to the measurement process will unavoidably lead to further conditional prediction about what will happen, what will become actual (even for a macroscopic system) if, . . . and so on. A never ending dilemma.

3. DESCRIPTION VERSUS EXPLANATION

Before going on to analyse some of the conceptual problems raised by the theory, we consider it appropriate to state that we take a realistic position and in particular that we assume that the regularities of nature which we recognize and which we try to account for by our theories refer to something which exists objectively, independently of any observer. This position may be considered very naive, it is surely expressed in too sketchy terms, but we hope the reader will grasp what we have in mind. In a sense our assumption corresponds to the adoption of the principle of inference to the best explanation, i.e. to the idea that the best explanation of the reason why we are able to make successful predictions about some regularities we guess and we verify (or better we find not to be falsified) in our experiments, consists in assuming that there are objective phenomena exhibiting objective regularities.

We are, then, interested in building a theoretical scheme to accomodate the regularities we have discovered. Such a scheme, to be taken seriously as a conceptual framework for understanding nature (we are not committing ourselves to its being 'true', so we accept that it could and actually should be improved and/or changed, but we are simply requiring that it is able to let us know something about something which is out there) must contain, and must be judged (see Maxwell (1988) and Redhead (1987)) on the basis of two kinds of criteria: the empirical ones and some general nonempirical criteria. There is no doubt that the empirical criteria, as Galileo taught us, play an absolutely primary role; a theory which does not meet them must be rejected. When this is granted, the adoption of non-empirical criteria leads to what one could call an interpretation of the theory.

The lowest level of interpretation consists in assuming that only the empirical criteria have to be taken into account; this leads to the so-called minimal instrumentalist interpretation, see Redhead (1987). The theory is reduced to the rules establishing the connection between preparation and measurement procedures. In so doing one can be said to accept the regularities as brute facts: all statements going beyond them are regarded as metaphysical speculations. This attitude leads either to strict instrumentalism or to positivistic and anti-realistic positions.

Obviously, even the minimal interpretation can be enriched by the adoption of a specific type of nonempirical criteria, in particular criteria of economy, simplicity, mathematical beauty etc., which allow to choose between conflicting theories. However, if a fundamentally instrumentalist attitude is taken, the choice is only a matter of convenience: the idea of a theory being 'more satisfactory' than another in any sense which goes beyond this level cannot be entertained.

We do not consider such a position satisfactory. If one wants to proceed further and to be allowed to consider the theory as something which does not simply 'describe' our experiences but serves the purpose of 'explaining' how it happens that we are able to discover regularities in natural phenomena, one has to resort to nonempirical criteria, or, in other words, one has to commit oneself to an interpretation in the proper sense (i.e. going beyond the minimal instrumentalist one). It seems to us that, when these criteria are chosen in an appropriate way and the scheme satisfies them, the theory can be considered as contributing to some extent to an 'understanding' of natural phenomena.

We are well aware that this point of view can be considered as too naive and that surely it would deserve a deeper and more rigorous analysis, and we also know that many scientists and philsophers do not share it. However, we cannot avoid stressing that, after all, all variants of the OIQM are interpretations in this second sense, i.e. they have recourse to nonempirical criteria. It seems, then, a useful task to identify some criteria on the adoption of which one can expect a large consensus and to see whether the OIQM meets them.

4. NONEMPIRICAL CRITERIA

Without pretending to be exhaustive we now list some nonempirical criteria whose consideration will allow us to review the conceptual problems of the formalism and which (together with others) should, in our opinion, be taken into account when judging the 'explanatory power' of a theory.

C1. The first fundamental criterion, which cannot be given up, is that of the internal consistency of the scheme, i.e. the requirement that the axioms of the theory are not contradictory. In the case of quantum mechanics the crucial problem, as already stressed, arises in connection with assumption (f). At the present stage we cannot yet say that wave packet reduction conflicts directly with the other assumptions on which the theory is based; the occurrence of a contradiction depends on what one assumes about the nature of the systems involved in the measurement process. However one can slightly modify the question: is the postulate of wave packet reduction compatible with the assumption that the measurement process itself be a genuine quantum process, governed by the linear evolution equation of the theory?

C2. This criterion, as well as the following one, have already been mentioned in Section 1. What does the theory pretend to describe: physical systems or laboratory experiments? Furthermore:

C3. What is the role of the measurement process and of the observer in the theory?

Basically, the two questions we have just formulated raise the problem of identifying the referents (or better the reference class) of the theory. As appropriately pointed out by Bunge (1973) one can consider two kinds of expressions occurring in physical writings: those making reference only to physical systems such as '*the value of the property P equals P*' and those, such as '*observer Z has found the value P for the property P*', in which two referents are involved, i.e. a system and an observer. Both kinds of sentences may be perfectly legitimate, but the crucial and philosophically significant problems derive from the fact that, since the referent may be a physical system or a subject, or some combination of the two, one can give different answers to the question: What is the physical theory about? The possible different answers correspond to the adoption of different theses about science, like the realistic, the subjectivist or the dualist one, etc.

C4. What is the unifying power of the theoretical scheme, i.e. can it be applied to a class of phenomena larger than the one for which it has been elaborated? Does it account for phenomena on differnt scales? Can it be used to build up an acceptable account for the evolution of the universe?

C5. In the case in which one is compelled to recognize that the theory cannot rule all phenomena, is its domain of applicability clearly defined by its precise mathematical structure or does it turn out to be to some or to a large extent arbitrary and shifty?

5. A FIRST COMPARISON OF QUANTUM THEORY WITH NONEMPIRICAL CRITERIA

We will now look at the quantum scheme from the point of view of some of the previously listed criteria.

C1. As far as this criterion is concerned we will not spend time in discussing it, since it is analyzed in great detail in Ghirardi and Rimini (1990). We then simply state that the answer to the final question raised under C1 is plainly: 'No!', i.e. the postulate of wave packet reduction contradicts the idea that the measurement process is governed by quantum mechanics itself. The conceptual reasons for this difficulty have already been mentioned in Section 2. Formally, the incompatibility derives from the fact that, while Schrödinger's evolution is linear and deterministic, wave packet reduction exhibits nonlinear and stochastic features.

C2. As we have already stressed, in its very formulation the theory speaks of the probabilities of the outcomes, conditional on the fact that specific measurement procedures are actually performed. We stress again that the acceptance of this position has been, in a certain sense, imposed on science by the recognition that the superposition principle holds for microsystems. Such a principle requires us to accept that being in a superposition of two states corresponding, e.g. to the two properties $A = a_k$, and $A = a_j$, i.e.:

$$(3) \qquad |\Psi, t> = \alpha |a_k > + \beta |a_j >$$

is not logically equivalent or compatible with the statement that

$$(4) \qquad either \ A = a_k \ or \ A = a_j, \ holds.$$

Therefore only a precise measurement procedure can make property A actual; without such a procedure A is indeterminate. We are then led to the conclusion that the theory is essentially about laboratory experiments.

C3. For what concerns the role of the observer in the theory, let us start by remarking that the original attempt by Schrödinger to interpret the modulus square $|\Psi(x)|^2$ of the wave function in configuration space as a density of matter and/or of charge, can be regarded as a serious attempt to give and objective meaning to the wave function, independently of any act of observation and of any conscious observer.

As is well known, however, if Schrödinger's linear dynamics is considered to be universally valid, such an interpretation cannot be consistently maintained, due to the unrestricted validity which one is then attributing to the superposition principle. Thus, an electron can very well be in the superposition.

(5) $|\Psi, t> = |$Being Here $> + |$Being There $>$

but then, how does it happen that it is always found *either* Here *or* There when its position is detected?

Thus, Born won the struggle with his probabilistic interpretation: the quantity $|\Psi(x)|^2$ does not represent a density, but a probability density. Probability of what? It could not be the probability of *being* Here or There, since from the occurrence of interference phenomena we know that state (5) describes a situation which is physically different from the one which would correspond to the assumption that Being Here *or* Being There is true. Thus the probability must be referred not to *being* but to *being found*. At this point the crucial role of the measurement process and of the observer has become an essential ingredient of the theory.

It is appropriate to point out that one can identify at least two 'orthodox positions' about the measurement process and the role of the observer.

– The strict Copenhagen view recognizes that, in any measurement, object, apparatus and observer become entangled (see Weber's contribution), and, as a consequence, they completely lose their identity. This unbreakable wholeness of the 'global physical system' is peculiar to the quantum nature of physical processes; a theory of measurement would attempt to distinguish subject from object, and would therefore destroy this fundamental characteristic of the quantum world. Concluding: no genuine quantum account of the measurement process must be given and the statements of the theory about measurement outcomes must employ the concepts and the language of classical physics which acquire the status of logical prerequisites for the theory itself.

– The von Neumann view is rather different: the consideration of a measurement process requires suspending the fundamental dynamical postulate of the theory (i.e. Schrödinger's equation) and adopting in its place the postulate of wave packet reduction. The choice of when and how the linear evolution law has to be given up is to a remarkable extent arbitrary and can be shifted at will up to the point of considering it as occurring only as a consequence of the conscious perception by an observer, see Wigner (1962).

Concerning the specific role played by the observer within such a view we
cannot do better than quote Bunge (1973):

The active role this account assigns to the conscious observer in determining the outcome of a
measurement is best brought home by the following imaginary procedure, which may be called
the *mensura interrupta* technique. You set up an experimental arrangement to measure a given
magnitude of an object of a certain kind and operate the device all the way but abstain from taking
the final reading. After a while you flip a coin: if heads you look at the pointer and register
its position; if tails you walk out of the laboratory. Being a subjectivist, you are unwilling to
distinguish the *physical* fact that the pointer came to rest at a given position, from the *psychical*
fact that you take or do not take cognizance of such a physical fact: what is more, you refuse to
believe that there is such a thing as an autonomous physical fact. Then you are bound to conclude
that the outcome of the measurement, i.e. the value of the magnitude concerned, depends on the
observer's consciousness. Assume further that you abide by the operationist tenet that calculated
values are possible measurement values: then you will conclude that the conscious observer is an
essential member of the quantum theory and, in general, that Man can no longer be ignored by
physics.

6. A DIGRESSION: LOOKING FOR DETERMINISTIC COMPLETIONS OF THE FORMALISM

In spite of the successes and the dominant role of the OIQM, a part of the
scientific community (including scientists of the level of Einstein) went on
trying hard to save an objective and realistic interpretation of the formalism, in
particular one which would not make the observer and the act of observation
to play such a fundamental role. The attempt to achieve this result has been
based on the search for a deterministic completion of quantum mechanics by
the consideration of hidden variables. The program consisted in trying to
build up a theory which would allow an ignorance interpretation for quantum
indeterminism. Such a theory should contain hidden parameters (besides or
alternatively to the wave function) whose precise knowledge would make
definite the values of all conceivable observables: it is the ignorance of these
parameters which compels us to make only statistical predictions. The program
had a quite peculiar story: von Neumann 'proved' an impossibility theorem for
such theories, but Bohm and others exhibited explicit examples of them. They
were right, von Neumann had made use of logically unnecessary assumptions
in his proof.

It is important to stress that, as previously mentioned, the supporters of the
OIQM had succeeded in convincing the scientific community that it is the very
occurrence of indeterminism which requires to attribute a peculiar and central
role to the observer. Thus, people motivated by their dissatisfaction for this
subjectivistic position, have been naturally led to try to restore determinism.

With reference to the program we discuss in the second part of the paper, it may be interesting to remark that, up to very recent times, nobody has considered it worthwhile to explore a different line of research, i.e. trying to build up a theory which would not require the making of a systematic reference to measurement procedures and observers, but would nevertheless retain indeterminism as a firm feature characterizing natural phenomena.

It is also important to recall that J.S. Bell (1964) has proved, with complete generality, that any hidden variable theory must be nonlocal. Only the derivation of this fundamental result has made possible the full appreciation of the revolutionary implications of quantum mechanics.

7. THE OTHER NONEMPIRICAL CRITERIA

C4. The recognition that the reduction of the wave packet occurring in a measurement cannot be consistently described by the linear evolution equation of the theory implies the acceptance that in such a process systems are involved which somehow do not obey quantum mechanical laws. More generally, one has to face the problem of the quantum description of macroscopic 'classical' objects. As is well known, various derivations of the 'classical limit' of quantum mechanics have been presented. It has, however, to be remarked that these derivations, which involve relating classical variables to the mean values of quantum observables, have a precise meaning only when one considers the ensemble level of description of physical phenomena. By contrast, when the individual level of description is taken into account, they contradict experience and common sense. What meaning can in fact be attributed to a superposition of two states of a macro-object corresponding to its being in two remote spatial regions? Moreover, what physical significance can be attached, in the case of an individual system, to the mean value of its position in such a state? J.S. Bell (1987) correctly shared Schrödinger's opinion that quantum mechanics is unable 'to account for the definiteness, the particularity of the world of experience'.

Once more, in order not to be misunderstood, we have to state that we are well aware of the fact that many scientists would not consider the previous arguments as cogent, on the basis of the fact that, from a practical point of view, it turns out to be extremely difficult, in the case of a macroscopic object, to distinguish a linear superposition from a statistical mixture. Such a distinction requires, in fact, correlation experiments involving all the constituents of the macroscopic system. This is true; nevertheless, in our opinion, to be content with such a solution amounts simply to recognizing that (as everybody knows), *"quantum mechanics is just fine FAPP (for all practical purposes)"*, Bell (1990), and to go back to an essentially instrumentalist position.

One should also, as required by C4, consider whether quantum mechanics can be used to build up an acceptable picture of the evolution of our universe. This question too raises fundamental difficulties. In fact, as we have seen, within the OIQM a physical system only has a state insofar as it is subjected to preparation and measurement procedures which are external or additional to the system under consideration. 'Who prepares the universe and who measures it?' then become really embarrassing questions. Another way of being compelled to realize that we are facing a crucial problem derives from the consideration of the program of combining quantum mechanics with general relativity. In order to quantize gravity, the space-time itself needs to be given a quantum state which, in turn, requires devices external to space-time. A deep puzzle.

C5. The difficulties met by the OIQM with this criterion are, in our opinion, the most crucial ones. After having been compelled (by the impossibility of fitting wave packet reduction within the scheme) to accept that there are nonquantum systems, it turns out to be impossible to identify, on the basis of the theory itself, the different classes of physical systems or processes which lie within or outside its domain of applicability. The splitting is basically shifty. Questions like: 'what qualifies a physical process as a measurement, what is quantum and what is classical, when has the linear evolution law to be suspended?' do not admit any definite answer.

It is interesting to remark that, in a certain sense, the debate on the measurement problem in quantum mechanics can be taken as an emblematic illustration of this fundamental arbitrariness and shiftness of the boundary between the two classes of phenomena. Different authors in different contexts have placed the borderline at different levels. Bohr, in his debate with Einstein, was lead various times to restrict the 'family' of objects which allowed a classical treatment (in particular to include in the quantum description shutters, diaphgrams etc.); von Neumann could be asserted to have, in a sense, 'axiomatised' the shiftiness trying to transform it from a drawback into a merit of the theory; Wigner has chosen to place the splitting at the level of consciousness (which in turn is a shifty concept), and so on. It is appropriate to remark that one cannot relate the splitting simply to the number of particles which are involved in the process under consideration since there are macroscopic objects which exhibit a genuine quantum behaviour.

Obviously, concerning this problem, too, different positions have been taken. The most radical ones are those which simply avoid it by denying the occurrence of wave packet reduction, so that all phenomena are genuinely quantum. We will not discuss such proposals, the most well known being the so-called Many Universes Theory. They are considered in Ghirardi and Rimini (1990). We do not believe that such a line can be consistently followed.

We conclude this analysis by stating that we consider the intrinsic lack of precision about the distinction between micro and macro, quantum and classical, as the most serious limitation of the theory, the one which forbids us, or renders extremely difficult for us to base on it an articulate, systematic and coherent picture of natural phenomena, i.e. what A. Shimony (1989) calls a "physical worldview".

PART II

This second part is devoted to a concise description of a recent attempt to overcome the difficulties discussed in Part I. Such an attempt is admittedly of the heretic type, i.e. it accepts that Schrödinger's equation is not always right. From the point of view of the conceptual analysis developed in this paper, the actual interest of the proposal does not derive from its specific technical aspects, but from the fact that it constitutes an explicit proof that it is possible to devise a theory which, even though it accepts fully the indeterminism of natural phenomena (actually, in the model, indeterminism is even increased) it nevertheless allows an objective description (i.e. one that does not require us to assign a central role to the observer) of macroscopic physical events.

A few remarks. First of all we consider it appropriate to point out immediately that the 'heretic' aspects of the model concern exclusively its dynamics; the formal mathematical apparatus, as well as the language, are precisely those of standard quantum mechanics. Secondly, the theory, in contrast to the hidden variable approach, maintains as a fundamental and firm point that the wave function yields the complete description of reality. The model does not add variables but, by introducing mathematically precise modifications of the standard dynamics, allows us to get rid of the embarassing (at the macroscopic level) ambiguities of the theory and to avoid the ill defined division of the world into parts, system and apparatus, as well as the necessity of suspending its fundamental dynamical equation during measurement processes.

8. MODIFYING THE DYNAMICS: HINTS AND PROBLEMS

Let us recall the conclusion of the previous critical analysis: quantum theory, on the one hand, is unable to describe wave packet reduction but, on the other, it needs it to account for the actualization of specific properties for macro-objects, in agreement with our experience with individual processes involving them. If, to find a way out of such a dilemma, one accepts considering the possibility of modifying the dynamics of the standard theory (whose characteristic features are those of being linear and deterministic), with the aim of making it compatible with wave packet reduction (which, as we have repeatedly remarked, exhibits nonlinear and stochastic aspects), the most natural line to be followed seems to be that of introducing nonlinear and stochastic terms into the dynamical equation. Such modifications are assumed to describe universal natural mechanisms governing all physical processes.

In recent years, several interesting attempts at a dynamic description of the reduction mechanism have been made, see de Broglie (1960), Pearle (1976) and Gisin (1984a,b). Many of them have considered the possibility of nonlinear and stochastic modifications of the standard theory. However, such programs did not lead, up to recent times, to a real breakthrough, due to some crucial problems they had to face.

First of all, since the dynamic modifications which one wants to consider (and which must be assumed to occur at the level of the microconstituents of any physical system, to have a fundamental character) should, in a sense, induce the actualization of some of the incompatible potentialities summarized by the state vector, one must face the question: which specific properties should one require to be dynamically and spontaneously actualized? Such a problem is usually referred to as the one of the choice of the preferred basis, see Pearle (1989): no explicit answer is given for it in the papers cited.

The second problem derives from the fact that the program seems to have to satisfy divergent instances: on one side the dynamical modifications should have a practically negligible effect for all microscopic systems (due to the extremely high degree of validity of standard quantum predictions for them); on the other, they should be able (this is called the 'trigger problem' in Pearle (1989)) to induce an extremely rapid suppression of macroscopic superpositions. In fact, modern technological developments allow us to complete a measurement in very short times (of the order of nanoseconds).

Concerning the first point an important hint comes from the recognition that the most disturbing feature of the superposition principle (i.e. of the linear nature of the theory) derives from its allowing the occurrence of states like the one of Equation (5), in which the specifications 'Being here' and 'Being there' refer to a macroscopic object. As repeatedly stressed by Bell (1986) any theory, to be considered a serious alternative to the standard one *"would allow electrons to enjoy the cloudiness of waves"* but, simultaneously should imply *"that tables and chairs and ourselves and black marks on photographs"* be *"rather definitely in one place rather than another"*. This consideration point towards the identification of the position as the preferred basis and therefore towards the consideration of spontaneous, stochastic and nonlinear processes leading to localizations. It turns out that this idea, as simple and natural as it is, allows us to devise a dynamics leading at the same time to the solution of the second problem. To be precise, the nonstandard additional dynamical mechanism which is considered by the model automathically amplifies as the number of constituents of a physical system increases. Thus one can choose the parameters characterizing it in such a way that an actual microsystem is essentially unaffected, while for macroscopic ones, relevant effects occur, leading exactly to what one is looking for.

9. THE MODEL: QUANTUM MECHANICS WITH SPONTANEOUS LOCALIZATION

We are now ready to discuss the model, proposed by Ghirardi, Rimini and Weber (1986) (see also (1987), (1988)) referred to as quantum mechanics with spontaneous localization (QMSL), which represents a mathematically precise and successful attempt to achieve the two divergent aims discussed in the previous section, i.e. those of leaving the physics of microscopic systems essentially unaltered but at the same time of forbidding the occurrence of superpositions of macroscopically distinguishable states.

The model, in its original and simpler version Ghirardi, Rimini and Weber (1986) and Bell (1987) is based on the following assumption (compare the considerations of the last part of the previous section): *the state vector of any physical system, besides evolving according to the standard Schrödinger equation, is subjected, at randomly distributed times, with an appropriate mean frequency, to spontaneous localization processes affecting the elementary constituents of the system itself. Such processes are assumed to correspond to some fundamental mechanism of nature.*

Obviously, one must make precise the specific features of such a mechanism, in particular one has to give an answer to the following questions:

- What are the specific temporal features of the spontaneous processes, i.e. WHEN do they take place?
- HOW does the wave function of the system change as a consequence of the occurrence of a localization process?
- At which places, i.e. WHERE is the particle localized?

For the moment we will confine our considerations to the system of one particle in one dimension, whose wave function in configuration space will be denoted by $\Psi(q)$. The mathematically precise answers to the previous questions are the following:

WHEN? As already stated, the times at which the spontaneous localization processes occur are randomly distributed; the only necessary further specification is that their distribution is of the Poisson type. The mean frequency for their occurrence will be denoted by λ, so that λdt represents the probability of a spontaneous process occurring in the time interval dt.

HOW? Suppose a spontaneous localization (we will also call it a hitting) occurs around the point x. Then the wave function, which before the hitting is $\Psi(q)$, changes suddenly according to

$$(6) \qquad \Psi(q) \Rightarrow \Psi_x(q) = \Phi_x(q) \,/\, \|\Phi_x(q)\|$$

where

$$(7) \qquad \Phi_x(q) = \left[\frac{\alpha}{\pi}\right]^{1/4} e^{-\frac{1}{2}\alpha(q-x)^2} \Psi(q) \,.$$

In Equation (6) we have denoted, as usual, by the symbol $\| \ \|$ the norm of a state. Note that $1/\sqrt{\alpha}$ has the dimensions of a length and represents the accuracy of the localization.

WHERE? One has to make precise how the position x of the localization is chosen. The corresponding assumption is quite natural and parallels the one concerning position measurements in standard quantum theory: the localizations take place with higher probability at the points at which the particle has a larger probability density of being found according to Born's recipe. To be precise: the probability density of the process taking place at x is given by $\|\Phi_x\|^2$.

It is important to stress that both the specification of WHEN as well as the one of WHERE the localizations take place introduce stochastic elements, and that the rule about WHERE it occurs introduces nonlinear feature in the theory. The model contains two parameters, i.e. the mean frequency λ and the spatial accuracy $1/\sqrt{\alpha}$ of the localizations. If the model is taken as describing fundamental natural mechanisms such parameters acquire the status of new constants of nature.

Let us now try to make clear, still confining our considerations to the case of one particle, while the spontaneous processes, when they occur, tend to

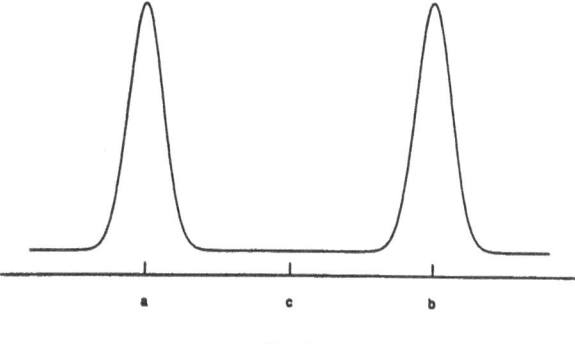

Fig. 1.

suppress linear superpositions of remote states. Suppose that the wave function be the one corresponding to the superposition (5), where, as shown in Figure 1, here means around a and there around b, with spreads smaller than $1/\sqrt{\alpha}$. Suppose a hitting occurs at $x \simeq a$: then the wave function after the process is sharply peaked around a. An analogous effect occurs when $x \simeq b$. Finally, the probability of a hitting taking place at a point x appreciably different from both a and b (e.g. at the middle point c of the figure), turns out to be vanishingly small since $\|\Phi_c(q)\| \simeq 0$.

Concluding, the occurrence of a localization, when the system is in a state like (5), transforms the state in one corresponding to either |Being around $a >$ or |Being around $b >$, i.e. it suppresses the linear superposition.

In the case of a physical system containing many (say N) constituents, according to the assumption on which QMSL is based, things go in exactly the same way, the only difference being that the spontaneous processes have to be referred to any individual particle. Thus one will consider hittings occurring for, e.g., the kth particle, whose effect is that of multiplying the wave function $\Psi(q_1, \ldots, q_N)$ times the factor $[\alpha/\pi]^{1/4} \exp[-(\alpha/2)(q_k - x)^2]$.

10. THE MAIN IMPLICATIONS OF QMSL

As already remarked QMSL contains two parameters, the frequency λ and the accuracy $1/\sqrt{\alpha}$ of the localizations. The most relevant feature of the theory consists in the fact that, provided α is chosen appropriately, in the case of an almost rigid body, the spontaneous processes essentially affect only the centre of mass motion. Moreover, the frequency of the localizations is amplified with the number N of constituents.

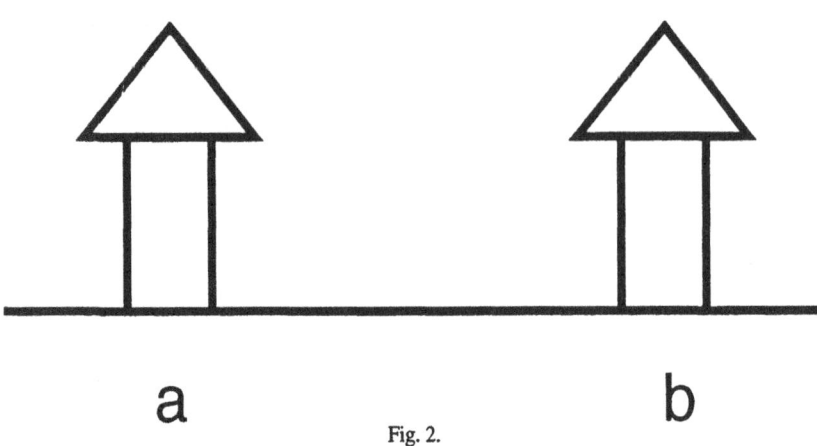

Fig. 2.

Formally:

(8) $\lambda_{N\text{-particles}} = N \, \lambda_{1\text{-particle}}$.

The above statements need some comments. The expression 'almost rigid body' denotes a body, like an insulating solid, for which the spreads of the positions of its lattice points around their equilibrium positions are smaller than $1/\sqrt{\alpha}$. The amplification of the mechanism for what concerns the dynamics of the centre of mass holds true in all cases. However, if the body is not almost rigid, the internal motion can be considerably affected.

Without going into technical details we can make clear the physical mechanism of amplification. Consider a superposition of type (5) for a macroscopic pointer of an apparatus (see Figure 2), and suppose that the separation between 'Here $\simeq a$' and 'There $\simeq b$' is larger than $1/\sqrt{\alpha}$. In such a case all the N particles of the pointer are around a in the state |Being here> and all of them are around b in the state |Being there>.

Localizing any one of the constituents then leads to a localization of the whole pointer around a and b.

We are now ready to make a specific choice of the parameters and to discuss the ensuing implications. Assume

(9) $\lambda \simeq 10^{-16} \sec^{-1}$; $[1/\sqrt{\alpha}] \simeq 10^{-5}$ cm .

The choice of λ implies that a microscopic system suffers a localization, on the average, every $10^8 - 10^9$ years! Thus, for example, an electron in a superposition like (5) will remain in such a state for about one tenth of the age of the universe. Moreover the internal state of any composite microscopic bound system (e.g. an atom) will not be appreciably affected, even when a localization occurs, since $1/\sqrt{\alpha}$ is much larger than atomic dimensions. In the case of a macroscopic system, however, the amplification mechanism (8) has remarkable effects; since in such a case N is of the order of Avogadro's number $(\simeq 10^{23})$, one has:

$$(10) \qquad \lambda_N \simeq 10^{23} \, 10^{-16} \sec^{-1} = 10^7 \sec^{-1} .$$

This means that a superposition like (5), for such a system, is spontaneously reduces either to $|$Being here$>$ or to $|$Being there$>$ in about 10^{-7} sec. This should make clear why the model is able to meet both divergent requirements mentioned in the previous section.

Two remarks are appropriate. First of all it has to be noted that the spatial localization mechanism also forbids, indirectly, the persistence of other types of superpositions of macroscopically distinguishable states. The simplest example is that of a superposition of two states of a macro-object corresponding to its being essentially in the same place but possessing different momenta. The localization mechanism is not immediately effective, but, as soon as the different velocities lead to different positions, it suppresses one of the two components, i.e. it picks up a state of definite velocity. Concerning the central problem of the identification of the borderline between micro- and macrophenomena, it has to be stressed that, according to the model, it is not simply the number of particles by itself that determines whether the superposition principle holds or not. The suppression of the superposition is governed by the number of particles which are displaced by a distance larger than $1/\sqrt{\alpha}$.

11. RECENT DEVELOPMENTS AND CONCLUSIONS

The implications of QMSL which have just been discussed make it legitimate to state that the theory contains the essential elements allowing a realistic and objective description of macroscopic phenomena, without however, contradicting all known facts about the microscopic world. It is important to stress that the proposal represents a genuinely scientific alternative to quantum mechanics and not simply a reinterpretation of it. QMSL has, in fact, different implications and therefore the choice between the two theories can, in principle, be experimentally decided.

The version of the theory we have presented still suffers, however, from some drawbacks, so that a considerable elaboration is necessary before it can be considered as a satisfactory theoretical scheme. The first fact to be taken into account is that QMSL cannot, as it stands, describe systems containing identical constituents, since the hitting process, when it affects a specific particle, destroys the symmetry properties of the wave function, an unacceptable consequence. This problem has found a satisfactory solution, see Pearle (1989) and Ghirardi, Pearle and Rimini (1990), from a combination of the ideas of Ghirardi, Rimini and Weber (1986) with the previous important work of Pearle (1976). The new formulation of the theory (referred to as continuous spontaneous localization model, CSL), besides allowing us to deal with systems containing identical particles, has also a remarkable mathematical beauty. The wave function no longer suffers the sudden discontinuous changes associated with the localizations; the evolution of the state vector is continuous and corresponds to a stochastic Markov process taking place in the Hilbert space. One could picture the situation by saying that the state vector undergoes a sort of Brownian motion, see also the analysis by Diosi (1987) and (1988).

The second problem which the theory has to face is more serious and much more difficult to solve. J.S. Bell in (1987) – a paper devoted to an exposition of QMSL – has immediately remarked that the theory possesses a sort of first-order relativistic invariance, and has been led by this fact to assert that: the QMSL model *"takes away the ground of my fear that any exact formulation of quantum mechanics must conflict with fundamental Lorentz invariance"*. However, to recognize that there could be a compatibility of the theory with relativity does not mean, by itself, either that a relativistic generalization of it is possible, or that it can be easily found. Bell himself had repeatedly stressed that he considered the finding of such a generalization absolutely crucial and extremely difficult. Actually, in his last paper (1990) on this subject, after having once again expressed his dissatisfaction with standard quantum mechanics he stated that, in his opinion, the only available interesting alternatives to it are the the de Broglie–Bohm pilot wave theory and QMSL. He says: *"The big question, in my opinion, is which, if either of these two precise pictures can be developed in a Lorentz invariant way"*, As far as this problem is concerned we mention that interesting steps have recently been made towards its solution, see Pearle (1990) and Ghirardi, Grassi and Pearle (1990a,b). The just quoted papers have made it plausible that CSL admits relativistic generalizations and have also brought new insight about some problems of epistemological relevance, in particular nonlocality, the cause–effect relation, and causality-and-chance in physical processes. We do not consider it appropriate to enter here into the details of these approaches.

Let us summarize our discussion by listing the main features of QMSL which are relevant from the point of view of the analysis presented in Part I of this paper.

Quantum mechanics with spontaneous localizations is a theory which is based on a unique dynamic principle which is assumed to govern all physical processes. The dynamics is such that in the case of microscopic systems it leads to predictions which agree, to a degree of accuracy which is well beyond the (present) experimental possibilities, with those of standard quantum mechanics. For such systems, the superposition principle has (practically) unrestricted validity: as a consequence they are allowed *"to fully enjoy the cloudiness of waves"*. Nevertheless, the unique dynamical principle, when macroscopic objects are considered, leads to an almost immediate suppression of linear superpositions of macroscopically distinguishable states. In particular, the measurement problem finds a fully satisfactory solution within the model in the sense that wave packet reduction with fixed pointer positions can be shown to follow by the dynamical equations describing the system–apparatus interaction. Analogously, the classical behaviour of macroscopic objects, without the occurrence of the embarassing superpositions of macroscopically distinguishable states follows from the unique dynamical principle governing the evolution of any physical system. According to QMSL: *"Schrödinger's cat is not both dead and alive for more than a split second"*, (Bell 1987).

The actualization of the potentialities occurs, within the model, according to the precise rules defined by its mathematical structure. As a consequence, the borderline between quantum and classical is precisely defined and the vagueness, the shiftiness about this point which characterizes the standard approach, disappears.

Finally, the theory allows us to adopt a realist macro-objectivist position about natural phenomena. Nature itself compels macro-objects to choose between macroscopically different potentialities, without the need for any act of observation or of any observer. The role of the observer and of his consciousness is brought back to the one which, in our opinion, is appropriate for science: the only observer which is essential is the inanimate apparatus which amplifies microscopic events into macroscopic ones. Obviously the experimenter can choose whether to perform a measurement or not, and can adjust the apparatus. In this sense, there is indeed a dependence on the mental processes of the conscious observer. But, once the experimental set-up is in place and functioning, it is totally irrelevant whether the observer decides to use or not the 'mensura interrupta' technique considered in Bunge (1973).

To conclude, we stress that, due to the impossibility (using the present experimental techniques), of subjecting the theory to crucial tests, its acceptance as a satisfactory scheme for the description of natural phenomena is, to a large extent, a matter of taste and depends on the epistemological position one is keen to take. However, as remarked by H.P. Stapp (1989) (who does not share our position), there is at least one point about the dynamic reduction models which cannot be denied and which makes their investigation interesting: *"These proposals show that one can certainly erect a coherent quantum ontology that generally conforms to ordinary ideas at the macrosopic level"*.

Dipartimento di Fisica Teorica
Università degli Studi di Trieste

REFERENCES

Bell, J.S.: 1964, 'On the Einstein-Podolsky-Rosen paradox', *Physics* **1**, 195–200.

Bell, J.S.: 1986, 'Six possible worlds of quantum mechanics', in *Proceedings of the Nobel Symposium 65: Possible Worlds in Arts and Sciences* (ed. by S. Allen).

Bell, J.S.: 1987, 'Are there quantum jumps?', in *Schrödinger: Centenary Celebration of a Polymath*. (ed. by C.W. Kilmister) Cambridge University Press, Cambridge, pp. 201–212.

Bell, J.S.: 1990, 'Against measurement', in *Sixty-two Years of Uncertainty* (ed. by A.I. Miller) Plenum Press, New York and London, pp. 17-31.

Bunge, M.: 1973, *Philosophy of Physics*, D. Reidel Publ. Co., Dordrecht, Holland.

Busch, P.: 1991, 'Macroscopic quantum systems and the objectification problem', in *Symposium on the Foundations of Modern Physics 1990* (ed. by P.J. Lahti and P. Mittelstaedt), World Scientific, Singapore, pp. 62–76.

De Broglie, L.: 1960, *Nonlinear Wave Mechanics*, Elsevier, Amsterdam.

Diosi, L.: 1987, 'A universal master equation for the gravitational violation of quantum mechanics', *Physics Letters* **A120**, 377–381.

Diosi, L.: 1988, 'Continuous quantum measurements and Ito formalism', *Physics Letters* **A129**, 419–423.

Einstein, A., Podolsky, B., and Rosen, N.: 1935, 'Can quantum-mechanical description of physical reality be considered complete?', *Physical Review* **47**, 777–780.

Gisin, N.: 1984a, 'Quantum measurements and stochastic processes', *Physical Review Letters* **52**, 1657–1660.

Gisin, N.: 1984b, 'Gisin responds', *Physical Review Letters* **53**, 1776–1777.

Ghirardi, G.C., Grassi, R., and Pearle, P.: 1990, 'Relativistic dynamical reduction models: general framework and examples', *Foundations of Physics* **20**, 1271–1316.

Ghirardi, G.C., Grassi, R., and Pearle, P.: 1991, 'Relativistic dynamical reduction models and nonlocality', in *Symposium on the Foundations of Modern Physics 1990* (ed. by P.J. Lahti and P. Mittelstaedt), World Scientific, Singapore, pp. 109–123.

Ghirardi, G.C., Pearle, P., and Rimini, A.: 1990, 'Markov processes in Hilbert space and continuous spontaneous localization of systems of identical particles', *Physical Review* **A42**, 78–89.

Ghirardi, G.C., Rimini, A., and Weber, T.: 1986, 'Unified dynamics for microscopic and macroscopic systems', *Physical Review* **D34**, 470–491.

Ghirardi, G.C., Rimini, A., and Weber, T.: 1987, 'Disentanglement of quantum wave functions: answer to 'Comment on "Unified dynamics for microscopic and macroscopic systems"', *Physical Review* **D36**, 3287–3289.

Ghirardi, G.C., Rimini, A., and Weber, T.: 1988, 'The puzzling entanglement of Schrödinger's wave function', *Foundations of Physics* **18**, 1–27.

Ghirardi, G.C. and Rimini, A.: 1990, 'Old and new ideas in the theory of quantum measurement', in *Sixty-two Years of Uncertainty* (ed. by A.I. Miller), Plenum Press, New York and London, pp. 167–191.

Gottfried, K.: 1989, 'Does quantum mechanics describe the 'collapse' of the wave function?', preprint, Cornell University, Ithaca N.Y.

Maxwell, N.: 1988, 'Quantum propensiton theory: a testable resolution of the wave/particle dilemma', *British Journal of Philosophy of Science* **38**, 1–50.

Pearle, P.: 1976, 'Reduction of the state vector by a nonlinear Schrödinger equation', *Physical Review* **D13**, 857–868.

Pearle, P.: 1989, 'Combining stochastic dynamical reduction with spontaneous localization', *Physical Review* **A39**, 2277–2289.

Pearle, P.: 1990, 'Toward a relativistic theory of state vector reduction', in *Sixty-two Years of Uncertainty* (ed. by A.I. Miller), Plenum Press, New York and London, pp. 193–214.

Redhead, M.: 1987, *Incompleteness, Nonlocality and Realism: a Prolegomenon to the Philosophy of Quantum Mechanics*, Clarendon Press, Oxford.

Shimony, A.: 1989, 'Search for a worldview which can accomodate our knowledge of microphysics', in *Philosophical Consequences of Quantum Theory: Reflections on Bell's Theorem* (ed. by J.T. Cushing and E. McMullin), University of Notre Dame Press, Notre Dame, Indiana, pp. 25–37.

Squires, E.: 1986, *The Mystery of the Quantum Worlds*, Adam Hilger Ltd., Bristol and Boston.

Stapp, H.P.: 1989, 'Quantum nonlocality and the description of nature', in *Philosophical Consequences of Quantum Theory: Reflections on Bell's Theorem* (ed. by J.T. Cushing and E. McMullin), University of Notre Dame Press, Notre Dame, Indiana, pp. 154–174.

Wigner, E.P.: 1962, 'Remarks on the Mind–body question', in *The Scientist Speculates* (ed. by R. Good), Heinemann, London, pp. 284–302.

TULLIO WEBER

INDETERMINISM, NONSEPARABILITY AND THE EINSTEIN-PODOLSKY-ROSEN PARADOX

> *[Quantum Mechanics has] two powerful
> bodies of fact in its favour, and only one
> thing against it. First, in its favour are all
> the marvellous agreements that the theory
> has had with every experimental result to
> date. Second [...] it is a theory of aston-
> ishing and profound mathematical beauty.
> The one thing that can be said against it is
> that it makes absolutely no sense!*
>
> Roger Penrose

1. INTRODUCTION

In spite of the marvellous successes of quantum mechanics in explaining microscopic phenomena occurring at nuclear, atomic and molecular levels, this
theory still today presents embarrassing conceptual problems. Fundamentally,
they are indeterminism and nonseparability, and the related problems connected
with the quantum description of macroscopic objects and measurement theory.

In this paper we will discuss the first two of the above problematic aspects
of quantum mechanics.

2. INDETERMINISM

Just in order to clarify the terms of the problem, let us start by considering
classical mechanics.

Classical mechanics is a deterministic theory, perfectly compatible with the
existence of an external reality, i.e. of a reality independent of the scientist
who is studying it, in the sense that he can act on but not create it. Within
the framework of such a theory one can safely admit the existence of a world

199

G. Corsi et al. (eds), Bridging the Gap: Philosophy, Mathematics, and Physics, 199–209.
© *1993 Kluwer Academic Publishers.*

made up of objects (which we will call systems) having intrinsic and objective attributes, in terms of which it can be comprehended. These attributes (or properties) are in general variable in time and the laws of classical mechanics governing their evolution are deterministic in the sense that, given the values of all the parameters characterizing mathematically the properties of a system in its initial situation, in principle one is able to determine the state of affairs of the system at all subsequent times.

Probability comes into this description only in a subjective way, it it is only connected with our ignorance of details. As a consequence, it can change even if no interaction occurs with the system it refers to. To make this point quite clear let us take a very trivial example. Suppose we have two classical objects, having the intrinsic properties of being one of mass m_1 and the other of mass m_2. The two objects are put in two identical boxes, A and B. If the details of this operation are not known and one asks where mass m_1 is, the only answer which can be given is that there is a probability $1/2$ that the object having mass m_1 is in box A, and a probability $1/2$ that it is in box B. If we open box A and find inside the object with mass m_2, the probability of finding m_1 in B instantly becomes 1. This happens without any interaction with box B or object m_1, but there is nothing strange in this fact: it is only our knowledge of the situation which changes when we open box A.

Let us now consider quantum mechanics. A central role in quantum mechanics is played by the concept of *state* of a system, which can be defined saying that it corresponds to *the maximum possible specification of a system, which cannot be supplemented without contradiction*. The mathematical counterpart of this statement is the following: the dynamical observables of the system are represented by Hermitian operators in an abstract Hilbert space and their possible values are the eigenvalues of these operators; the state is represented by a vector in such a space and is an eigenvector of a set of commuting Hermitian operators (only observables corresponding to commuting operators can have definite values at the same time); the specification characterizing the state corresponds to the known eigenvalues of the set of commuting operators; it is maximal when the knowledge of the eigenvalue of any other operator commuting with those of the set gives no more information about the system, i.e. when such a new operator is a function of the preceding ones; in such a case the set is said to be complete and the specification contained in the state cannot be supplemented. Therefore, the state is represented by a vector which is an eigenvector of a complete set of commuting Hermitian operators.

The Schrödinger equation, governing the time evolution of the state, is deterministic in exactly the same way as the equations of classical mechanics, i.e. it allows us to determine the undisturbed state of the system at all times, given the state at $t = 0$.

However, as far as the specifications which the definition of state refers to are concerned, can we say that they correspond to objective properties of the related system?

Let us examine this point, taking the simplest example we can, i.e. considering the spin of a particle like the electron. As is well known, the spin is a sort of intrinsic vector and we can measure its components along any given direction. In the case of the electron, the experimental evidence tells us that, whatever be the direction along which we measure such a component, we can obtain only the values $+\hbar/2$ or $-\hbar/2$. Let us then consider an electron for which we know *for certain (i.e. with probability 1)* that a measurement of the component of spin along the z-direction (which we shall indicate by s_z) will give the result $+\hbar/2$, if performed. This is a specification about the spin of the electron, and quantum mechanics tells us that it is indeed a maximum specification, uniquely identifying the spin state of the electron. Moreover, in such a case *we are allowed to say, without any contradiction, that the electron can be considered to possess some property assuring that the result of a possible measurement of* s_z *will be* $+\hbar/2$. Then the answer to our starting question is in the affirmative sense.

However, as already remarked, quantum mechanics tells us that for such an electron we can measure also the x-component of the spin, s_x, and that, if we do it, we obtain for sure either $+\hbar/2$ or $-\hbar/2$. Suppose we make such an experiment and we obtain the result $+\hbar/2$. It is then natural to ask: if we take another electron with $s_z = +\hbar/2$ and we also perform on it the measurement of s_x, shall we again obtain the result $+\hbar/2$? The answer is negative: the result $-\hbar/2$ can also be obtained. If the measurement is repeated many times, we find that the statistical frequencies of the results approach, for increasing number of experiments, increasingly closely to $1/2$, both for $s_x = +\hbar/2$ and $s_x = -\hbar/2$. We can then ask another question: Is it possible, at least conceptually, to assume that those individual electrons for which the result $s_x = +\hbar/2 \, (-\hbar/2)$ was obtained, had, even before the measurement, some property guaranteeing that the result $s_x = +\hbar/2 \, (-\hbar/2)$ would have been obtained? The quantum scheme tells us that this is not allowed. The reason is the following. If a property exists, telling us that fifty per cent of the electrons of a beam with $s_z = +\hbar/2$ have $s_x = +\hbar/2$ and the remaining fifty per cent have $s_x = -\hbar/2$, then one of the rules of quantum mechanics identifies uniquely the state of these electrons: it corresponds to a statistical operator which in principle is definitely different from the state in which all electrons have $s_z = +\hbar/2$, as it gives different physical results for measurements of spin in arbitrary directions.

The conclusion we can draw is thus the following: if no property exists guaranteeing that for electrons with $s_z = +\hbar/2$, s_x will be $+\hbar/2$ or $-\hbar/2$, and in repeated measurements the two results are obtained with statistical frequencies approaching $1/2$, then we can only say that for one electron in the state corresponding to $s_z = +\hbar/2$ there is a probability $1/2$ of obtaining $s_x = +\hbar/2$ in a measurement, and a probability $1/2$ of obtaining $s_x = -\hbar/2$.

This is a very general situation in quantum mechanics, not related only to the spin: there is an objective indefiniteness and we are obliged to admit that quantum systems possess some *intrinsic probabilities*, which are not related to our ignorance and are attributes of each system, but which cannot be measured on one system, only. This fact has a very important consequence:

Systems prepared identically and with the maximum accuracy permitted by the theoretical scheme, when subjected to identical measurements, can in general give different results.

This is the essence of *quantum indeterminism*. It compels us to give up the principle that different effects arise from different causes.

There is also another important aspect of the previous considerations which must be stressed. We saw that neither the value $s_x = +\hbar/2$, nor $s_x = -\hbar/2$ can be attributed to an electron in the state $s_z = +\hbar/2$. However, in quantum mechanics there is a postulate which has been added to its rules to link physics and mathematical formalism: it is the so-called wave-packet reduction. In the case of spin, it assumes that after a measurement of s_x the state of the electron jumps from that corresponding to $s_z = +\hbar/2$ to the one in which s_x is equal to $+\hbar/2$ or $-\hbar/2$, according to the result which has been found. This means that the electron after the measurement possesses a physical property which was not present before. In general one can say that experiments not only display existing properties but actually create attributes not present before the measurement. As P. Jordan wrote (Jammer 1974): *"We ourselves produce the results of measurement"*.

The intrinsic indeterminism of quantum mechanics has been considered quite differently by scientists. Einstein (Jauch 1968) rejected it as a lack of completeness (in the sense which will be clarified in what follows) of the theory:

I am in fact firmly convinced that the essential statistical character of contemporary quantum theory is solely to be ascribed to the fact that this theory operates with an incomplete description of physical systems.

Pauli (Jauch 1968), on the contrary, was satisfied with the new scheme, considering the indeterminism as an acceptable feature of the theory:

The new epistemological situation underlying Quantum Mechanics is satisfactory, both from the standpoint of physics and from the broader standpoint of human knowledge in general.

Schrödinger (1957) was even more satisfied, considering this aspect of quantum mechanics as a promising one:

Once we have discarded our rooted predilection for absolute causality, we shall succeed in overcoming the difficulties.

Opinions on quantum indeterminism are divergent, obviously depending on the philosophical background. If one assumes a realistic point of view, being convinced of the existence of an external reality which cannot be created by the observer, then quantum indeterminism is very difficult to swallow. If, on the contrary, one accepts positivistic or instrumentalist philosophies, then all difficulties are dissolved. The words of Bohr (Jammer 1974) in this connection are illuminating:

There is no quantum world. There is only an abstract quantum mechanical description. It is wrong to think that the task of physics is to find how Nature is. Physics concerns what we can say about Nature.

3. NONSEPARABILITY

Nonseparability is a feature of quantum mechanics related to the description of composite systems. Let us consider two types of quantum systems, A and B, forming a unique system $(A + B)$. Sticking to the rules of quantum mechanics, it is very easy to prove the following

THEOREM (Ghirardi *et al.* 1977) *A necessary and sufficient condition in order that there exists a complete set* $\{O\}$ *of commuting observables of A (or B) having probability 1 of giving a definite result if measured, is that the composite system $(A + B)$ be described by a factorized state, i.e. by a state product of two states, one representing A and the other B:*

$$(3.1) \qquad |\Psi_{AB}> = |\Psi_A > \cdot |\Psi_B > .$$

In such a case $|\Psi_A >$ (or $|\Psi_B >$) is the eigenstate of $\{O\}$ pertaining to the definite result. Only if this situation is verified can one foresee with certainty the result of a measurement of the set $\{O\}$ on A (or B), even without making it.

This theorem has important consequences for the possibility of describing the component systems as individual entities. To clarify this point, let us consider two ensembles of systems A and B, described by states $|\Psi_A >$ and $|\Psi_B >$, respectively, and let the two ensembles interact for a finite time. Suppose that before the interaction the overall ensemble made by systems A and B is described by the product $|\Psi_A > \cdot |\Psi_B >$; in general, after the interaction the two types of systems will change state, and a transition will occur:

(3.2) $\qquad |\Psi_A > \cdot |\Psi_B > \Rightarrow |\varphi_A > \cdot |\varphi_B >$,

the change depending on the states before the interaction.

If the initial state of systems A is $|\Psi'_A >$, the same mechanism gives

(3.3) $\qquad |\Psi'_A > \cdot |\Psi_B > \Rightarrow |\varphi'_A > \cdot |\varphi'_B >$.

Since the superposition principle holds in quantum mechanics, suppose the state $|\Psi_A > + |\Psi'_A >$ to be a physically realizable state, eigenstate of some complete set $\{O\}$ of commuting observables (physical examples could be very easily found). Then, if we prepare systems A in this new state and let him interact with systems B still in state $|\Psi_B >$, because of the linearity of quantum mechanics the change will be

(3.4) $\qquad (|\Psi_A > + |\Psi'_A >) \cdot |\Psi_B > \Rightarrow |\varphi_A > \cdot |\varphi_B > + |\varphi'_A > \cdot |\varphi'_B >$.

The state at the right is no longer a factorized state. Owing to the theorem we have stated, this means that after the interaction no set of observables of A (or B) has probability 1 of having definite values, i.e. it is no longer possible to attribute to A (or B) definite, though possibly unknown, values for a complete set of observables.

Suppose now that after the interaction systems A and systems B separate, maintaining, however, the non-factorized overall state (physical examples of this behaviour are easy to find, as we shall see in what follows). Then we arrive at the surprising conclusion that *the now isolated systems $A(B)$ have completely lost their individuality*, i.e. that we can no longer attribute to them any state of their own; only the overall systems have well defined properties. This feature of non factorized states has been called by Schrödinger the *entanglement* of quantum wave functions and corresponds to what is usually indicated as nonseparability.

Just in order to illustrate the above formal discussion and to go a little bit further, let us take the simple example of two particles wiht spin 1/2, considering only the spin part of the wave functions. We denote by $|i, +z >$ and $|i, -z >$ the states of particle i having spin $+\hbar/2$ and $-\hbar/2$ along the z-axis, respectively. For the overall system of particles $(1 + 2)$ let us consider the state $|1, +z > \cdot |2, -z >$. This state represents a system $(1 + 2)$ in which particle 1 possesses the property of having spin $+\hbar/2$ along the z-axis, while particle 2 certainly has the spin in the opposite direction. Then we can say that in the situation described by the above state the two particles present definite individual properties.

Let us now consider the so-called singlet state

$$(3.5) \quad |S> = \frac{1}{\sqrt{2}} \{|1, +z > \cdot |2, -z > - |1, -z > \cdot |2, +z >\}.$$

In this case the total spin of the overall system $(1 + 2)$ has a definite value, which is 0, i.e. there are overall observables having precise values, while there is no spin observable of system 1 (or 2) for which we can foresee the result of a measurement with probability 1. This is the essence of nonseparability: we cannot attribute precise values to the spin of the single constituents, only system $(1 + 2)$ has a definite spin, and, if particles 1 and 2 move away, the spin part of the wave function always remaining of the form (3.5), the situation remains the same, even if 1 and 2 are mutually remote and no longer interacting.

However, there is a way to recover definite properties for the individual systems. Suppose we perform, for example, a measurement of the spin of particle 1 along the z-direction. In such a case the postulate of wave packet reduction tells us that, if the result is (e.g.) $+\hbar/2$, the state $|S >$ reduces to $|1, +z > \cdot |2, -z >$, which is a factorized state. Then measurement had produced what Schrödinger called a *damned quantum jump*, instantaneously giving individual properties to systems 1 and 2. Moreover, what is even more astonishing, *one recovers properties for particle 2 acting only on system 1!*

Schrödinger commented this feature, writing (Schrödinger 1935):

It is rather discomforting that the theory should allow a system to be steered or piloted into one or the other type of state at the experimenter's mercy in spite of his having no access to it.

The uneasiness one can prove at this point is twofold. On one side *the possibility of describing the world by dividing it into small parts and completely describing each part – a procedure which is often regarded as essential for the progress of science* (Sudbery 1986) – is no longer possible in quantum mechanics. On the other, the rules of quantum mechanics imply a *spooky action at a distance*, as Einstein characterized (Born 1971) the strong nonlocality of the theory.

4. THE EINSTEIN--PODOLSKY--ROSEN PARADOX

The impossibility of attributing objective properties to individual physical systems and the nonlocality of the theory, naturally induce us to imagine that quantum mechanics can conflict with the assumption of the existence of an external reality, and led Einstein, Podolsky and Rosen to formulate their famous paradox (Einstein *et al.* 1935) (known simply as EPR). We will discuss here a variant of the original argument, i.e. Bohm's spin version (Bohm 1951), which is now commonly used.

If one wants to judge the correctness of a physical theory by confronting it with general assumptions about reality, one must be very cautious, assuming only principles which are as objective and neutral as possible with respect to the philosophical positions one can adopt. Thus Einstein, Podolsky and Rosen avoid a 'comprehensive definition of reality' and assume a quite restrictive and 'reasonable' (as they say) criterion, which we will call the 'Principle of Reality': *If, without in any way disturbing a system, we can predict with certainty (i.e., with probability equal to unity) the value of a physical quantity, then there exists an element of physical reality corresponding to this physical quantity.*

Einstein, Podolsky and Rosen comment on this statement in the following way:

It seems to us that this criterion, while far from exhausting all possible ways of recognizing a physical reality, at least provides us with one such way .... Regarded not as a necessary but merely as a sufficient condition of reality, this criterion is in agreement with classical as well as quantum-mechanical ideas of reality.

As a matter of fact, we already accepted as non contradictory this assumption when we discussed indeterminism. There we called 'property' what here is more precisely called 'element of physical reality'. This criterion seems indeed to be the weakest form of realism one can adopt: if, for example, we know with certainty that a measurement of the spin along a given direction for an electron not interacting with us will give the result $+\hbar/2$, then it is very hard not to admit that the electron has some property corresponding to this definite result. If it is not allowed to speak of such a property, we cannot speak of an electron either, and the only way out is the position of Bohr already quoted: "There is no quantum world".

Another assumption, which has not been made explicit in the EPR paper, but which is tacitly admitted in their work, is the so called *Einstein's locality requirement*, which can be stated as follows (D'Espagnat 1989):

If a physical system remains, during a certain time, mechanically (including electromagnetically, etc.) isolated from other systems, then the evolution of its properties during this whole time interval cannot be influenced by operations carried out on other systems.

Formulated in this way, the requirement appears almost self-evident. However, it has a precise physical meaning, which is the following: an operation made at some point in space cannot modify the features of a system being at a distance d before the time d/c, which is the time necessary for the light to cover the distance d. Thus, the elements of physical reality in a system cannot be influenced during this time; in particular they cannot be influenced instantaneously.

The EPR paradox confronts these assumptions, which appear almost unavoidable for any normal-minded physicist, with the description given by quantum mechanics. The question, for EPR, is whether this theory provides a complete description of systems which it is applied to. Thus it is essential to clarify what we require in order that a theory can be considered 'complete'. In this case also EPR make an assumption which appears unavoidable:

Every element of the physical reality must have a counterpart in the physical theory.

In other words, a complete theory must contain the formal counterparts corresponding to the physical properties possessed by the systems it pretends to describe. In the opposite case, the theory is clearly incomplete and is not able to give a good description of reality, not even in the restricted sense of EPR.

The EPR analysis has the aim of studying the compatibility of the two general principles, of reality and of locality, with the completeness of quantum mechanics. To this end, let us consider a composite system containing two spin $1/2$ particles, described, as far as the spin part of the wave function is concerned, by the singlet state $|S>$ given by (3.5). Suppose that at time $t = t_0$ the composite system decays and the two component particles go in opposite directions, $|S>$ always being their overall spin state (experimental situations of this kind can be realized). Let us now consider a time $t_1 > t_0$, at which particles 1 and 2 can be considered completely separated and no longer interacting. Then at a time $t_2 > t_1$ a measurement of the spin of particle 2 on, say, the z-direction is performed. Suppose the result $-\hbar/2$ has been found. Then wave packet reduction tells us that the state immediately after the measurement is $|1, +z > |2, -z >$. This is a factorized state: using the principle of reality we can assert that from time t_2 on the element of physical reality 'particle 1 has spin up' has been gained. This has been accomplished without disturbing in any way system 1, which was isolated from 2 already from time $t_1 < t_2$. Thus the principle of locality tells us that from time t_1 to time t_2 the evolution of the properties of system 1 cannot have been influenced by what happened to particle 2. In fact, if at time t_2 the distance between 1 and 2 was r, some information, due to the measurement at time t_2, could have been reached system 1 at time $t_2 + r/c$ at best. But since we know that at time t_2 the element of physical reality 'particle 1 has spin up' is present, this means that it was present also at time t_1. However, at time t_1 the overall state was

$|S>$, which is a non factorized state implying that particle 1 has spin neither up nor down.

There is an evident contradiction, which can be interpreted in two ways. For Einstein and his coworkers quantum mechanics is simply not complete, it does not contain the formal (i.e. mathematical) counterpart of the element of physical reality 'particle 1 has spin up in the z-direction before time t_2'. In fact, state $|S>$, which is the state describing systems 1 and 2 before time t_2, does not contain this information. The situation is clear, there is no paradox, the EPR *Gedankenexperiment* simply provides a proof of the incompleteness of quantum mechanics.

If, on the contrary, one assumes that quantum mechanics is a fundamental theory, and as such complete, then it must be recognized that there is an unescapable contradiction with the two principles of reality and locality stated before. Since these two principles seem almost unavoidable, a paradox arises.

The paradox is a purely quantum one, being based essentially on nonseparability and on the wave packet reduction. To stress these non classical features it seems appropriate to quote the nice example devised by J. Bell (Bell 1987).

Bell has an eccentric friend, Professor Bertlmann, who always wears two socks of different colours. One cannot predict the colour he will have on a foot a given day, but one knows for certain that if one sock is pink, the other will not be pink. Thus, observing only the first sock, one immediately obtains information on the second one. This is a classical experiment in which there are two non-interacting systems (the socks) and in which information about one of the two systems can be gained making an experiment only on the other one, as in EPR. There is no paradox here, since the situation is totally different. In the case of Professor Bertlmann, it is only our knowledge which increases, nothing more happens: we are perfectly entitled to think that the second sock was not pink even before our observation. In the EPR case it is our experiment on particle 2 which 'creates' for particle 1 an element of physical reality which was not present before the observation.

This instantaneous action at a distance poses conceptual problems but cannot be used for practical purposes, e.g. to send signals with superluminal velocity, as has been proposed many times. The idea is more or less the following. Suppose we have, as in the EPR argument, a composite quantum system $U + V$, the two subsystems U and V lying in widely separated spatial regions and being non interacting. In these regions we put two apparatuses A and B devised to perform measurements on U and V, respectively. One might then suppose that it is possible to arrange the experimental set-up in such a way that there are physical effects on the apparatus B, due to its interaction with V, which are different according as the measurement of U at A has taken place or not. If this is the case, one can choose the times of the two measurements in such a way that the measurement at B shows that something happened in A before any signal could arrive at B from A.

This is indeed impossible. In fact, the rules of quantum mechanics imply (Ghirardi *et al*. 1980) that no physical difference at B can arise, whether U has interacted with A or not.

If not practical, the conceptual and epistemological importance of the EPR argument is enormous, and it still represents a real challenge for those physicists who trust in the existence of an objective external world, but cannot, obviously, reject quantum mechanics.

The embarrassment and the doubts of scientists, who have at their disposal and accept a scheme which appears perfect from the predictive point of view, but who, at the same time, cannot deny that its logical foundations are very weak, are well represented by the words of R.P. Feynmann (Feynmann 1982): "I cannot define the real problem, therefore I suspect there's no real problem, but I'm not sure there's no real problem".

Dipartimento di Fisica Teorica

REFERENCES

Bell, J.S.: 1987, *Speakable and Unspeakable in Quantum Mechanics; Collected Papers on Quantum Philosophy*, Cambridge University Press, Cambridge, p. 139.
Bohm, D.: 1951, *Quantum Theory*, Prentice-Hall, New York.
Born, M.: 1971, *The Born–Einstein letters* (ed. by M. Born), Macmillan, London, p. 158.
D'Espagnat, B.: 1989, *Conceptual Foundations of Quantum Mechanics*, Addison Wesley, Reading, Mass., p. 81.
Einstein, A., Podolsky, B., and Rosen, N.: 1935, *Phys. Rev.* 47, 777.
Feynmann, R.P.: 1982, *Int. J. Theor. Phys.* 21, 471.
Ghirardi, G.C., Rimini, A., Weber, T., and Omero, C.: 1977, *Il Nuovo Cimento B* 39, 130.
Ghirardi, G.C., Rimini, A., and Weber, T.: 1980, *Lettere al Nuovo Cimento* 27, 293.
Jammer, M.: 1974, *The Philosophy of Quantum Mechanics*, Wiley, New York, pp. 161, 204.
Jauch, J.M.: 1968, *Foundations of Quantum Mechanics*, Addison Wesley, Reading, Mass., p. 111.
Schrödinger, E.: 1935, *Proc. Cambridge Phil. Soc.* 31, 555.
Schrödinger, E.: 1957, 'What is a law of nature?', in *Science, Theory and Man*, Dover, New York, p. 133.
Sudbery, A.: 1986, *Quantum Mechanics and the Particles of Nature*, Cambridge University Press, Cambridge, p. 212.

GIANNI CASSINELLI AND PEKKA J. LAHTI

SIGMA-CONVEX STRUCTURES OF THE SETS OF STATES AND PROBABILITY MEASURES IN QUANTUM MECHANICS

1. INTRODUCTION. BASIC NOTATION AND TERMINOLOGY

In quantum mechanics each observable of a physical system defines a mapping from the set of states of the system to the set of probability measures on the value space of the observable. This mapping is σ-convex (in all conceivable ways) and it has some natural continuity properties. The paper aims to investigate the properties of this mapping in detail. In that the usual Hilbert space formulation of quantum mechanics will be applied. The notations and terminology will be fixed next. In that we follow rather closely the monographs of Beltrametti and Cassinelli (1981) and of Davies (1976). On the basic results of the Hilbert space operator theory we rely on Reed and Simon (1971).

The description of a physical system S will be based on a complex, separable Hilbert space $\mathcal{H}$ with the inner product $< \cdot | \cdot >$. Let $\mathcal{L}(\mathcal{H})$ denote the set of bounded linear operators on $\mathcal{H}$, and let $\mathcal{L}(\mathcal{H})^+$ be its subset of positive elements. Any *observable* of S will be represented as (and identified with) a normalized positive-operator-valued (POV) measure $E : \mathcal{F} \to \mathcal{L}(\mathcal{H})^+$ on a measurable space $(\Omega, \mathcal{F})$, the value space of E. The defining properties of E are: (i) $E(X) \geq E(\emptyset) = O$ for all $X \in \mathcal{F}$; (ii) if $(X_i) \subset \mathcal{F}$ is a disjoint sequence, then $E(\cup X_i) = \Sigma E(X_i)$ (the series converging in the weak operator topology of $\mathcal{L}(\mathcal{H})$) and (iii) $E(\Omega) = I$, the identity operator on $\mathcal{H}$. If, in addition, E is multiplicative, i.e. $E(X \cap Y) = E(X) E(Y)$ for all $X, Y \in \mathcal{F}$, then E is, in fact, a projection-operator-valued (PV) measure. According to the spectral theorem E is then associated with a unique self-adjoint operator A in $\mathcal{H}$ if and only if the value space $(\Omega, \mathcal{F})$ is a subspace of the real Borel space $(\mathfrak{R}, \mathcal{B}(\mathfrak{R}))$. In that case we write $E = E^A$. Let $\mathcal{T}(\mathcal{H})$ denote the set of trace class operators on $\mathcal{H}$. Any *state* of S is represented as (and identified with) a positive trace class operator T of trace (norm) one. The set of states is denoted as $\mathcal{T}(\mathcal{H})_1^+ = \{T \in \mathcal{T}(\mathcal{H}) : T \geq 0, \text{tr}[T] = 1\}$.

It is a basic result of quantum mechanics that any pair (E, T) of an observable E and a state T defines a probability measure on the value space of E:

211

G. Corsi et al. (eds), Bridging the Gap: Philosophy, Mathematics, and Physics, 211–224.
© *1993 Kluwer Academic Publishers.*

(1) $E_T : \mathcal{F} \to [0,1]$, $X \mapsto E_T(X) \doteq \mathrm{tr}[TE(X)]$.

The *Born interpretation* of this measure is that the number $E_T(X)$ is the probability that a measurement of the observable E on the system $\mathcal{S}$ in the state T leads to a result in the value set X. It is another basic result of quantum mechanics that the above representations of observables and states are the most general ones compatible with the probability structure of the theory.

For a fixed observable E, the induced probability measures E_T, $T \in \mathcal{T}(\mathcal{H})_1^+$, are Kolmogorovian, i.e., $E_T \in \mathcal{M}(\Omega)_1^+$ – the set of probability measures on $(\Omega, \mathcal{F})$. In spite of this partial compatibility of the probability structure of quantum mechanics with the classical probability theory, the family of the induced probability measures $\{E_T : T \in \mathcal{T}(\mathcal{H})_1^+\} \subset \mathcal{M}(\Omega)_1^+$ shows some nonclassical features. It is the intention of this paper to investigate these features. In fact, we shall study the properties of the mapping $T \mapsto E_T$ from the set of states $\mathcal{T}(\mathcal{H})_1^+$ of the system $\mathcal{S}$ to the set of probability measures $\mathcal{M}(\Omega)_1^+$ on the value space $(\Omega, \mathcal{F})$ of the observable E. In that we shall proceed as follows: in Section 2 and 3 the σ-convex structures of the sets $\mathcal{T}(\mathcal{H})_1^+$ and $\mathcal{M}(\Omega)_1^+$ will be studied; Section 4 contains a study of the continuity properties of the mapping $T \mapsto E_T$; Section 5 gives an important convergence theorem for σ-convex combinations of states, whereas Section 6 studies the range of this mapping, in general and in particular.

2. σ-CONVEX STRUCTURES OF $\mathcal{T}(\mathcal{H})_1^+$

2.1. Convexity

The set of states $\mathcal{T}(\mathcal{H})_1^+ = \{T \in \mathcal{T}(\mathcal{H}) : T \geq O, \mathrm{tr}[T] = 1\}$ is a *convex* subset of the vector space $\mathcal{T}(\mathcal{H})$ of trace class operators on $\mathcal{H}$. Indeed, if T_1 and T_2 are states, then $wT_1 + (1 - w)T_2$ is also a state for any 'weight factor' $0 \leq w \leq 1$. The extremal elements of this convex set are known to be its idempotent elements, or equivalently, the one-dimensional projection operators on $\mathcal{H}$:

(2) $\mathrm{Ex}(\mathcal{T}(\mathcal{H})_1^+) \doteq \{T \in \mathcal{T}(\mathcal{H})_1^+ : \text{if}$

$\qquad T = wT_1 + (1 - w)T_2 , \ 0 < w < 1 ,$

$\qquad \text{then } T = T_1 = T_2\}$

$\qquad = \{T \in \mathcal{H})_1^+ : T = T^2\}$

$\qquad = \{P[\varphi] : \varphi \in \mathcal{H}, \ <\varphi|\varphi> = 1\} .$

Here $P[\varphi]$ is the projection operator defined through $P[\varphi]\psi = <\varphi|\psi>$ $\varphi, \psi \in \mathcal{H}$. Any convex set is closed under finite convex combinations of its elements: if $(T_1, \ldots, T_n) \subset T(\mathcal{H})_1^+$, $(w_1, \ldots, w_n) \subset [0, 1]$, $\sum_{i=1}^{n} w_i = 1$, then $\sum_{i=1}^{n} w_i T_i \in T(\mathcal{H})_1^+$. In the case of $T(\mathcal{H})_1^+$ there are several natural ways (topologies) to extend this notion to sequences of states, i.e., to define the *σ-convex structures* on $T(\mathcal{H})_1^+$. We shall now study some of them.

2.2. Topologies

The set $\mathcal{L}(\mathcal{H})$ is a (complex) Banach space with respect to the operator norm $\|A\| \doteq \sup\{< A\varphi|A\varphi >^{1/2} : \varphi \in \mathcal{H}, < \varphi|\varphi >= 1\}$. In addition to this (uniform) *operator topology*, the $\| \cdot \|$-topology, the set $\mathcal{L}(\mathcal{H})$ is equipped also, e.g., with the *weak operator topology*, i.e., the topology generated by the seminorms $A \mapsto |\text{tr}[AF]|$, $F \in \mathcal{F}(\mathcal{H})$ – the finite rank operators on $\mathcal{H}$. (See, e.g., Brezis 1985, or Reed and Simon 1971).

   The set $T(\mathcal{H})$ is a vector subspace, in fact a *-ideal of $\mathcal{L}(\mathcal{H})$, and $\text{tr}[\cdot]$ is a linear functional on $T(\mathcal{H})$. The formula $\|T\|_1 \doteq \text{tr}[|T|] (= \text{tr}[(T^*T)^{1/2}])$ defines a norm, *the trace norm*, on $T(\mathcal{H})$ with respect to which $T(\mathcal{H})$ is a (complex) Banach space. Since $\|T\| \leq \|T\|_1$ for any $T \in T(\mathcal{H})$, the $\| \cdot \|_1$-*topology* is finer than the $\| \cdot \|$-*topology*, i.e., the identity mapping

(3) $\iota : (T(\mathcal{H}), \| \cdot \|_1) \mapsto (T(\mathcal{H}), \| \cdot \|)$

is continuous. We recall that since $T(\mathcal{H})$ is not $\| \cdot \|$-closed in $\mathcal{L}(\mathcal{H})$ (its closure being the set of compact operators), the operator norm does not define a Banach norm on $T(\mathcal{H})$. The restriction of the weak operator topology to $T(\mathcal{H})$ defines the *weak operator topology* on $T(\mathcal{H})$. It may, again, be defined as the topology generated by the seminorms $T \mapsto |\text{tr}[TF]|$, $F \in \mathcal{F}(\mathcal{H})$, and it is coarser than any of the two norm topologies on $T(\mathcal{H})$. In order to write down the *weak topology* on $T(\mathcal{H})$, i.e., the topology of $T(\mathcal{H})$ generated by all of its (*sup*-norm bounded) linear functionals, we recall that $(\mathcal{L}(\mathcal{H}), \| \cdot \|)$ can be identified with the dual of the Banach space $(T(\mathcal{H}), \| \cdot \|_1)$ (via the mapping $A \mapsto f_A$, $A \in \mathcal{L}(\mathcal{H})$, with $f_A(T) = \text{tr}[TA]$ for all $T \in T(\mathcal{H})$). The weak topology of $T(\mathcal{H})$ is then exactly the one generated by the seminorms $T \mapsto |\text{tr}[TA]|$, $A \in \mathcal{L}(\mathcal{H})$. (Since $\mathcal{F}(\mathcal{H}) \subset \mathcal{L}(\mathcal{H})$, the weak topology is finer than the weak operator topology on $T(\mathcal{H})$). As always, the weak topology is coarser than the (relevant) norm topology. Hence, the identity mappings

(4) $\iota : (T(\mathcal{H}), \| \cdot \|_1) \mapsto (T(\mathcal{H}), \mathcal{L}(\mathcal{H}))$,

and

(5) $\iota : (T(\mathcal{H}), \mathcal{L}(\mathcal{H})) \rightarrow (T(\mathcal{H}), \mathcal{F}(\mathcal{H}))$

are continuous. We emphasize that $\operatorname{tr}[\cdot] : T(\mathcal{H}) \to \mathbf{C}$ is continuous with respect to the trace norm topology and the weak topology, but it is not continuous with respect to the operator topology nor with respect to the weak operator topology.

2.3. Sigma-Convexity

We are now ready to investigate the σ-convex structures of $T(\mathcal{H})_1^+$ with respect to the four 'natural' topologies of $T(\mathcal{H})$ described above. Let $(T_i)_{i=1}^\infty \subset T(\mathcal{H})$ be a sequence of states, and let $(w_i)_{i=1}^\infty \subset [0,1]$ be a sequence of weights. It turns out that the series $\sum_{i=1}^\infty w_i T_i$ converges in each of the above topologies, and that the sum is always the same. At first glance, this may appear strange, since it was just pointed out that the trace is not continuous with respect to the operator topology, and, *a fortiori*, with respect to the weak operator topology.

Let $(T_i)_{i=1}^\infty \subset T(\mathcal{H})_1^+$ and $(w_i)_{i=1}^\infty \subset [0,1]$, with $\sum_{i=1}^\infty w_i = 1$, be given. Denote $s_n = \sum_{i=1}^n w_i T_i$. Then for any $n, m = 1, 2, \ldots, n \geq m$, $\|s_n - s_m\|_1 = \|\sum_{i=m}^n w_i T_i\|_1 \leq \sum_{i=m}^n w_i$, which shows that (s_n) is a Cauchy-sequence in $(T(\mathcal{H}), \|\cdot\|_1)$. Let $T = \lim s_n = \sum_{i=1}^\infty w_i T_i$ (with respect to the trace norm). Since trace is $\|\cdot\|_1$-continuous we get $\operatorname{tr}[T] = \operatorname{tr}[\lim s_n] = \lim \operatorname{tr}[s_n] = \sum_{i=1}^\infty w_i = 1$. But, due to (4) and (5), we also have $\lim <\varphi|s_n\varphi> = <\varphi|T\varphi>$ for each $\varphi \in \mathcal{H}$ so that $T \geq O$ (since $s_n \geq O$ for all n). Hence $T = \lim s_n = \sum_{i=1}^\infty w_i T_i \in T(\mathcal{H})_1^+$, i.e., $T(\mathcal{H})_1^+$ is *σ-convex* with respect to the trace norm topology.

The above rather standard proof of the σ-convexity of $T(\mathcal{H})_1^+$ (with respect to the $\|\cdot\|_1$-topology) already shows to what extent this result can be generalized (with respect to coarser topologies). Let τ be a topology on $T(\mathcal{H})$ which is coarser than its trace norm topology, so that the identity mapping $\iota : (T(\mathcal{H}), \|\cdot\|_1) \to (T(\mathcal{H}), \tau)$ is again continuous. Let (S_i) be any convergent sequence of $(T(\mathcal{H}), \|\cdot\|_1)$ and let $S = \|\cdot\|_1 - \lim S_i$. Then $(\iota(S_i)) = (S_i)$ is a convergent sequence in $(T(\mathcal{H}), \tau)$, too, and $S = \iota(S) = \iota(\|\cdot\|_1 - \lim S_i) = \tau - \lim \iota(S_i) = \tau - \lim S_i$. In particular, with the above notations,

$$(6) \qquad T = \|\cdot\|_1 - \lim s_n = \sum_{i=1}^\infty w_i T_i$$

$$= \tau - \lim s_n = \sum_{i=1}^\infty w_i T_i .$$

Thus, if τ is Hausdorff, then T is the unique τ-limit of (s_n), and $T \in T(\mathcal{H})_1^+$, since also $T = \| \cdot \|_1 - \lim s_n$. Clearly, the above-referred topologies on $T(\mathcal{H})$ are Hausdorff.

2.4. Krein-Milman Property

The set of states $T(\mathcal{H})_1^+$ has a rich subset of extreme elements $\mathrm{Ex}(T(\mathcal{H})_1^+)$, the pure states. Let $\mathrm{conv}(\mathrm{Ex}(T(\mathcal{H})_1^+))$ and $\sigma_\tau - \mathrm{conv}(\mathrm{Ex}(T(\mathcal{H})_1^+))$ denote the convex and σ_τ-convex hulls of $\mathrm{Ex}(T(\mathcal{H})_1^+)$. Then

(7) $\qquad \sigma_\tau - \mathrm{conv}(\mathrm{Ex}(T(\mathcal{H})_1^+)) = \sigma_{\|\cdot\|_1} - \mathrm{conv}(\mathrm{Ex}(T(\mathcal{H})_1^+))$

for any Hausdorff topology τ which is coarser than the trace norm topology. Furthermore,

(8) $\qquad \sigma_\tau - \mathrm{conv}(\mathrm{Ex}(T(\mathcal{H})_1^+)) = \sigma_{\|\cdot\|_1} - \mathrm{conv}(\mathrm{Ex}(T(\mathcal{H})_1^+)) \subset T(\mathcal{H})_1^+$,

and

(9) $\qquad \mathrm{conv}(\mathrm{Ex}(T(\mathcal{H})_1^+)) \subset \sigma_\tau$

$\qquad\qquad -\mathrm{conv}(\mathrm{Ex}(T(\mathcal{H}_1^+)) \subset \overline{\mathrm{conv}(\mathrm{Ex}(T(\mathcal{H})_1^+))}^\tau \subset \overline{T(\mathcal{H})_1^+}^\tau$.

If the last set-inclusion in (9) is, in fact, an equality, we say – following Dieudonne (1970) – that the *Krein–Milman property* holds for τ. We discuss next to what extent these set-inclusions are proper and to what extent the closure-operations are necessary.

We recall first that the set of states $T(\mathcal{H})_1^+$ is $\| \cdot \|_1$-closed in $T(\mathcal{H})$. Indeed, if $(T_i) \subset T(\mathcal{H})_1^+$ is such that $\| \cdot \|_1 - \lim T_i = T$, then $T \in T(\mathcal{H})_1^+$, since trace is continuous in the trace norm (so that $\mathrm{tr}[\lim T_i] = \lim \mathrm{tr}[T_i] = 1$) and the trace norm topology is finer than the weak operator topology (so that $\lim < \varphi | T_i \varphi > = < \varphi | T \varphi >$ for each $\varphi \in \mathcal{H}$ showing that $T \geq O$). As a norm-closed convex subset of $T(\mathcal{H})$ the set of states $T(\mathcal{H})_1^+$ is also weak-closed in $T(\mathcal{H})$ (Brezis, 1985). Thus, briefly,

(10) $\qquad \overline{T(\mathcal{H})_1^+}^{\|\cdot\|_1} = T(\mathcal{H})_1^+$,

and

(11) $\qquad \overline{T(\mathcal{H})_1^+}^w = T(\mathcal{H})_1^+$.

In order to appreciate the nontriviality of the latter result one should recall that $\overline{T(\mathcal{H})_{=1}}^w = T(\mathcal{H})_{\leq 1}$ (since $T(\mathcal{H})$ is infinite dimensional). (See, e.g., Brezis 1985, or Reed and Simon 1971). Since $T(\mathcal{H})_1^+$ is not compact in the norm, weak, and weak operator topologies, one cannot apply the Krein–Milman theorem (Royden, 1968) to show that the Krein–Milman property holds for the topologies in question. In spite of that, the spectral theorem gives that property for $T(\mathcal{H})_1^+$, and, in fact, somewhat more.

Consider a state $T \in \mathcal{T}(\mathcal{H})_1^+$, and let $T = \sum_i \lambda_i P_i$ be its spectral decomposition (as a self-adjoint compact operator). Since any spectral projection P_i is finite-dimensional, we may also write $T = \sum_i \lambda_i P[\varphi_i]$, where (φ_i) is a (finite or infinite) orthonormal sequence (and in the case of infinite sequence the series converges in the operator norm). Each λ_i is an eigenvalue of T, and it appears in the series $\dim(\ker(T - \lambda_i I)$-times. Since $T \geq 0$, each $\lambda_i \geq 0$. Moreover, one can show that $\|T\|_1 = \sum \lambda_i$, which then gives that $T \in \sigma_{\|\cdot\|} - \mathrm{conv}(\mathrm{Ex}(\mathcal{T}(\mathcal{H})_1^+))$. In fact, this also shows that the series $T = \sum \lambda_i P[\varphi_i]$ converges in the trace norm, too. Thus, we get:

$$(12) \qquad \sigma_\tau - \mathrm{conv}(\mathrm{Ex}(\mathcal{T}(\mathcal{H})_1^+)) = \mathcal{T}(\mathcal{H})_1^+$$

for any Hausdorff topology τ coarser than (or equal to) the trace norm topology. Moreover, the Krein–Milman property holds for the trace norm and weak topologies:

$$(13) \qquad \sigma - \mathrm{conv}(\mathrm{Ex}(\mathcal{T}(\mathcal{H})_1^+)) = \overline{\mathrm{conv}(\mathrm{Ex}(\mathcal{T}(\mathcal{H})_1^+))}^{\|\cdot\|_1, w} = \mathcal{T}(\mathcal{H})_1^+ .$$

2.5. Nonunique Decomposability

The result (12) shows that any state $T \in \mathcal{T}(\mathcal{H})_1^+$ can be expressed as a σ-convex combination of some pure states $P[\varphi] \in \mathrm{Ex}(\mathcal{T}(\mathcal{H})_1^+)$. The question then arises whether such a decomposition is unique. As it is well-known, the answer to this question is negative. Indeed, for any $T \in \mathcal{T}(\mathcal{H})_1^+$,

$$(14) \qquad T = wP[\varphi] + (1-w)T'$$

for some $0 < w < 1$ and $T' \in \mathcal{T}(\mathcal{H})_1^+$ if and only if $\varphi \in \mathrm{cl}(\mathcal{R}(T^{1/2}))$ (Hadjisavvas 1981). Thus, if T is not pure, then T admits continuously many decompositions into pure states. A consequence of this *nonunique decomposability* of mixed states in quantum mechanics is that quantum probabilities do not, in general, admit an epistemic ignorance interpretation.

2.6. Finite-Dimensional Case

In the finite-dimensional Hilbert space $\mathcal{H}$ the sets $\mathcal{L}(\mathcal{H})$ and $\mathcal{T}(\mathcal{H})$ coincide. Also the different topologies (operator, weak operator, trace norm, and weak) are the same. Moreover, the convex hull of the set of pure states $\mathrm{Ex}(\mathcal{T}(\mathcal{H})_1^+)$ exhaust the set of states $\mathcal{T}(\mathcal{H})$. The nonunique decomposability of mixed states, of course, remains.

3. σ-CONVEX STRUCTURES OF $\mathcal{M}(\Omega)_1^+$

3.1. General

In the definition of an observable E its value space was taken to be a measurable space $(\Omega, \mathcal{F})$ with no further specification. From nowon, we assume that Ω is a compact separable metrisable space and $\mathcal{F} = \mathcal{B}(\Omega)$ is the natural Borel σ-algebra of subsets of Ω. The assumption on the compactness of Ω implies no loss in generality as concerns the problems analysed here.

Let $\mathcal{M}(\Omega)$ denote the vector space of complex measures on Ω, i.e., σ-additive mappings $\mu : \mathcal{B}(\Omega) \to \mathbf{C}$. Let $\mathcal{M}(\Omega)_1^+$ denote the set of *probability measures* on Ω, i.e., the positive ($\mu(X) \geq 0$ for each $X \in \mathcal{B}(\Omega)$) normalized ($\mu(\Omega) = 1$) elements of $\mathcal{M}(\Omega)$. Clearly, each E_T, $T \in \mathcal{T}(\mathcal{H})_1^+$, is such. It is also evident that the set $\mathcal{M}(\Omega)_1^+$ is *convex*. To determine its σ-convex structures we need to consider, again, the topologies on $\mathcal{M}(\Omega)$.

3.2. Topologies

Let $\mu \in \mathcal{M}(\Omega)$. Its *total variation* $|\mu|$ is uniquely characterized as the smallest of all positive measures $\nu \in \mathcal{M}(\Omega)^+$ for which $|\mu(X)| \leq \nu(X)$ for all $X \in \mathcal{B}(\Omega)$ (Rudin, 1966). The formula $\|\mu\|_1 = |\mu|(\Omega)$ then gives rise to a norm on $\mathcal{M}(\Omega)$, the *total variation norm*, with respect to which $(\Omega, \|\cdot\|_1)$ is a (complex) Banach space. We shall also consider $\mathcal{M}(\Omega)$ endowed with the *vague* topology. According to the Riesz-Markov theorem $(\mathcal{M}(\Omega), \|\cdot\|_1)$ can be identified with the dual of the Banach space $(\mathcal{C}(\Omega), \|\cdot\|_\infty)$ of continuous functions with the *sup*-norm. (The formula $\mu(f) = \int_\Omega f_{d\mu}$, $\mu \in \mathcal{M}(\Omega)$, $f \in \mathcal{C}(\Omega)$, gives this identification). The vague topology on $\mathcal{M}(\Omega)$ is the weak*-topology on $\mathcal{C}(\Omega)^* \simeq \mathcal{M}(\Omega)$, i.e., the topology generated by the seminorms $\mu \mapsto |\mu(f)|$ $f \in \mathcal{C}(\Omega)$ (Brezis 1985, Reed and Simon 1971). Equivalently, it can be given by the seminorms $\mu \mapsto |\mu(X)|$, $X \in \mathcal{B}(\Omega)$. The vague topology is coarser than the norm topology, i.e., the identity mapping

$$(15) \qquad \iota : (\mathcal{M}(\Omega), \|\cdot\|_1) \mapsto (\mathcal{M}(\Omega), vague)$$

is continuous.

3.3. Sigma-Convexity

As it was the case with $\mathcal{T}(\mathcal{H})_1^+$, the set $\mathcal{M}(\Omega)_1^+$ is σ-convex with respect to the $\|\cdot\|_1$-topology as well as with any other coarser Hausdorff topology on $\mathcal{M}(\Omega)$. Indeed, let $(\mu_i) \subset \mathcal{M}(\Omega)_1^+$ and $(w_i) \subset [0, 1]$, with $\sum_{i=1}^{\infty} w_i = 1$, be given, and define $s_n = \sum_{i=1}^{n} w_i \mu_i$ for each $n = 1, 2, \ldots$. Then $\|s_n - s_m\|_1 = \|\sum_{i=m}^{n} w_i \mu_i\|_1 \leq \sum_{m}^{n} w_i$ for each $m \leq n$ showing that (s_n) is a Cauchy-sequence in $(\mathcal{M}(\Omega), \|\cdot\|_1)$. Let $\mu = \lim s_n$ be its $\|\cdot\|_1$-limit. But due to (15) we also have that $\mu = vague - \lim s_n$, i.e., $\mu(X) = \lim s_n(X)$ for each $X \in \mathcal{B}(\Omega)$. This shows that μ is positive and normalized, i.e., $\mu \in \mathcal{M}(\Omega)_1^+$. Due to the continuity of the identity mapping $\iota : (\mathcal{M}(\Omega), \|\cdot\|_1) \to (\mathcal{M}(\Omega), vague)$, the σ-convexity of $\mathcal{M}(\Omega)_1^+$ with respect to the vague topology is guaranteed, as well. To summarise,

$$(16) \qquad \mu = \|\cdot\|_1 - \lim s_n = \sum_{i=1}^{\infty} w_i \mu_i$$

$$= vague - \lim s_n = \sum_{i=1}^{\infty} w_i \mu_i .$$

3.4. Krein–Milman Property

The set of probability measures $\mathcal{M}(\Omega)_1^+$ has, again, a rich subset of extremal points $\mathrm{Ex}(\mathcal{M}(\Omega)_1^+)$, the set of point measures μ_ω, $\omega \in \Omega$, on Ω. Let $\mathrm{conv}(\mathrm{Ex}(\mathcal{M}(\Omega)_1^+))$ and $\sigma_\tau - \mathrm{conv}(\mathrm{Ex}(\mathcal{M}(\Omega)_1^+))$ denote the convex and σ_τ-convex hulls of $\mathrm{Ex}(\mathcal{M}(\Omega)_1^+)$ (with τ denoting either the norm or vague topology on $\mathcal{M}(\Omega)$). Then

$$(17) \qquad \mathrm{conv}(\mathrm{Ex}(\mathcal{M}(\Omega)_1^+)) \subset \sigma_{vague} - \mathrm{conv}(\mathrm{Ex}(\mathcal{M}(\Omega)_1^+))$$

$$= \sigma_{\|\cdot\|_1} - \mathrm{conv}(\mathrm{Ex}(\mathcal{M}(\Omega)_1^+)) \subset \mathcal{M}(\Omega)_1^+ .$$

Clearly, the set $\mathcal{M}(\Omega)_1^+$ is, again, norm-closed in $\mathcal{M}(\Omega)$, so that (as a convex set) it is weak-closed in $\mathcal{M}(\Omega)$. But, in fact, it is also vaguely closed. To see this, consider a net $\{\mu_\alpha\} \subset \mathcal{M}(\Omega)_1^+$ converging (vaguely) to $\mu \in \mathcal{M}(\Omega)$. But, for each $X \in \mathcal{B}(\Omega)$, the mapping $\mu \to \mu(X)$ is a continuous linear form on $\mathcal{M}(\Omega)$ (with the vague topology), so that $\mu(X) = (\lim_\alpha \mu_\alpha)(X) = \lim_\alpha \mu_\alpha(X) \geq 0$ and, thus, $\mu(\Omega) = \|\mu\|_1 = \lim_\alpha \mu_\alpha(\Omega) = 1$ for all $X \in \mathcal{B}(\Omega)$, i.e. $\mu \in \mathcal{M}(\Omega)_1^+$. According to the Banach–Alaoglu theorem, the unit ball of $\mathcal{M}(\Omega)$ is vague-compact (Brezis 1985). As a vague-closed subset of a vague-compact set, the set $\mathcal{M}(\Omega)_1^+$ is also vague-compact. The Krein–Milman theorem then says that the set of extremal points of $\mathcal{M}(\Omega)_1^+$ is vague-dense in $\mathcal{M}(\Omega)_1^+$, i.e.,

$$(18) \qquad \overline{\mathrm{conv}\,\mathrm{Ex}(\mathcal{M}(\Omega)_1^+)}^{vague} = \mathcal{M}(\Omega)_1^+ .$$

We note further that the σ-convex hull of $\mathrm{Ex}(\mathcal{M}(\Omega)_1^+)$ is always norm-closed but, in general, not vague-closed. Thus, in general, $\sigma - \mathrm{conv}(\mathrm{Ex}(\mathcal{M}(\Omega)_1^+))$ is a proper subset of $\mathcal{M}(\Omega)_1^+$.

As an example, consider the value space $\Omega = [0,1]$. Since no diffuse measure on Ω is a σ-convex combination of its point measures, the set-inclusion $\sigma - \text{conv}(\text{Ex}(\mathcal{M}(\Omega)_1^+)) \subset \mathcal{M}(\Omega)_1^+$ is now proper. As another example, consider a discrete value space $\Omega = \{\omega_1, \omega_2, \ldots\}$. Then, for any $\mu \in \mathcal{M}(\Omega)_1^+$, $\mu(X) = \sum_{\omega_i \in X} \mu(\{\omega_i\}) \mu_{\omega_i}$ for each $X \in \mathcal{B}(\Omega)$, so that now the σ-convex hulls of the set of point measures exhaust the set of probability measures. Finally, if the set Ω is finite, then (and only then) the vague-topology is the same than the norm-topology on $\mathcal{M}(\Omega)$. Clearly, then $\text{conv}(\text{Ex}(\mathcal{M}(\Omega)_1^+)) = \mathcal{M}(\Omega)_1^+$.

3.5. The Unique Decomposability

If the value space Ω is countable, then any $\mu \in \mathcal{M}(\Omega)_1^+$ can be expressed in a unique way as a σ-convex combination of the point measures μ_ω, $\omega \in \text{supp}(\mu)$. In general, the relation (18) guarantees that each μ can be expressed as a vague limit of a net (μ_α) of point measures. The (unique) expression $\mu = \sum w_i \mu_\omega$, which is valid for all $\mu \in \sigma - \text{conv}(\text{Ex}(\mathcal{M}(\Omega)_1^+))$, has a generalization as an integral representation, the so-called barycentric representation, for each $\mu \in \mathcal{M}(\Omega)_1^+$. The unique decomposability of the probability measures into the pointmeasures is, of course, a characteristics of simplexes. However, it seems that the ignorance interpretation can be carried out in a meaningful way only for the elements of $\sigma - \text{conv}(\text{Ex}(\mathcal{M}(\Omega)_1^+))$.

4. CONTINUITY OF THE MAPPING $T \mapsto E_T$

The mapping $T(\mathcal{H})_1^+ \to \mathcal{M}(\Omega)_1^+$, $T \mapsto E_T$, induced by the observable $E : \mathcal{B}(\Omega) \to \mathcal{L}(\mathcal{H})^+$ is convex:

$$(19) \qquad E_{wT_1+(1-w)T_2} = wE_{T_1} + (1-w)E_{T_2}$$

for each $T_1, T_2 \in T(\mathcal{H})_1^+$, and for all $w \in [0,1]$. But the set of states $T(\mathcal{H})_1^+$ as well as the set of probability measures $\mathcal{M}(\Omega)_1^+$ are σ-convex. Hence it is natural to ask whether the mapping $T \mapsto E_T$ is σ-convex, too.

Let $T_s(\mathcal{H})$ and $\mathcal{M}_\Re(\Omega)$ be the set of self-adjoint trace class operators and the set of finite real measures on $\mathcal{H}$ and Ω, respectively. We recall that $T(\mathcal{H}) = T_s(\mathcal{H}) \oplus \iota T(\mathcal{H})$ and $\mathcal{M}(\Omega) = \mathcal{M}_\Re(\Omega) \oplus \iota \mathcal{M}_\Re(\Omega)$. Consider the mapping

$$(20) \qquad V : T_s(\mathcal{H}) \to \mathcal{M}_\Re(\Omega), \, T \mapsto V(T)$$

defined through the formula

(21) $V(T)(X) = \text{tr}[TE(X)]$, $X \in \mathcal{B}(\Omega)$.

Clearly, $V(T) \in \mathcal{M}_\Re(\Omega)$ for all $T \in \mathcal{T}_s(\mathcal{H})$, and V is linear.

Consider now $\mathcal{T}_s(\mathcal{H})$ equipped with its weak topology, i.e., the one generated by the seminorms $T \mapsto |\text{tr}[TA]|$, $A \in \mathcal{L}_s(\mathcal{H})$, and $\mathcal{M}_\Re(\Omega)$ equipped with its total variation norm $\| \cdot \|_1$. For any $T \in \mathcal{T}_s(\mathcal{H})$ let P_+ and P_- denote its spectral projections on the positive and negative parts of its spectrum so that $T = T_+ - T_- = P_+T - P_-T$. Let $V(T)_+$ and $V(T)_-$ denote the positive and negative parts of the measure $V(T)$. They are the smallest of the positive measures that allow $V(T)$ to be decomposed as $V(T) = \mu_1 - \mu_2$. Hence $V(T)_\pm \leq V(T_\pm)$ and

(22) $\|V(T)\|_1 = |V(T)|(\Omega)|$

$$= V(T)_+(\Omega) + V(T)_-(\Omega)$$

$$\leq V(T_+)(\Omega) + V(T_-)(\Omega)$$

$$= \text{tr}[TP^+] + \text{tr}[TP^-] = |\text{tr}[TP^+]| + |\text{tr}[TP^-]| .$$

But this simply means that V is continuous (with respect to the given topologies.).

Since any $T \in \mathcal{T}(\mathcal{H})$ and $\mu \in \mathcal{M}(\Omega)$ can uniquely be written as $T = T_1 + \iota T_2$, with $T_1, T_2 \in \mathcal{T}_s(\mathcal{H})$, and $\mu = \mu_1 + \iota\mu_2$, with $\mu_1, \mu_2 \in \mathcal{T}_s(\mathcal{H})$, we may define $\hat{V}(T) = V(T_1) + \iota V(T_2)$ so that $\hat{V}$ is a continuous linear mapping from $(\mathcal{T}(\mathcal{H}), weak)$ to $(\mathcal{M}(\Omega), \| \cdot \|_1)$. We denote this mapping as V, too. Together with the results of Sections 2 and 3 we may conclude that all the mappings in the following diagram are continuous:

(23) $(\mathcal{T}(\mathcal{H}), \| \cdot \|_1) \to_\iota (\mathcal{T}(\mathcal{H}), weak) \to_V (\mathcal{M}(\Omega), \| \cdot \|_1)$

$$\to_\iota (\mathcal{M}(\Omega), vague) .$$

This result shows, in particular, that the mapping V is σ-convex in all conceivable ways, i.e.,

(24) $V(\sum w_i T_i) = \sum w_i V(T_i)$

in any Hausdorff topologies of $\mathcal{T}(\mathcal{H})$ and $\mathcal{M}(\Omega)$ which are coarser (or equal to) their norm topologies.

5. INTEGRATION VS. SUMMATION

As an important application of the above results we give an analog of the monotone convergence theorem which shows the conditions under which the integration and summation order can be interchanged in the case of σ-convex combinations of states. Thus, let $T = \sum_{i=1}^\infty w_i T_i$ be a σ-convex combination of the states T_i with the weights $0 \leq w_i \leq 1$, and let $f : \Re \to \Re$ be a Borel function. Then f is $V(T)$-integrable if and only if f is $V(T_i)$-integrable for each $i = 1, 2, \ldots$ and the series $\sum_{i=1}^{\infty} \int |f| \, dV(T_i)$ converges. In that case

$$(24) \qquad \int f \, dV(T) = \sum w_i \int f \, dV(T_i).$$

In order to proof this result, we denote $V(T)_n = \sum_{i=1}^{n} w_i V(T_i)$, and observe that $(V(T)_n)$ is a monotone increasing sequence of finite positive measures converging (vaguely) to $V(T)$. But the vague convergence $(V(T)_n \to V(T)$ implies the setwise convergence $(V(T)_n(X) \to V(T)(X)$ for all $X \in \mathcal{B}(\Omega))$. Hence an analog (Royden 1968) of the monotone convergence theorem implies that for any positive measurable function f

$$\int f \, dV(T) = \lim_{n \to \infty} \int f \, dV(T)_n.$$

Moreover, an analog (Royden 1968) of the Lebesgue theorem shows that if a measurable function f is $V(T)_n$-integrable for any n, $V(T)$-integrable, and $\lim_n \int |f| \, dV(T)_n = \int |f| \, dV(T)$, then

$$\lim_{n \to \infty} \int f \, dV(T)_n = \int f \, dV(T).$$

Taking together, these two results amount to show the aforementioned result.

As an illustration of the above result, we recall that the *expectation* of an observable E in the state T is defined as the expectation of the probability measure $V(T)$, i.e., $\mathrm{Exp}(E, T) \doteq \int \iota \, dV(T)$. Here ι denotes the identity mapping on $\Re$. Then, if $T = \sum w_i T_i$, then $\mathrm{Exp}(E, T)$ is finite exactly when all $\mathrm{Exp}(E, T_i)$ are finite and the series $\sum w_i \int |\iota| \, dV(T_i)$ converges. In that case, $\mathrm{Exp}(E, T) = \sum w_i \mathrm{Exp}(E, T_i)$. (See also Cassinelli and Olivieri 1984.)

6. THE RANGE OF V

The mapping $V : T \mapsto V(T) \doteq E_T$ has shown to be σ-convex from the set of states $\mathcal{T}(\mathcal{H})_1^+$ to the set of probability measures $\mathcal{M}(\Omega)_1^+$ on the value space of the observable $E : \mathcal{B}(\Omega) \to \mathcal{L}(\mathcal{H})^+$. On the other hand, the set of states is highly nonsimplicial (each nonpure state admitting uncountably many decompositions into its pure components) whereas the set of probability measures is, in fact, a simplex (where each probability measure has a unique, in general integral representation in terms of the point measures). The fact that V *preserves the σ-convex structures* may thus appear contradictory. In order to get a better feeling on this important mapping we shall now consider the structure of the image of the set of states under V. We denote it as $\mathcal{R}(V)$, so that $\mathcal{R}(V) = \{V(T) : T \in T(\mathcal{H})_1^+\}$.

For simplicity, we shall assume next that E is, in fact, a PV measure on a (bounded) Borel subspace of the real line $\Re$. Let A be the corresponding self-adjoint operator, and let $\mathrm{sp}(A) = \mathrm{sp}_p(A) \cup \mathrm{sp}_c(A)$ denote its spectrum, decomposed into the set of eigenvalues $\mathrm{sp}_p(A)$ and the continuous part $\mathrm{sp}_c(A) = \mathrm{sp}(A) \setminus \mathrm{sp}_p(A)$. Then $\Omega = \mathrm{supp}(E) = \mathrm{sp}(A)$. Of course, this is the usual case in quantum mechanics.

We note first that the set $\mathcal{R}(V) \subset \mathcal{M}(\Omega)_1^+$ is σ-convex. Secondly, one may immediately verify that $V(T) \in \mathrm{Ex}(\mathcal{M}(\Omega)_1^+)$, i.e., $V(T) = \mu_\omega$ for some $\omega \in \Omega$, if and only if $\omega \in \mathrm{sp}_p(A)$ and $T = P[\varphi]$ is a corresponding eigenvector (i.e., $A\varphi = \omega\varphi$). But, since E is a PV measure, one then readily observes that

$$(25) \qquad \mathrm{Ex}(\mathcal{R}(V)) = \mathcal{R}(V) \cap \mathrm{Ex}(\mathcal{M}(\Omega)_1^+) \,.$$

In particular, we stress that if a unit vector $\varphi \in \mathcal{H}$ is not an eigenvector of A then $V(P[\varphi]) \in \mathcal{M}(\Omega)_1^+ \setminus \mathrm{Ex}(\mathcal{M}(\Omega)_1^+)$. Thus V is never a *pure* mapping.

Consider a unit vector $\varphi \in \mathcal{H}$, and define $\psi_X = e^{\iota E(X)} \varphi$ for all $X \in \mathcal{B}(\Omega)$. Then, in general, $P[\varphi] \neq P[\psi_X]$, but, clearly, always $V(P[\varphi]) = V(P[\psi_X])$. Hence, V is never *injective*. Below we shall observe the conditions under which V is surjective.

Consider the case where the value space of E consists solely of the eigenvalues of A : $\Omega = \mathrm{sp}_p(A) = \{\omega_1, \omega_2, \ldots\}$. Then $\mathcal{M}(\Omega)_1^+ = \sigma - \mathrm{conv}(\mathrm{Ex}(\mathcal{M}(\Omega)_1^+))$, and, clearly, also $\mathcal{R}(V) = \mathcal{M}(\Omega)_1^+$. In that case V is *surjective* and the image set $\mathcal{R}(V)$ is simplex so that each $V(T)$ can uniquely be decomposed as $\sum w_i P[\varphi_i]$ for some eigenvectors φ_i of A and for some weights $0 < w_i < 1$. But since the mapping V is not injective, the property of unique decomposability of $V(T)$ cannot be transferred to that of T.

Consider next the case $\Omega = \mathrm{sp}_p(A) \cup \{\omega\}$. The point ω is not an eigenvalue of A, but it is an accumulation point of $\mathrm{sp}_p(A)$. Clearly, $\mu_\omega \in \mathcal{M}(\Omega)_1^+ = \sigma - \mathrm{conv}(\mathrm{Ex}(\mathcal{M}(\Omega)_1^+))$, but $\mu_\omega \notin \mathcal{R}(V)$ since $\mathcal{R}(V)$ is neither norm nor vague closed. The Krein–Milman theorem cannot hold now for $\mathcal{R}(V)$. Finally, assume that $\Omega = \mathrm{sp}_c(A)$ so that $\mathrm{Ex}(\mathcal{R}(V))$ is empty. Then, clearly, $\mathcal{R}(V)$ is a subset of $\mathcal{M}(\Omega)_1^+ \setminus \sigma - \mathrm{conv}(\mathrm{Ex}(\mathcal{M}(\Omega)_1^+))$, and, again, no Krein–Milman property holds for $\mathcal{R}(V)$.

To close our discussion, we consider the case of an injective mapping $V : T \mapsto V(T) \doteq E_T$. As pointed out above, such a mapping cannot arise from a PV measure. Since the injectivity of V simply means that different states T_1 and T_2 always determine different probability measures $V(T_1) = E_{T_1}$ and $V(T_2) = E_{T_2}$ we observe that the *injectivity* of V is, in fact, the same than the *informational completeness* of the associated observable E. Indeed, an observable E is said to be informationally complete if the condition $E_{T_1} = E_{T_2}$ implies that $T_1 = T_2$ for all $T_1, T_2 \in T(\mathcal{H})_1^+$. Such observables are important, e.g., in the theory of joint measurements of complementary observables (Busch and Lahti 1989).

Consider now the mapping V induced by an informationally complete POV measure $E : B(\Omega) \to \mathcal{L}(\mathcal{H})^+$. Since now $V(T) = wV(T_1) + (1 - w) V(T_2)$ if and only if $T = wT_1 + (1 - w) T_2$ one observes that the extremal points of $\mathcal{R}(V)$ are exactly those probability measures which arise from the pure states, i.e.,

$$(26) \qquad \text{Ex}(\mathcal{R}(V)) = \{V(P[\varphi] : \varphi \in \mathcal{H}, \|\varphi\| = 1\} .$$

Contrary to Equation (25), the set $\mathcal{R}(V) \cap \text{Ex}(\mathcal{M}(\Omega)_1^+)$ is now empty. Indeed, if this were not the case, there would be a point $\omega \in \Omega$ and a vector $\varphi \in \mathcal{H}$ such that $E(\{\omega\}) \varphi = \varphi$. But this would imply that $E(\{\omega\}) E(X) \varphi = E(X) E(\{\omega\}) \varphi$ for all $X \in B(\Omega)$ – a fact which would contradict the total noncommutativity of an informationally complete E (Busch and Lahti 1989). We observe also that

$$(27) \qquad \mathcal{R}(V) = \sigma - \text{conv}(\text{Ex}(\mathcal{R}(V))) ,$$

i.e., the Krein–Milman property hold for $\mathcal{R}(V)$. This shows also that the mapping V can never be surjective. In fact, as a convex set, $\mathcal{R}(V)$ has the same structure than $T(\mathcal{H})_1^+$, so that, in particular, $\mathcal{R}(V)$ is now highly nonsimplicial – in spite of the fact that $\mathcal{R}(V)$ is a subset of $\mathcal{M}(\Omega)_1^+$ where, e.g., no superposition principle holds true.

GIANNI CASSINELLI
Dipartimento di Fisica, Università di Genova,
Istituto Nazionale di Fisica Nucleare, sez. di Genova

PEKKA J. LAHTI
I.N.F.N. Sez. di Genova,
and
Dept. of Physical Sciences,
University of Turku, Finland

REFERENCES

Beltrametti, E. and Cassinelli, G.: 1981, *The Logic of Quantum Mechanics*, Addison-Wesley, Reading-Massachusetts.

Brezis, H.: 1985, *Analyse Fonctionelle*, Masson, Paris.

Busch, P. and Lahti, P.: 1989, 'The determination of the past and the future of a physical system in quantum mechanics', *Foundations of Physics* 19, 633–678.

Cassinelli, G. and Olivieri, G.: 1984, 'The statistics of unbounded observables in Hilbert-space quantum mechanics', *Il Nuovo Cimento* 84B, 43–52.

Davies, E.B.: 1976, *Quantum Theory of Open Systems*, Academic Press, London.

Dieudonne, G.: 1970, *Treatise on Analysis, II*, Academic Press, New York.

Hadjisavvas, N.: 1981, 'Properties of mixtures of non-orthogonal states', *Lett. Math. Phys.* 5, 327–332.

Reed, M. and Simon, B.: 1971, *Methods of Modern Mathematical Physics, I: Functional Analysis*, Academic Press, New York.

Royden, H.L.: 1968, *Real Analysis*, Second Ed., MacMillan, New York.

Rudin, W.: 1966, *Real and Complex Analysis*, McGraw Hill, New York.

GIANPIERO CATTANEO

THE 'LOGICAL' APPROACH TO AXIOMATIC QUANTUM THEORY

1. THE FORMAL LANGUAGE OF STATISTICAL THEORIES

A *statistical theory* (ST) is based on a formalized structure in which peculiar classes of objects are interpreted as representing the *primitive physical concepts* of *states*, consisting of classes of *preparations* of individual samples of the physical entity under well defined and repeatable conditions; *observables* or *physical measurable quantities* (or *magnitudes*); a *probability* function.

In order to avoid formal complications, but without any loss in correctness, we give a *formal pre-realization* of the language of a statistical theory, based on usual abstract set theory and class logic. To be precise, the *specific alphabet* of the *formal language* of a statistical theory consists of the following nonlogical undefined objects:

(a1) S is non-empty set, whose elements are denoted by $w, v, \ldots$, with indices if necessary;

(a2) $\mathcal{O}$ is a non-empty set, whose elements are denoted by $A, B, \ldots$, with indices if necessary;

(a3) $\mathcal{P}$ is a mapping
$$P \: : \: S \times \mathcal{O} \times \mathcal{B}(I\!\!R) \mapsto [0, 1]$$
(where $\mathcal{B}(I\!\!R)$ is the set of all Borel subsets of the real line $I\!\!R$).

This alphabet, specific of any statistical theory (in which there is no reference to any concrete mathematical structure, e.g., the one involved in the Hilbert spaces theory or in the measure spaces theory), will be denoted by
$$\mathbf{L}_{ST} \equiv \langle S, \mathcal{O}, \mathcal{P} \rangle \: .$$

The *rules of interpretation*, which transform this formal language into a language about an empirical domain of physical objects, are the following:

(RI-1)   The elements of the set S are interpreted as describing *states* of the physical entity, the set S is the *phase* or *state-space*;

(RI-2)   the elements of the set $\mathcal{O}$ are interpreted as *physical quantities* (or *magnitudes*) which can be measured on the physical entity, the set $\mathcal{O}$ is the set of *observables*;

225

G. Corsi et al. (eds), Bridging the Gap: Philosophy, Mathematics, and Physics, 225–260.
© 1993 Kluwer Academic Publishers.

(RI-3) for any triple $(w, A, \Delta) \in \mathcal{S} \times \mathcal{O} \times \mathcal{B}(\mathbb{R})$, the real number

$$P(w; A, \Delta) \in [0, 1]$$

is interpreted as the *probability* that for the physical entity prepared in the state w a measurement of the physical quantity A gives a value contained in the set Δ of real numbers.

We list some *derived definitions* which will be very useful in the sequel.

(d1) A *yes–no device* is any pair (A, Δ), consisting of an observable $A \in \mathcal{O}$ and a real Borel subset $\Delta \in \mathcal{B}(\mathbb{R})$.

(d2) The collection of all yes–no devices is denoted by $\mathcal{Y}$; i.e., $\mathcal{Y} := \{(A, \Delta) \; : \; A \in \mathcal{O}, \; \Delta \in \mathcal{B}(\mathbb{R})\}$. Yes–no devices are denoted by $r, s, t, \ldots$, with indices if necessary, in the sequel.

Quoting Beltrametti and Cassinelli (1981):

Consider now the ordered pairs (A, Δ) with $A \in \mathcal{O}$, $\Delta \in \mathcal{B}(\mathbb{R})$. From the physical point of view, we may picture the pair (A, Δ) as the *experimental device* that is obtained from the instrument used to measure the physical quantity A by attaching to the reading scale a window that isolates the numerical subset Δ; according as the pointer of the instrument does or does not appear within the window, one may answer 'yes' or 'no' to the *question*

Is the measured result of A contained in Δ?

(d3) An *elementary* (or *yes–no) experiment* is any pair (w, r) consisting of a state $w \in \mathcal{S}$ and a yes–no device $r \in \mathcal{Y}$;

(d4) The collection of all elementary experiments will be denoted by $\mathbf{Y}$.

According to Ludwig (1971):

we have (...) to return (...) to experimental situations which everybody, physicist or layman, might examine as objectively given events. Now what kind of experimental situation should be selected as a starting point? (...) We find a *preparing part* which, via a microscopic channel, can act on a *signal part* (a part, that is, where the *effects* are produced). In the course of an experiment, the signal part will either respond or not. The preparing part, for example, might be an accelerator with the target placed in the beam; and the signal part might be a counter. (...) The preparing part w shall be an apparatus, objectively given and technically describable and the same is required of the signal part r together with its response ('yes') or lack of response ('no').

The macroscopic arrangement of the measuring part can interact with individual samples of the physical system in such a way "that a direct, objectively traceable [macroscopic alternative] effect occurs or does not occur (e.g., counter signal, a cloud-chamber-track, blackening of a photographic plate, etc.)" (Ludwig 1971); "the presence of the effect is conventionally taken as the answer 'yes' whereas its absence is 'no'" (Mielnik 1976).

Another empirical fact is the existence of so called measuring instruments, which are capable of undergoing macroscopically observable changes due to ('triggered by') their interaction with single systems. The simplest type of measuring instrument is one on which just a single change may be triggered. (...) One usually defines the result of a single measurement to be 'yes' if the effect occurs, and 'no' if the effect does not occur" (Kraus 1983).

According to Bohr the external conditions of an experiment have to be described in the language of classical physics. In the case usually considered this classical description specifies:

(i) The construction and application of a *preparing instrument* which 'emits' (produces, selects) single samples [of the physical entity].

(ii) The construction, application and reading of a *measuring instrument* which is applied to these samples. In particular, the response of the measuring instrument is a classical (objective, observer-independent) event which leads to a unique measured [yes–no] value for every single sample.

Being thus based operationally on the same kind of objective facts and events which are already familiar from classical physics, this interpretation (...) avoids the introduction of any subjective elements (like knowledge of observers, or human consciousness) into the theory (Kraus 1986).

2. THE SENTENIAL LOGIC OF STATISTICAL THEORIES

A typical *simple proposition* of a statistical theory is expressed by *theoretical sentences* (Bub 1973) or *elementary statements* (van Fraassen 1974) (or *exact question*) of the form

$$\mathrm{val}(A) \in \Delta := \text{'the value of the physical magnitude } A \in \mathcal{O} \text{ lies in the}$$
$$\text{Borel set } \Delta \in \mathcal{B}(\mathbb{R}) \text{ of real numbers'.}$$

These elementary statements are tested by yes–no devices, according to the one-to-one and onto correspondence pictured by the following table.

Propositional Logic	Statistical Theory
Elementary Statements	Yes–No Devices
$\mathrm{val}(A) \in \Delta$	(A, Δ)

This agrees with the following assertion of Gudder (1970): "A *proposition (event, [exact] question)* may be thought of physically as a statement concerning the system [i.e., $\mathrm{val}(A) \in \Delta$] for which there is an experiment [i.e., (A, Δ)] that can verify whether the statement is true or false".

Owing to the above identification, in the sequel we will denote such elementary statements by the same variable signs $r, s, t, \ldots$, used to denote yes–no devices. Starting from these elementary statements, using the standard connectives of the everyday language, one constructs *complex sentences* such as '$r \underline{o} s$' (or), '$r \& s$' (and), '$\neg r$' (not), and so on. "For example, 'the electron has

x coordinate between 2 and 3 cm *and* x momentum component between 1 and 2 g cm sec^{-1} might be a proposition" (Gudder 1970). In discussing the so-called paradoxes of quantum physics and referring to the two-slit experiment, we have occasion to speak about the fact that

- 'the particle has passed through the slit 1 *or* through the slit 2', described by a complex proposition of the form '$(Q, \Delta_1) \underline{o} (Q, \Delta_2)$';

or

- 'the spin of the particle along the z direction is up *and* along the x direction is up', described by a complex proposition of the form '$(S_z, \Delta^\uparrow) \& (S_x, \Delta^\uparrow)$';

these statements have been the source of a lot a discussions.

Hence, underlying to any statistical theory is a *propositional language* $\mathbf{PL_{ST}}$ which includes two components:

Alphabet of $\mathbf{PL_{ST}}$

    $r, s, t, \ldots$ *propositional variables* (or *sentential letters*), with indices if necessary (which can be interpreted as simple propositions of the form (A, Δ)); the collection of all sentential letters will be denoted by $P_{ST}^{(a)}$.

    $\underline{o}$, & binary connectives and $\neg$ unary connective;

    (,) brackets.

Rules of Formation for $\mathbf{PL_{ST}}$

1   Any propositional variable r is a wff;
2   If α, β are wffs, then so are $\alpha \underline{o} \beta$, $\alpha \& \beta$, and $\neg \alpha$;
3   Only those strings generated from rules 1 and 2 are wffs, whose collection will be denoted by P_{ST}.

Normal conventions on bracketing apply.

We can summarize this sentential logic by the following structure

$$(2.1) \qquad \mathbf{PL_{ST}} \equiv \langle P_{ST}, \underline{o}, \&, \neg \rangle \,.$$

which is closed with respect to unary operator $\neg$ and binary operators &, $\underline{o}$. Note that the sentential logic $\mathbf{PL_{ST}}$ "is the logic only of the set of elementary statements" (van Fraassen 1974), and not the classical logical system by means of which theorems of the statistical theory are derived from axioms.

3. THE EMPIRICAL SEMANTIC OF STATISTICAL SENTENTIAL LOGICS

Once introduce the sentential logic of a statistical theory, one has to face the problem that 'sentences are statements which have the property of being sometimes *true* or *false* or, in some cases, also *indeterminate*'. With respect to this problem, first of all, one has to face the problem of how the truth or falsity or indeterminacy is assigned to elementary statements from $P_{ST}^{(a)}$. We claim that the following methaphysical assumption (Ludwig's *pre-decision*) must be made in order to avoid misconceptions and misunderstandings.

(s1)   *The truth values of elementary statements of the sentential logic under-*
lying any physical theory (and so, in particular, any statistical theory)
must be introduced making use of the **only** *notions available in the*
formal structure of the theory.

In agreement with Ludwig (1971),

I hasten to add that nobody, of course, is obliged to accept that pre-decision. One might have, for whatever reasons, a quite different conception (...). Now it is not my intention to speak ill of any different view, but only to explain my *own* approach and to present it as one possible way of proceeding.

Anyway, if someone would treat about sentences of the form 'it is true that the particle has passed through the slit 1' or of the form 'it is true that it has passed through the slit 2', trying to connect the truth of these sentences with the truth of the sentence 'the particle has passed through the two-slit experiment', then he should specify which truth-value criterion is used to state this connection.

In the case of a statistical theory, we introduce the following two binary predicate signs, 'true' T and 'false' F, realized as 2-argument relations states $w \in S$ and elementary statements $r = (A, \Delta) \in P_{ST}^{(a)}$, according to pre-decision (s1).

(3.1) $(w, r)\, T$ iff $P(w, r) = P(w, A, \Delta) = 1$

(3.2) $(w, r)\, F$ iff $P(w, r) = P(w, A, \Delta) = 0$.

Whether or not the statement r is 'true' (resp., 'false') depends on the state of the system; precisely, an elementary statement r is 'true' (resp., 'false') in the state w iff in this state the effect tested by the yes–no device bijectively associated to r occurs (resp., does not occur) with certainty, i.e., probability 1 (resp., 0). From (3.1) and (3.2) we can derive a further 2-argument relation

(3.3) (w, r) iff neither $(w, r)\, T$ nor $(w, r)\, F$
          iff $P(w, r) \neq 1, 0$.

Quoting von Weizsäcker (1958):

From the standpoint of formal logic, [statistical theory] employs a multi-valued concept of truth. Here, besides the truth-values 'true' and 'false' a statement can have the truth-value 'indeterminate' with such and such a probability [i.e., $P(w, r) \neq 1$ or 0] to be true.

At any rate, according to van Fraassen (1974):

Much philosophical puzzelment has been occasioned by the use of the term 'third truth value'. This term is unfortunate; the case can be stated much more perspicuosly by saying that certain statements are neither true nor false, rather than that they have a third truth value, the value *indeterminacy*. A value assignment may assign T, F, or U to a sentence; or perhaps 1,0, or $\frac{1}{2}$; but this assignment is only a marker for the corresponding class of sentences.

In any statistical theory, these assignments can be formalized by a set of mappings

$$\mathcal{V}_{\mathrm{ST}} := \{v_w \ : \ w \in \mathcal{S}\}$$

where, for any fixed state $w \in \mathcal{S}$, the mapping

$$v_w \ : \ P_{\mathrm{ST}}^{(a)} \mapsto \{T, F, U\}$$

called also *valuation-function* induced by state w, is defined as follows:

$$v_w(r) := \left\{ \begin{array}{l} T \ \text{iff} \ P(w,r) = 1 \\ F \ \text{iff} \ P(w,r) = 0 \\ U \ \text{iff} \ P(w,r) \neq 1,0 \end{array} \right.$$

We also introduce two sets of states, for any fixed elementary statement $r \in \mathcal{Y}$:
– The *certainly-true* domain of statement r, denoted by $S_{\mathrm{T}}(r)$, and defined as follows

$$S_{\mathrm{T}}(r) := \{w_1 \in \mathcal{S} \ : \ P(w_1, r) = P(w_1, A, \Delta) = 1\} \, .$$

– The *certainly-false* domain of statement r, denoted by $S_{\mathrm{F}}(r)$, and defined as follows

$$S_{\mathrm{F}}(r) := \{w_0 \in \mathcal{S} \ : \ P(w_0, r) = P(w_0, A, \Delta) = 0\} \, .$$

Hence, states $w_1 \in S_{\mathrm{T}}(r)$ (resp., $w_0 \in S_{\mathrm{F}}(r)$) are the ones in which the elementary statement $r = (A, \Delta)$ is 'true' (resp., 'false'). Thus, making use of the above definitions, we introduce the mapping

$$h \ : \ P_{\mathrm{ST}}^{(a)} \mapsto \mathcal{P}(\mathcal{S}) \times \mathcal{P}(\mathcal{S})$$

assigning to any elementary statement $r \in P_{\mathrm{ST}}^{(a)}$ the pair of subsets of states ($\mathcal{P}(\mathcal{S})$ being the power set of $\mathcal{S}$),

$$h(r) := (S_{\mathrm{T}}(r), S_{\mathrm{F}}(r))$$

called *proposition* associated to r, and consisting of the set of states in which r is 'true' and the set of states in which r is 'false'.

This being stated, one has to face the second, very delicate, problem that the truth values of a complex sentence must depend on the truth values of constituent elementary statements. This problem consists in the following *semantical requirement*.

(s2)   *To find a suitable set of specific axioms for the statistical theory in such a way that an algebraic structure, or* **semantical structure**, *can be deduced from the involved axiomatic theory.*

Concretely, this requirement means that it should be possible to determine from the theory

(i)   an algebraic structure (the *Lindenbaum–Tarski* algebra of the sentential logic) on a suitable subset $\mathcal{L}(S) \subseteq \mathcal{P}(S) \times \mathcal{P}(S)$ of the kind

$$\langle \mathcal{L}(S), \nabla, \Delta, - \rangle$$

where $\nabla, \Delta \; : \; \mathcal{L}(S) \times \mathcal{L}(S) \mapsto \mathcal{L}(S)$, and $- : \; \mathcal{L}(S) \mapsto \mathcal{L}(S)$; such that propositions of the form $(S_T(r), S_F(r))$, for r running on $P_{ST}^{(a)}$, are elements of $\mathcal{L}(S)$;

(ii)   an extension of the mapping h, denoted by h too,

$$h \; : \; P_{ST} \mapsto \mathcal{L}(S)$$

such that

$$h(r\underline{\text{o}}s) \; = h(r) \nabla h(s) \; = (S_T(r), \; S_F(r)) \nabla (S_T(s), \; S_F(s))$$
$$h(r\&s) = h(r) \Delta h(s) \; = (S_T(r), \; S_F(r)) \Delta (S_T(s), \; S_F(s))$$
$$h(\neg r) \;\; = -h(r) \qquad = -(S_T(r), \; S_F(r)) \; .$$

As a summary of the above discussion, we quote van Fraassen (1974):

Implicit in the work of Carnap and many of his contemporaries in philosophy of science is the following picture of a *physical theory*:

– It is ideally constructed by adding axioms with empirical content [specific axioms plus the intended interpretation] to a formalized system of logic and mathematics, say *Principia Mathematica*. The latter consists of *standard logic* plus *axioms for sets*.

In that sense one might say that the current picture of a physical theory implied that its (eventual) formalization must be 'based' on standard logic. [See also Ludwig (1971) "One of the most important pre-decision is connected with an axiomatic theory by choosing as logical rules the laws of classical logic".] (...) But from our point of view a *logic* of [statistical theories] is simply an attempt to give a systematic account of the semantic relations [i.e., the (ii) of (s2)] among the elementary statements of that theory. And these semantic relations are to be *deduced from* the theory [semantical pre-decision (s1)]. (...) [This logic] is not meant to be the basis for a formalization of the theory (...)

– The relation between statistical theories] and [corresponding sentential logics] is that the former provides the semantics for the latter.

4. MACKEY'S APPROACH TO QUANTUM THEORY AS STATISTICAL THEORY

So far we have only considered the formal language of a statistical theory, without any reference to the choice of a possible system of specific axioms. In this section we will sketch Mackey's approach to axiomatic quantum theory (Mackey 1963).

The first assumption is that any quantum mechanical entity is described by a statistical theory characterized by the following specific axioms, which qualify Mackey's approach to quantum physics as an axiomatic statistical theory, say $(\mathcal{QT})_M$, different from other statistical theories based on the same formal language.

Axiom I. (i)

$$(\forall w \in \mathcal{S}), \ (\forall A \in \mathcal{O}),$$

$$P(w, A, \emptyset) = 0, \ P(w, A, \mathbb{R}) = 1$$

Axiom I. (ii)

$$(\forall w \in \mathcal{S}), \ (\forall A \in \mathcal{O}), \ (\forall \{\Delta_n \in \mathcal{B}(\mathbb{R}) \ : \ i \neq j, \ \Delta_i \cap \Delta_j = \emptyset\})$$

$$P(w, A, \cup_n \Delta_n) = \sum_{n=1}^{\infty} P(w, A, \Delta_n)$$

Axiom II. (i) If for every state $w \in \mathcal{S}$ and every real Borel set $\Delta \in \mathcal{B}(\mathbb{R})$ we have that

$$P(w, A, \Delta) = P(w, A', \Delta) \ \text{then} \ A = A' \, .$$

Axiom II. (ii) If for every observable $A \in \mathcal{O}$ and every real Borel set $\Delta \in \mathcal{B}(\mathbb{R})$ we have that

$$P(w, A, \Delta) = P(w', A, \Delta) \ \text{then} \ w = w' \, .$$

Axiom III. Let $A \in \mathcal{O}$ be an observable and let $f : \mathbb{R} \mapsto \mathbb{R}$ be any Borel function. Then an observable $B \in \mathcal{O}$ exists such that

$$(\forall w \in \mathcal{S}), \ (\forall \Delta \in \mathcal{B}(\mathbb{R})),$$

$$P(w, B, \Delta) = P(w, A, f^{-1}(\Delta))$$

It follows from Axiom II(i) that B is uniquely determined by A and f and we will denote it by $f(A)$.

Axiom IV. For every countable family of states $\{w_n\} \subseteq S$ and every corresponding family of non negative real numbers $\{\lambda_n\} \subseteq \mathbb{R}_+$, such that $\sum \lambda_n = 1$, a state $w \in S$ exists such that

$$(\forall A \in \mathcal{O}),\ (\forall \Delta \in \mathcal{B}(\mathbb{R})),$$

$$P(w, A, \Delta) = \sum_{n=1}^{\infty} \lambda_n P(w_n, A, \Delta).$$

Axiom V. For every sequence of yes–no devices $\{(A_n, \Delta_n)\} \subseteq \mathcal{Y}$ which satisfies the condition $(\forall w \in S)$, $(\forall i, j \text{ with } i \neq j)$,

$$P(w, A_i, \Delta_i) + P(w, A_j, \Delta_j) \leq 1$$

a yes–no device $(A, \Delta) \in \mathcal{Y}$ exists such that $(\forall w \in S)$,

$$P(w, A, \Delta) = \sum_{n=1}^{\infty} P(w, A_n, \Delta_n).$$

In particular, in Mackey's quantum theory, $(\mathcal{QT})_M$, the following holds.

• *For any fixed state $w \in S$ and any fixed observable $A \in \mathcal{O}$, the mapping*

$$\mu_{(w,A)} : \mathcal{B}(\mathbb{R}) \mapsto [0,1],\ \Delta \to \mu_{(w,A)}(\Delta) := P(w, A, \Delta)$$

is a probability measure on the σ-algebra of all Borel sets of real line. "The distribution of an observable A in the state w (or under w) is the probability measure $\mu_{(w,A)}$." (Gudder 1965).

By making use of these probability measures, Axioms II can be restated in the following manner.

Axiom II. (i) Let $A_1, A_2 \in \mathcal{O}$, then

$$\forall w \in S,\ \mu_{(w,A_1)} = \mu_{(w,A_2)}\ \text{implies}\ A_1 = A_2.$$

Axiom II. (ii) Let $w_1, w_2 \in S$, then

$$\forall A \in \mathcal{O},\ \mu_{(w_1,A)} = \mu_{(w_2,A)}\ \text{implies}\ w_1 = w_2.$$

Taking into account interpretation rules (RI-1)–(RI-3) of Section 1.

– Axiom II(i) says that two observables, to be different, must have different probability distributions in at least one state. *(Indistinguishability principle of observables)*.

– Axiom II(ii) say that two states, to be different, must assign different probability distributions to at least one observable. *(Indistinguishability principle of states)*.

5. EVENTS, PROBABILITY MEASURES AND
EVENT-VALUED MEASURES

Quoting Mackey: "Because of Axiom III, (...) if $f_\lambda(x) := \lambda$, where λ is some real number, then $f_\lambda(A)$ is independent of A and will be called the *constant observable* with value λ or simply the observable λ" (Mackey 1963). To be precise, we have that for every $w \in S$ and every $\Delta \in \mathcal{B}(\mathbb{R})$

$$P(w, f_\lambda(A), \Delta) = P(w, A, f_\lambda^{-1}(\Delta)) = \begin{cases} 1 & \text{if } \lambda \in \Delta \\ 0 & \text{otherwise} \end{cases}$$

which is manifestly independent from both w and A. In particular we will consider the *null observable* and the *identity observable*

$$\mathbb{O} := f_0(A), \quad \mathbb{I} := f_1(A).$$

These observables are such that

$$P(w, \mathbb{O}, \Delta) = \begin{cases} 1 & \text{iff } 0 \in \Delta \\ 0 & \text{otherwise} \end{cases} \quad P(w, \mathbb{I}, \Delta) = \begin{cases} 1 & \text{iff } 1 \in \Delta \\ 0 & \text{otherwise} \end{cases}$$

This result leads to the following property, which holds for every $w \in S$,

$$\mu_{(w,\mathbb{O})}(\{0\}) = 1 \quad \mu_{(w,\mathbb{I})}(\{1\}) = 1$$

i.e., in every state $w \in S$ the measure $\mu_{(w,\mathbb{O})}$ (resp., $\mu_{(w,\mathbb{I})}$) is concentrated in 0 (resp., 1).

Note that, once introduced the real function $(\mathrm{id})^n(x) := x^n$, expression such as $A^2 := (\mathrm{id})^2(A)$, $A^3 + A := (\mathrm{id}^3 + \mathrm{id})(A)$ and

(5.1) $\mathbb{I} - A := (f_1 - \mathrm{id})(A)$

all make sense whenever A is an observable. The following is straightforward.

PROPOSITION 5.1. *Let A be an observable. Then the following are equivalent statements*

(i) $A = A^2$; *i.e., the observable is idempotent.*

(ii) $\forall w \in S$, $\mu_{(w,A)}(\{0,1\}) = 1$; *i.e., in every state $w \in S$ the measure $\mu_{(w,A)}$ is concentrated in the points 0 and 1.*

(iii) $A = \chi_{\{1\}}(A)$; χ_Δ *being the characteristic function of the real Borel set $\Delta \in \mathcal{B}(\mathbb{R})$ defined as follows*

$$\chi_\Delta(x) := \begin{cases} 1 & \text{iff } x \in \Delta \\ 0 & \text{iff } x \notin \Delta \end{cases}$$

This result allows us to introduce the following definition.

DEFINITION 5.1. Any observable which satisfies one of the conditions of

Proposition 5.1 is said to be an *event* (or Mackey's *question* or Ludwig's *decision effect*). The set of all events will be denoted by $\mathcal{E}$ in the sequel; in symbols

$$\mathcal{E} = \{\chi_\Delta(A) \ : \ A \in \mathcal{O}, \ \Delta \in \mathcal{B}(\mathbb{R})\} \ .$$

We now state a preliminary result on events.

PROPOSITION 5.2. *In Mackey's quantum theory we have that*

(i) $\mathbb{O}$ *and* $\mathbb{I}$ *are events, in this context called the absurd event and the certain event, respectively. In particular, for every observable* $A \in \mathcal{O}$,

$$\mathbb{O} = \chi_\emptyset(A) \ and \ \mathbb{I} = \chi_\mathbb{R}(A) \ ;$$

(ii) *for every event* $\chi_\Delta(A) \in \mathcal{E}$, *the observable* $\mathbb{I} - \chi_\Delta(A)$ *is an event, too, called the opposite or inverse event of* $\chi_\Delta(A)$, *and*

$$\mathbb{I} - \chi_\Delta(A) = \chi_{(\mathbb{R}/\Delta)}(A) \ .$$

From now on we denote by E the arbitrary event from $\mathcal{E}$ and if $E \in \mathcal{E}$, with $E = \chi_\Delta(A)$, we set $E' := \mathbb{I} - \chi_\Delta(A)$.

5.1. *State–Event–Probability (SEVP) Structure*

Mackey's approach to axiomatic quantum theory, described by any triple *state–observable–probability* (SOP)

$$\mathbf{L}_{\text{STM}} \equiv \langle \mathcal{S}, \mathcal{O}, \mathcal{P} \rangle$$

satisfying Axioms I, II, III, IV, and V, induces a *state–event–probability* (SEVP) structure consisting of the triple

$$\mathbf{L}_{\text{SEVP}} \equiv \langle \mathcal{S}, \mathcal{E}, p \rangle$$

where $\mathcal{S}$ is the non-empty set of all *states*, $\mathcal{E}$ is the non-empty set of all *events* and the *probability function* is the mapping

$$p \ : \ \mathcal{S} \times \mathcal{E} \mapsto [0, 1]$$

defined, for arbitrary $E = \chi_\Delta(A) \in \mathcal{E}$, as follows

(5.2) $p(w, E) := P(w, E, \{1\}) = P(w, \chi_\Delta(A), \{1\}) = P(w, A, \Delta)$

and satisfying the following.

Axiom 1. There exists two events $\mathbb{O}$ and $\mathbb{I}$, the *absurd* and the *certain* event respectively, such that

$$\forall w \in \mathcal{S} \ , \ p(w, \mathbb{O}) = 0 \ , \ p(w, \mathbb{I}) = 1 \ .$$

Axiom 2-$\mathcal{E}$. Let $E_1, E_2 \in \mathcal{E}$; then,

$$\forall w \in S, \ p(w, E_1) = p(w, E_2) \ \text{implies} \ E_1 = E_2 \ ;$$

(Indistinguishability principle of events).

Axiom 2-S. Let $w_1, w_2 \in S$; then,

$$\forall E \in \mathcal{E}, \ p(w_1, E) = p(w_2, E) \ \text{implies} \ w_1 = w_2 \ ;$$

(Indistinguishability principle of states).

Axiom 3. $\forall E \in \mathcal{E}, \exists E' \in \mathcal{E}$ (the *inverse* of E) such that

$$\forall w \in S, \ p(w, E) + p(w, E') = 1 \ .$$

Axiom 4-S. For every countable family of states $\{w_n\} \subseteq S$ and every corresponding family of non negative real numbers $\{\lambda_n\} \subseteq \mathbb{R}_+$, such that $\sum \lambda_n = 1$, a state $w \in S$ exists such that

$$\forall E \in \mathcal{E}, \ p(w, E) = \sum_{n=1}^{\infty} \lambda_n \, p(w_n, E) \ .$$

Axiom 5. For every sequence of events $\{E_n\} \subseteq \mathcal{E}$ which satisfies the condition $(\forall w \in S), (\forall i, j \text{ with } i \neq j)$,

$$p(w, E_i) + p(w, E_j) \leq 1$$

an event $E \in \mathcal{E}$ exists such that

$$\forall w \in S, \ p(w, E) = \sum_{n=1}^{\infty} p(w, E_n) \ .$$

Now, the SEVP structure $\mathbf{L}_{\text{SEVP}}$ so obtained from any SOP, in its turn, induces the following structure, usually called the *quantum logical* (QL) structure of the SEVP [based on the primitive notion of *event*, according to Gudder approach to quantum physics (Gudder 1970)]

$$\mathbf{L}_{\text{QL}} \equiv \langle \mathcal{E}, \mathcal{S}(\mathcal{E}), \mathcal{O}(\mathcal{E}) \rangle$$

where $\mathcal{E}$ is the non-empty set of all *events* (primitive notion of the theory), $\mathcal{S}(\mathcal{E})$ is a family of *probability measures* on $\mathcal{E}$ and $\mathcal{O}(\mathcal{E})$ is a family of $\mathcal{E}$-*valued measures* satisfying the following

Axiom QL$_1$. *The set of all events $\mathcal{E}$ has a structure*

$$\langle \mathcal{E}, \emptyset, \mathbb{I}, \leq, ' \rangle$$

of orthomodular σ-orthoposet with respect to
(1-i) The partial order relation $\leq$ on $\mathcal{E}$

(po) $E_1 \leq E_2$ iff $\forall w \in S$, $p(w, E_1) \leq p(w, E_2)$.

(The g.l.b. and l.u.b. in $\langle \mathcal{E}, \leq \rangle$, if they exist, will be denoted by $\wedge$ and $\vee$ respectively. This poset is bounded by the least element $\mathbb{O}$ and the greatest element $\mathbb{I}$, i.e., for all $E \in \mathcal{E}$, $\mathbb{O} \leq E \leq \mathbb{I}$.)

(2-ii) *The (standard) orthocomplementation mapping* $'$: $\mathcal{E} \mapsto \mathcal{E}$, $E \to E' = \mathbb{I} - E$

(oc) $\forall w \in S$, $p(w, E') = 1 - p(w, E)$.

(In particular, the 'non-contradiction' law $\forall E$, $E \wedge E' = \mathbb{O}$ and the 'excluded-middle' law $\forall E$, $E \vee E' = \mathbb{I}$ are derived using Axiom 5.)

(a) For any state $w \in S$, the mapping

$$p_w : \mathcal{E} \mapsto [0, 1]$$

defined, for arbitrary $E \in \mathcal{E}$, as

$$p_w(E) = p(w, E)$$

is a probability measure on $\mathcal{E}$, i.e.,

(pm-1) $p_w(\mathbb{O}) = 0$, $p_w(\mathbb{I}) = 1$

(pm-2) $\forall \{E_n : i \neq j, E_i \perp E_j\} \subseteq \mathcal{E}$,

$$p_w(\vee E_n) = \sum_{n=0}^{\infty} p_w(E_n) .$$

Axiom QL$_2$. *The set of all such probability measures on $\mathcal{E}$*

$$S(\mathcal{E}) := \{p_w : \mathcal{E} \mapsto [0, 1] \mid w \in S\}$$

(2-i) *is order-determining, i.e., for all $E_1, E_2 \in \mathcal{E}$*

if $p_w(E_1) \leq p_w(E_2)$ for any $p_w \in S(\mathcal{E})$, then $E_1 \leq E_2$;

(2-ii) *is σ-convex, i.e., for all $\{p_n\} \subseteq S(\mathcal{E})$ and all $\{\lambda_n\} \subseteq \mathbb{R}_+$ such that $\sum \lambda_n = 1$, we get that $\sum \lambda_n p_n \in S(\mathcal{E})$.*

• $S(\mathcal{E})$ is in a one-to-one and onto correspondence with the family of all states S according to the graph

$$S \equiv S(\mathcal{E})$$

$$w \longleftrightarrow p_w$$

(b) For any observable $A \in \mathcal{O}$, the mapping

$$E_A : \mathcal{B}(\mathbb{R}) \mapsto \mathcal{E}$$

defined, for arbitrary $\Delta \in \mathcal{B}(\mathbb{R})$, as

$$E_A(\Delta) := \chi_\Delta(A)$$

is an $\mathcal{E}$-valued measure, i.e.,

(vm-1) $E_A(\emptyset) = \Phi$, $E_A(\mathbb{R}) = \mathbb{I}$

(vm-2) $\Delta_1 \cap \Delta_2 = \emptyset$ implies $E_A(\Delta_1) \perp E_A(\Delta_2)$

(vm-3) $\forall \{\Delta_n : i \neq j, \Delta_i \cap \Delta_j = \emptyset\} \subseteq \mathcal{B}(\mathbb{R})$

$$E_A(\cup \Delta_n) = \vee E_A(\Delta_n) .$$

Axiom QL₃. *The set of all $\mathcal{E}$-valued measured*

$$\mathcal{O}(\mathcal{E}) := \{E_A : \mathcal{B}(\mathbb{R}) \mapsto \mathcal{E} \mid A \in \mathcal{O}\}$$

(3-i) *is a surjective family, i.e.*

$$(\forall E \in \mathcal{E}), (\exists A \in \mathcal{O}), (\exists \Delta \in \mathcal{B}(\mathbb{R})), \text{ s.t. } E_A(\Delta) = E;$$

(3-ii) *is closed with respect to the product for real Borel functions, i.e., for all $E_A \in \mathcal{O}(\mathcal{E})$ and all $f : \mathbb{R} \mapsto \mathbb{R}$, real Borel function, we have that $E_A \circ f^{-1} \in \mathcal{O}(\mathcal{E})$.*

• $\mathcal{O}(\mathcal{E})$ is in a one-to-one and onto correspondence with the family of all observables $\mathcal{O}$ according to the graph

$$\mathcal{O} \equiv \mathcal{O}(\mathcal{E})$$

$$A \longleftrightarrow (E_A : \mathcal{B}(\mathbb{R}) \mapsto \mathcal{E})$$

THEOREM 5.1. **'Representation Theorem'** (Maczynski 1973, Beltrametti and Cassinelli 1976). *From any triple*

$$\langle \mathcal{E}, \mathcal{S}(\mathcal{E}), \mathcal{O}(\mathcal{E}) \rangle$$

where $\mathcal{E}$ is an abstract orthomodular orthocomplemented σ-orthoposet, $\mathcal{S}(\mathcal{E})$ is an order-determining, σ-convex family of probability measures on $\mathcal{E}$, and $\mathcal{O}(\mathcal{E})$ is a surjective family of $\mathcal{E}$-valued measures, the triple

$$\langle \mathcal{S}, \mathcal{O}, \mathcal{P} \rangle$$

where $S = \mathcal{S}(\mathcal{E})$, $\mathcal{O} = \mathcal{O}(\mathcal{E})$ and $P : \mathcal{S}(\mathcal{E}) \times \mathcal{O}(\mathcal{E}) \times \mathcal{B}(\mathbb{R}) \mapsto [0,1]$ is the probability function defined as follows, $(\forall p_w \in \mathcal{S}(\mathcal{E}))$, $(\forall E_A \in \mathcal{O}(\mathcal{E}))$, $(\forall \Delta \in \mathcal{B}(\mathbb{R}))$,

$$P(p_w, E_A, \Delta) := (p_w \circ E_a)(\Delta)$$

is an SOP structure, i.e., it satisfies Mackey's axioms I, II, III, IV and V. Moreover, there is a full equivalence between SOP and QL structures.

6. M-PROPOSITIONS AND EVENTS

Let us consider the set $\mathcal{Y} = \mathcal{O} \times \mathcal{B}(\mathbb{R})$ of all *elementary statements*. Following Maczynski (1973), and Beltrametti and Cassinelli (1976), we introduce the following definitions.

(d5)   The relation $\cong$ on $\mathcal{Y}$ defined as

$$(A_1, \Delta_1) \cong (A_2, \Delta_2) \text{ iff}$$

$$(\forall w \in \mathcal{S}), \ P(w, A_1, \Delta_1) = P(w, A_2, \Delta_2)$$

is an equivalence relation, called the *M-equivalence* on elementary statements.

(d6)   For any elementary statement (A, Δ), we denote by $[(A, \Delta)]_{\cong}$ the equivalence class generated by (A, Δ) and we call it an *M-proposition*.

(d7)   The set of all M-propositions (i.e., the quotient set $\mathcal{Y}/\cong$) will be denoted by $\mathcal{L}_M$.

(d8) $(A_1, \emptyset) \cong (A_2, \emptyset)$ holds; the equivalence class generated by any element of the form $(A, \emptyset)$ is called the *absurd* M-proposition.

$(A_1, \mathbb{R}) \cong (A_2, \mathbb{R})$ holds; the equivalence class generated by any element of the form $(A, \mathbb{R})$ is called the *certain* M-proposition.

(d9) $(A_1, \Delta_1) \cong (A_2, \Delta_2)$ implies $(A_1, \mathbb{R}/\Delta_1) \cong (A_2, \mathbb{R}/\Delta_2)$; for every M-proposition $[(A, \Delta)]_{\cong}$, the M-proposition $[(A, \mathbb{R}/\Delta)]_{\cong}$ is well defined and is called the *opposite* or *inverse* of $[(A, \Delta)]_{\cong}$.

The following result allows us to identify M-propositions and events.

PROPOSITION 6.1. *Let* (A_1, Δ_1) *and* (A_2, Δ_2) *be two elementary statements and* $E_{A_1}(\Delta_1)$ *and* $E_{A_2}(\Delta_2)$ *the corresponding events. Then*

$$(A_1, \Delta_1) \cong (A_2, \Delta_2) \ \text{implies} \ E_{A_1}(\Delta_1) = E_{A_2}(\Delta_2) \,.$$

Hence, the mapping

$$\mathcal{L}_M \mapsto \mathcal{E}, \ [(A, \Delta)]_{\cong} \to E_A(\Delta)$$

is well defined and is one-to-one and onto.

We summarize the results up to now by the following diagram

elementary statements	$(A, \Delta) \in \mathcal{Y}$

$$\downarrow$$

M-propositions	$[(A, \Delta)]_{\cong} \in \mathcal{L}_M = (\mathcal{Y}/\cong) \longleftrightarrow \mathcal{E} \ni E_A(\Delta)$	events

In the above identification between M-propositions and events we have, in particular, that

$$[(A, \emptyset)]_{\cong} \quad \longleftrightarrow \quad \mathbb{0}$$
$$[(A, \mathbb{R})]_{\cong} \quad \longleftrightarrow \quad \mathbb{I}$$
$$[(A, \mathbb{R}/\Delta)]_{\cong} \quad \longleftrightarrow \quad \mathbb{I} - E_A(\Delta) \, .$$

In other words, the absurd M-proposition is identified with the absurd event, the certain M-proposition with the certain event, and the opposite of an M-proposition with the opposite of the corresponding event.

7. THE ORTHOFRAME BASED ON THE PHASE SPACE

In this section we consider the *event-probability measures* (on $\mathcal{E}$) structure

$$\langle \mathcal{E}, S(\mathcal{E}) \rangle$$

consisting of an orthomodular orthoposet $\mathcal{E}$ (the set of all *events*) and a separating, σ-convex family $S(\mathcal{E})$ of probability measures on $\mathcal{E}$ (in this section, for the sake of simplicity, $S(\mathcal{E})$ is denoted by S and, owing to the above identification, is called the *state space*). The state space can be endowed with a *physical separation* relation, written $\perp$, defined for arbitrary $p_1, p_2 \in S$ as follows

$$p_1 \perp p_2 \text{ iff } \exists E \in \mathcal{E}, \ p_1(E) = 1 \text{ and } p_2(E) = 0 \, .$$

"Two states are in relation $\perp$ iff an event exists whic physically 'separates' them" (Aerts 1983). In the context of the structure $(S, \perp)$ we will take into account a further axiom about Mackey's axiomatic quantum physics.

Axiom VIII. If E is any event different from $\mathbb{0}$ a state $p \in S$ exists such that $p(E) = 1$.

The separation relation is a *preclusivity* (Cattaneo and Nisticò 1989) (or *orthogonality* according to Foulis and Randall 1972) *relation*, i.e., it satisfies the conditions

(og-1) $p_1 \perp p_2$ implies $p_2 \perp p_1$ (symmetric)

(og-2) $p_1 \perp p_2$ implies $p_1 \neq p_2$ (irreflexive)

The structure $(S, \perp)$, consisting of the nonempty set S of *semantical worlds* or *states* and the preclusivity (or orthogonality) relation $\perp$ of *physical separation*, is a *preclusivity space* (or orthogonality space).

Two subsets H and K of S are said to be *mutually preclusive*, written $H(\perp) K$, iff $h \perp k$ for all $h \in H$ and $k \in K$. If p is an element of S, the symbol

$p \perp H$ means that $\{p\}\,(\perp)\,H$. For every subset H of S, we call *preclusive orthocomplement* of H the subset $H^{\perp} := \{p \in S \;:\; p \perp H\}$ of S. In the sequel we set $H^{\perp\perp} = (H^{\perp})^{\perp}$, $H^{\perp\perp\perp} = (H^{\perp\perp})^{\perp}$, etc.

PROPOSITION 7.1. *The structure based on the power set $\mathcal{P}(S)$ of S*

$$\left\langle \mathcal{P}(S), \emptyset, S, \subseteq, \;^{\perp} \right\rangle$$

where $\subseteq$ is the usual set theoretic inclusion and $^{\perp}$ is the mapping

$$^{\perp} \;:\; \mathcal{P}(S) \mapsto \mathcal{P}(S), \; H \to H^{\perp}$$

is a complete lattice with intuitionistic-like orthocomplementation; that is for arbitrary $H, K \subseteq S$ we have that

(ioc-1) $H \subseteq H^{\perp\perp}$

(ioc-2) $H \subseteq K$ implies $K^{\perp} \subseteq H^{\perp}$

(ioc-3) $H \cap H^{\perp} = \emptyset$

(We recall that in a lattice with intuitionistic-like orthocomplementation, from conditions (ioc-1)–(ioc-3) the following results are valid: $H^{\perp} = H^{\perp\perp\perp}$ and $H^{\perp} \cap K^{\perp} = (H \cup K)^{\perp}$, the latter being a de Morgan law. In general neither the dual de Morgan law or the excluded middle law $H \cup H^{\perp} = S$ hold.)

A set $X \subseteq S$ is said to be an *exact* set (or an FR *perp-closed* set) iff $X = X^{\perp\perp}$. The set of all exact sets of $(S, \perp)$ is denoted by

$$\mathcal{L}(S, \perp) := \{X \subseteq S \;:\; X = X^{\perp\perp}\}$$

The trivial subsets $\emptyset$ and X are elements of $\mathcal{L}(X, \perp)$.

PROPOSITION 7.2. *With respect to the ordering of set theoretic inclusion the structure*

$$\left\langle \mathcal{L}(S, \perp), \emptyset, S, \subseteq, \;^{\perp} \right\rangle$$

is a complete lattice with the standard orthocomplementation

$$^{\perp} \;:\; \mathcal{L}(S, \perp) \mapsto \mathcal{L}(S, \perp), \; X \mapsto X^{\perp}.$$

In particular, for any family $\{X_j\} \subseteq \mathcal{L}(S, \perp)$ the greatest lower bound (g.l.b.), written ΔX_j, exists and it is

$$\Delta X_j = \cap X_j$$

i.e., it is the set theoretic intersection, and the least upper bound (l.u.b.), written ∇X_j, exists and it is

$$\nabla X_j = (\cup X_j)^{\perp\perp}$$

which contains, and in general does not coincide with the set theoretic union.

The binary relation of *orthogonality* induced from this orthocomplementation

$$X \perp Y \text{ iff } X \subseteq Y^{\perp} \text{ (iff } Y \subseteq X^{\perp})$$

is just the relation of mutual preclusion $X(\perp)Y$ introduced above. For any given $H \subseteq S$, we can construct the set

$$\bar{H} := \cap \{X \in \mathcal{L}(S, \perp) : H \subseteq X\}$$

which, as intersection of exact sets, is an exact set too; the exact set $\bar{H}$ is called the *closure* or *exact envelope* of H and trivially we get that $\bar{H} = H^{\perp\perp}$. Let H, K be two subsets of the preclusivity space $(S, \perp)$, then $\bar{H}^{\perp} = H^{\perp}$; moreover $H \perp K$ implies $\bar{H} \perp \bar{K}$.

The preclusivity space $(S, \perp)$ is said to be *discretely separating* iff any singleton $\{p\}$ is an exact set for all $p \in S$; in this case it is apparent that $\overline{\{p\}} = \{p\}$ and $\mathcal{L}(S, \perp)$ turns out to be an atomic (orthocomplemented complete) lattice whose atoms are just the singletons $\{p\}$.

For every fixed event $E \in \mathcal{E}$ we introduce the following two subsets of the state space:

- the *certainly-true* domain of event E, denoted by $X_T(E)$, and defined as follows

$$X_T(E) := \{p \in S : p(E) = 1\}$$

i.e., the set of all states (semantical worlds) in which the event E occurs with certainty (probability 1);

- the *certainly-false* domain of event E, denoted by $X_F(E)$, and defined as follows

$$X_F(E) := \{p \in S : p(E) = 0\}$$

i.e., the set of all states (semantical worlds) in which the event E does not occur with certainty (probability 0).

From the definition of the preclusivity relation of physical separation it follows immediately that for every event E we have that

$$X_T(E) \perp X_F(E) .$$

This allows us to introduce the mapping, called the *extensional mapping*, defined as follows

$$\text{ext} : \mathcal{E} \mapsto \mathcal{P}(S, \perp) \times \mathcal{P}(S, \perp), \ E \rightarrow (X_T(E), X_F(E))$$

whose *closure* is the mapping

$$\overline{\text{ext}} : \mathcal{E} \mapsto \mathcal{L}(S, \perp) \times \mathcal{L}(S, \perp), \ E \rightarrow (\overline{X_T(E)}, X_T(E)^{\perp})$$

which gives rise to the following diagram

$$E \in \mathcal{E} \;\xrightarrow{\text{ext}}\; \mathcal{P}(\mathcal{S},\perp) \times \mathcal{P}(\mathcal{S},\perp) \;\ni\; (X_T(E),\, X_F(E))$$

$$\searrow^{\overline{\text{ext}}} \qquad\qquad \bigcup | \qquad\qquad\qquad \downarrow \text{cl}$$

$$\mathcal{L}(\mathcal{S},\perp) \times \mathcal{L}(\mathcal{S},\perp) \;\ni\; (\overline{X_T(E)},\, X_T(E)^{\perp})$$

where $X_T(E) \subseteq \overline{X_T(E)}$, $X_F(E) \subseteq X_T(E)^{\perp}$ and $\overline{X_T(E)} \perp X_T(E)^{\perp}$. More-over, owing to Axiom VIII, for every event E different from $\emptyset$ we have that $X_T(E) \neq \emptyset$.

Taking into account the diagram of Section 6, connecting elementary state-ment of the form (A, Δ) with the set of all events (or all M-propositions), we can complete it in the following way.

$$\boxed{\text{elementary statements}}\;\; (A,\Delta) \in \mathcal{Y}$$

$$\downarrow \Phi$$

$$\boxed{\text{M-propositions}}\;\; [(A,\Delta)]_{\simeq} \in \mathcal{L}_M \;\xrightarrow{\Psi}\; \mathcal{E} \ni E_A(\Delta) \;\boxed{\text{events}}$$

$$\downarrow \qquad\qquad\qquad\qquad \downarrow \text{ext}$$

$$(\mathcal{S}_T(A,\Delta),\, \mathcal{S}_F(A,\Delta)) \;\longleftrightarrow\; \Big(X_T(E_A(\Delta)),\, X_F(E_A(\Delta))\Big)$$

$$\downarrow \text{cl}$$

$$\Big(\overline{X_T(E_A(\Delta))},\, X_T(E_A(\Delta))^{\perp}\Big)$$

We recall that the 'sentential logic' of a statistical theory, whose 'spe-cific alphabet' consists of the 'elementary statements' (or 'physical questions') 'val$A \in \Delta$', identified with the 'yes–no devices' $(A, \Delta) \in \mathcal{Y}$ they are tested by, has a structure

$$\langle \mathcal{Y}, \varrho, \&, \neg \rangle \;.$$

In the case of Mackey's axiomatic quantum theory as a statistical theory the orthoframe based on the state space gives rise to an algebraic structure

$$\Big\langle \mathcal{L}(\mathcal{S},\perp), \nabla, \Delta, \;^{\perp} \Big\rangle$$

which is a complete lattice with standard orthocomplementation. The mapping

$$h := (\text{cl} \circ \text{ext} \circ \Psi \circ \Phi) : \mathcal{Y} \mapsto \mathcal{L}(\mathcal{S},\perp)$$

defined by the rule

$$(A,\Delta) \to h(A,\Delta) := \Big(\overline{X_T(E_A(\Delta))},\, X_T(E_A(\Delta))^{\perp}\Big)$$

$$\updownarrow$$

$$\overline{X_T(E_A(\Delta))} \in \mathcal{L}(\mathcal{S},\perp)$$

could be a good candidate for a 'Lindenbaum–Tarski algebra' of the sentential logic of quantum theory based on the physical orthoframe $(\mathcal{S}, \perp)$ of all 'physical states' (or 'semantical worlds') endowed with the preclusivity relation of 'physical separation'. This conjecture shows a lot of problems with respect to the actual Mackey's axiomatization, which we now list.

- The set $X_T(E_A(\Delta))$ of all semantical worlds (physical states) in which sentence (A, Δ) occurs with certainty (probability 1) in general does not belong to $\mathcal{L}(\mathcal{S}, \perp)$; precisely, in general

$$X_T(E_A(\Delta)) \subseteq \overline{X_T(E_A(\Delta))} \in \mathcal{L}(\mathcal{S}, \perp) .$$

  Semantical worlds from $\overline{X_T(E_A(\Delta))} \setminus X_T(E_A(\Delta))$ could be considered as particular situations which make (A, Δ) 'true' without giving probability 1 to it.
- The set $X_F(E_A(\Delta))$ of all semantical worlds (physical states) in which sentence (A, Δ) does not occur with certainty (probability 0) in general does not belong to $\mathcal{L}(\mathcal{S}, \perp)$; i.e., in general

$$X_F(E_A(\Delta)) = X_T(E_A(\Delta)') \subseteq X_T(E_A(\Delta))^\perp .$$

  Semantical worlds from $X_T(E_A(\Delta))^\perp \setminus X_F(E_A(\Delta))$ could be considered as particular situations which make (A, Δ) 'false' without giving probability 0 to it.
- No weak form of distributivity (orthomodularity, modularity, distributivity) is inherited from the set of all events $\mathcal{E}$, which has a structure of orthomodular σ-orthoposet.

8. STATE-EFFECT-PROBABILITY (SEFP) STRUCTURE

The notion of *elementary experiment* as an experiment arrangement decomposable into a *preparing* part and a *measuring* part, can be applied to some physical situations which are not completely described by the set of all events of a SEVP structure; in general, these new physical situations oblige to take into account, as the starting point of an axiomatization, a class of measuring apparatuses testing *effects*; such a class turns out to be larger than the class of all events. A *state–effect–probability* (SEFP) structure is a triple

$$\mathbf{L}_{\text{SEFP}} \equiv \langle \mathcal{S}, \mathcal{F}, p \rangle$$

where

(a1) $\mathcal{S}$ is a not empty set, whose elements are denoted by $w, v, \ldots$, interpreted as *states*, realized by macroscopic apparatuses which prepare both single samples and ensembles of identical non interacting physical objects under well defined and repeatable conditions.

(a2) $\mathcal{F}$ is a not empty set, whose elements are denoted by $f, g, \ldots$, interpreted as *effects*, tested by dichotomic measuring macroscopic devices which, when interacting with a single sample of the physical entity, may or may not produce a certain definite macroscopic yes–no alternative. The occurrence of the alternative is taken as the answer 'yes' and its absence as the answer 'no'.

(a3) P is a mapping

$$p : \mathcal{S} \times \mathcal{F} \mapsto [0,1] .$$

For any pair $(w, f) \in \mathcal{S} \times \mathcal{F}$, the value

$$p(w, f) \in [0,1]$$

represents the probability of the occurrence of the 'yes' alternative for the effect f when the physical entity is prepared in w.

The basic Axioms of the SEFP structure are the following.

Axiom 1. There exists two effects $\emptyset$ and $\mathbb{I}$, the *absurd* and the *certain* effect, respectively, such that

$$\forall w \in \mathcal{S} , \ p(w, \emptyset) = 0 , \ p(w, \mathbb{I}) = 1 .$$

Axiom 2-$\mathcal{F}$. Let $f_1, f_2 \in \mathcal{F}$; then,

$$\forall w \in \mathcal{S} , \ p(w, f_1) = p(w, f_2) \ \text{implies} \ f_1 = f_2 ;$$

(Indistinguishability principle of effects).

Axiom 2-$\mathcal{S}$. Let $w_1, w_2 \in \mathcal{S}$; then,

$$\forall f \in \mathcal{F} , \ p(w_1, f) = p(w_2, f) \ \text{implies} \ w_1 = w_2 ;$$

(Indistinguishability principle of states).

Axiom 3. $\forall f \in \mathcal{F}, \ \exists f' \in \mathcal{F}$ (the *inverse* of f) such that

$$\forall w \in \mathcal{S} , \ p(w, f) + p(w, f') = 1 .$$

Axiom 4-$\mathcal{F}$. For every n-tuple of effects $\{f_1, \ldots, f_n\} \subseteq \mathcal{F}$ and every n-tuple of non negative real numbers $\{\lambda_1, \ldots, \lambda_n\} \subseteq \mathbb{R}_+$ such that $\sum_{j=1}^{n} \lambda_j = 1$, an effect $\sum_{j=1}^{n} \lambda_j f_j \in \mathcal{F}$ exists (the *convex combination* of the f_j with *weights* λ_j) such that

$$\forall w \in \mathcal{S} , \ p\left(w, \sum_{j=1}^{n} \lambda_j f_j\right) = \lambda_j p(w, f_j) .$$

Axiom 4-S. For every countable family of states $\{w_n\} \subseteq S$ and every corresponding family of non negative real numbers $\{\lambda_n\} \subseteq \mathbb{R}_+$, such that $\sum \lambda_n = 1$, a state $w \in S$ exists (the *mixture* of the w_n with *weights* λ_n) such that

$$\forall f \in \mathcal{F}, \ p(w, f) = \sum_{n=1}^{\infty} \lambda_n p(w_n, f) \, .$$

(Comparing with the Axioms of an SEVP structure outlined in Section 5.1 we see that in an SEFP structure, Axiom 5, which assures that the orthomplementation mapping is standard, does not hold; on the contrary in any SEFP structure Axiom 4-$\mathcal{F}$, which assures the convex combination of effects, holds.)

In any SEFP structure, owing to Axiom 2-$\mathcal{F}$ (indistinguishability of effects) and Axiom 3 (existence of an inverse) we have that for every effect f its inverse f' is unique. Hence, the following mapping, which associates to any effect its unique inverse, is well defined

$$' : \mathcal{F} \mapsto \mathcal{F}, \, f \to f' \, .$$

In particular $\mathbb{O} = \mathbb{I}'$ and $\mathbb{I} = \mathbb{O}'$. We extend now to effects the notions of certainty domains introduced in Section 7. Precisely, an effect f is *true* in a state w, written $(w, f)\, T$, iff $p(w, f) = 1$ and is *false* in a state w, written $(w, f)\, F$, iff $p(w, f) = 0$. The *certainly-yes* and the *certainly-no* domains of $f \in \mathcal{F}$ are defined as follows

$$S_1(f) = \{w \in S \ : \ p(w, f) = 1\}$$
$$S_0(f) = \{w \in S \ : \ p(w, f) = 0\} \, .$$

By a *question* (also *yes–no measurement*) one usually understands any physical arrangement which, when interacting with a microobject, may or may not produce a certain macroscopic *effect* interpreted as the answer 'yes'. Though the 'question' may be put to any *single* microobject, the answer becomes conclusive only if obtained for a great number of its independent replies. This leads to an abstract scheme where 'questions' idealize the macroscopic devices used to test the statistical ensembles of microsystems. (...) Given an ensemble w and a question f, one says that the answer 'yes' to the question f is *certain* for the individuals of w if 'yes' is obtained for the average fraction 1 of the individuals of that ensemble.
The ensembles w for which the answer 'yes' to the question f is certain will be told to form the *certainly-yes domain* of f (Mielnik 1976).

In any SEFP structure some phenomenological binary relations on the set of all effects $\mathcal{F}$ can be introduced.

(PR$_1$) Phenomenological partial order relation.

$$f \leq g \ \text{iff} \ \forall w \in S, \ p(w, f) \leq p(w, g) \, .$$

(PR$_2$) JP-phenomenological quasi-order relation.

$$f < g \quad \text{iff} \quad (w, f)\,T \text{ implies } (w, g)\,T$$
$$\text{iff} \quad S_1(f) \subseteq S_1(g)\,.$$

(PR$_3$) CN-phenomenological quasi-order relation.

$$f <_0 g \quad \text{iff} \quad (w, g)\,F \text{ implies } (w, f)\,F$$
$$\text{iff} \quad S_0(g) \subseteq S_0(f)\,.$$

Trivially $f \leq g$ implies $f < g$ and $f <_0 g$. With respect to the JP quasi-order relation, we quote the following interesting assertion of Mielnik (1976), since a very strong condition is assumed to hold in his approach, which is not required for SEFP structure:

The existence of statistical ensembles as the counterpart of $\mathcal{F}$ allows us to introduce a certain structure in $\mathcal{F}$ which is the most recognized element of geometry in quantum theory. (...) Given two questions f, g, we say that f is *more restrictive* than $g, f < g$, if the certainty of the answer 'yes' to the question f implies the certainty of 'yes' to the question g. Thus, $f < g$ if the 'certainly-yes' domain of f is contained in the 'certainly-yes' domain of g. [$f < g$ iff $S_1(f) \subseteq S_1(g)$] (...).

The relation $<$ is *reflexive* and *transitive*. The further properties of $<$ are associated with the *logical interpretation*. According to that interpretation the questions f represent the elements of an abstract '*logic*' which reflects the nature of the microsystem:

the relation $<$ is the *implication* of the logic.

Since in any logical system the pair of implications $f \Rightarrow g$ and $g \Rightarrow f$ means that 'f is equivalent g', one generally assumes that a similar property should hold in $\mathcal{F}$.

AXIOM ID (*Identity Axiom*).
Two questions f, g with identical 'certainly-yes' domains [i.e., $S_1(f) = S_1(g)$] are physically equivalent (i.e., cannot be distinguished by observing how they select any statistical ensemble). Formally,

$$f < g \text{ and } g < f \quad \Rightarrow \quad f = g\,.$$

In consequence, the relation $<$ introduces a partial order in $\mathcal{F}$.

On the contrary, in a SEFP structure the 'Identity Axiom' is *not* assumed to hold.

THEOREM 8.1. *The structure* $\langle \mathcal{F}, \mathbb{O}, \leq, ' \rangle$,

(i)   *is a poset bounded by the least element* $\mathbb{O}$ *and by the greatest element* $\mathbb{I}$;

(ii)   *the mapping* $' : \mathcal{F} \mapsto \mathcal{F}$, $f \to f'$ *is a fuzzy orthocomplementation* (Cattaneo and Nisticó 1989), *i.e.*,

(doc-1)   *For every* $f \in \mathcal{F}$, $f = f''$;

(doc-2)   *let* $f, g \in \mathcal{F}$; *then,* $f \leq g$ *implies* $g' \leq f'$;

(re)   *if* $f \leq f'$ *and* $g' \leq g$ *then* $f \leq g$.

The 'non-contradiction' law $(\forall f,\ f \wedge f' = \emptyset)$ *and the 'excluded-middle' law* $(\forall f,\ f \vee f' = \mathbb{I})$ *in general do not hold (since Axiom 5 of SEVP structure is not required to be satisfied here).*

Almost all the SEFP structures contain the *semi-transparent* effect $(\frac{1}{2}\mathbb{I})$ characterized by the property

$$\forall w \in \mathcal{S},\ P\left(w, (\tfrac{1}{2}\mathbb{I})\right) = \tfrac{1}{2}.$$

Because of Axiom 3

$$(\tfrac{1}{2}\mathbb{I}) = (\tfrac{1}{2}\mathbb{I})'$$

From this we get,

$$(\tfrac{1}{2}\mathbb{I}) \wedge (\tfrac{1}{2}\mathbb{I})' = (\tfrac{1}{2}\mathbb{I}) = (\tfrac{1}{2}\mathbb{I})' = (\tfrac{1}{2}\mathbb{I}) \vee (\tfrac{1}{2}\mathbb{I})'$$

Therefore,

$$(\tfrac{1}{2}\mathbb{I}) \wedge (\tfrac{1}{2}\mathbb{I})' \neq \emptyset \quad (\tfrac{1}{2}\mathbb{I}) \vee (\tfrac{1}{2}\mathbb{I})' \neq \mathbb{I}.$$

Moreover, it is not correct to conclude from the equalities above that $\emptyset = \mathbb{I}$. The structure of effects does *not* coalesce in a unique element. We quote the following remark by Mielnik which explains the different point of views:

In spite of its elegant generality, the idea of a 'question' as a *quite arbitrary macroscopic arrangement which produces a certain macroscopic alternative effect* is wrong. To illustrate this, consider a statistical ensemble of any objects and a macroscopic device which yields the answer 'yes' for an average of $\frac{1}{2}$ of them in a completely random way. A good approximation is a *semi-transparent* mirror in the path of a photon beam. No doubt, this is a certain macroscopic arrangement producing a macroscopic alternative effect:

either the photon reaches the screen 'yes' or it does not.

However, the arrangement *cannot be considered* one of 'questions'. *If it were, it would produce a sequence of catastrophes in the structure of 'quantum logic'.*

First of all, it would not be clear which device is the 'negative' of the semitransparent mirror a. By insisting on the purely verbal solution (just the interchange of 'yes' and 'no') one would conclude that a' is acting, in fact, identically as a: for it too gives the answer 'yes' in a completely random way for an average $\frac{1}{2}$ of the beam photon. Thus

$$a = a'.$$

This would further imply

$$\emptyset = a \wedge a' = a \wedge a = a = a \vee a = a \vee a' = \mathbf{I}$$

and so, *the whole structure of $\mathcal{F}$ would collapse* (Mielnik 1976).

Summarizing, the metatheoretical principles involved by SEFP structures are the following.

(MT$_1$) The idea of an *effect* as an objectively traceable alternative occurring in a quite arbitrary macroscopic arrangement.

(MT$_2$) The *indistinguishability principle* of effects: two effects, to be different, must have different probability distributions in at least one state. The Identity Axiom does *not* hold.

(MT$_3$) The *uniqueness* of the orthocomplement: the inverse of any effect is unique. The 'non-contradiction' law and the 'excluded-middle' law do *not* hold.

We give now some definitions according to the Jauch–Piron (JP) approach to quantum physics (QP) (Jauch 1971, Jauch and Piron 1969, Piron 1964, 1972, 1976a, 1976b, 1977, 1978, 1981) which can be introduced into SEFP structures (Cattaneo, Garola and Nisticó 1989).

(D1) The *absurd* or *impossible* effect φ and the *certain* effect $\mathbb{I}$;

(D2) $f < g$ iff $S_1(f) \subseteq S_1(g)$ (quasi-order relation!);

(D3) $f \sim g$ iff $f < g$ and $g < f$ (equivalently iff $S_1(f) = S_1(g)$);

(D4) a JP *proposition* is any equivalence class of effects with respect to $\sim$;

(D5) $\mathcal{L}(\mathcal{F}) = \mathcal{F}/\sim$ is the set of all JP propositions, whose elements are denoted by $a, b, c, \ldots$;

(D6) the *satisfaction domain* $S_1([f]_\sim)$ of a proposition $[f]_\sim$ is the certainly-yes domain of any effect in $[f]_\sim$.

Whenever a state w belongs to the satisfaction domain of a JP-proposition, we say that w is a state in which the JP-proposition (or the properties associated to it) is (are) 'true';

(D7) $0 := [\varphi]_\sim$ and $1 := [\mathbb{I}]_\sim$ are the *impossible* and the *certain* JP-propositions; in particular: $1 = \{\mathbb{I}\}$;

(D8) $\subseteq$ is the partial order relation induced on $\mathcal{L}(\mathcal{F})$ by the quasi-order relation $<$.

The semi-transparent effect $(\frac{1}{2}\mathbb{I})$ is such that $S_1(\frac{1}{2}\mathbb{I}) = S_1(\varphi) = \emptyset$ and so, it belongs to the impossible proposition $0 = [\varphi]_\sim$ generated by the absurd question; therefore, it verifies the same properties associated to the impossible proposition. Compare with Mielnik (1976):

One might reply, that the Axioms of quantum logic are exact, but they must be properly understood. The semi-transparent mirror is not a good example of a 'question' since it is not at all a measuring device: *it does not verify any physical property of the transmitted photons*. This is a good answer, but it means that the whole approach of 'quantum logic' should start from an information which is inverse to the usually given. *Not every arrangement producing a macroscopic alternative effect is a question* – the right information should read.

In a JP-proposition, the effects are characterized by having the same certainly-yes domain. In general, they have different certainly-no domains. Even if the effects f and g generate the same JP-proposition one cannot deduce that f' and g' generate the same JP-proposition, i.e., if $f \sim g$ (i.e.,

$S_1(f) = S_1(g)$) in general $f' \not\sim g'$ (i.e., $S_1(f') = S_0(f) \neq S_0(g) = S_1(g')$).

We do agree with the following assertion of Mielnik's:

It is not a logical impossibility to imagine a hypothetical physical world where to every domain of micro-objects with a certain specific property [i.e., $S_1([f]_\sim)$] there would be many possible 'complementing domain' [i.e., $S_0(f_1) \neq S_0(f_2) \neq \ldots$, for $f_1, f_2, \ldots \in [f]_\sim$] corresponding to many possible ways of being 'opposite' to that property. $(\ldots)$

For each of those devices [i.e., $f_1, f_2, \ldots$] the verbal negation operation (yes $\leftrightarrow$ no) could be easily performed leading to a sequence of devices $f'_1, f'_2, \ldots$ with different 'certainly-yes' domains $S_1(f'_1) \neq S_1(f'_2) \neq \ldots (\ldots)$ The devices $f_1, f_2, \ldots$ would be physically different, and even if we tried to neglect the difference $(\ldots)$, the negatives $f'_1, f'_2, \ldots$ could no longer be identified. $(\ldots)$ In consequence, *something would be broken in the assumed structure of $\mathcal{F}$: either the Identity Axiom or the uniqueness of the orthocomplement* (Mielnik 1976).

By contrast, in SEFP structures every arrangement producing a macroscopic alternative effect is a question, the Identity Axiom is broken (and substituted by the Indistinguishability principle of effects) and the Uniqueness of the orthocomplement holds (but without the non-contradiction law and the excluded-middle law).

9. COMPLETE SEFP STRUCTURES

We underline that the passage from *effects*, $\mathcal{F}$, to JP-propositions, $\mathcal{L} := \mathcal{F}/\sim$, has been done by the equivalence relation on effects (D3). Hence, one has the two posets:

$\langle \mathcal{F}, \emptyset, \leq '\rangle$ Poset with Fuzzy Orthocomplementation of all *Effects*

$$f \leq g \text{ iff } \forall w, P(w, f) \leq P(w, g)$$

$$f \to f' \text{ iff } \forall w, P(w, f') = 1 - P(w, f)$$

$\langle \mathcal{L}, [\emptyset]_\sim, \subseteq \rangle$ Poset of all *Propositions*

$$\mathbf{a} \subseteq \mathbf{b} \text{ iff } f \in \mathbf{a},\ g \in \mathbf{b},\ \begin{cases} S_1(f) \subseteq S_1(g) \\ P(w, f) = 1 \Rightarrow P(w, g) = 1 \end{cases}$$

We define the *extension* of any effect f as follows:

$$\text{ext}(f) := (S_1(f),\ S_0(f))$$

PROPOSITION 9.1. *The extension mapping satisfies the following:*

(i) $\text{ext}(\emptyset) = (\emptyset, \mathcal{S})$ *and* $\text{ext}(\mathbb{I}) = (\mathcal{S}, \emptyset)$.

(ii)   *For every* $f \in \mathcal{F}$, $\text{ext}(f') = (S_1(f), S_0(f))$.

(iii) *For every convex combination of effects* $\sum_{j=1}^{n} \lambda_j f_j$

$$\text{ext}\left(\sum_{j=1}^{n} \lambda_j f_j\right) = \left(\bigcap_{j=1}^{n} S_1(f_j), \bigcap_{j=1}^{n} S_0(f_j)\right).$$

In particular, for the product effect πf_j (i.e., the convex combination with $\forall \lambda_j = 1/n$)

$$\text{ext}(\pi f_j) = \left(\bigcap_{j=1}^{n} S_1(f_j), \bigcap_{j=1}^{n} S_0(f_j)\right).$$

DEFINITION 9.1. A SEFP structure is said to be *complete* iff the following Axioms hold:

Axiom JPc. For every set $\mathcal{G}$ of effects, an effect $\pi_{\mathcal{G}} f$ exists (the *product* of $\mathcal{G}$) such that

$$\text{ext}(\pi_{\mathcal{G}} f) = \left(\bigcap_{\mathcal{G}} S_1(f), \bigcap_{\mathcal{G}} S_0(f)\right).$$

Axiom CC. For every effect f, an effect $\nu(f)$ exists (the *necessity* of f) such that

(1) $S_1(\nu(f)) = S_1(f)$;

(2) $S_1(g) = S_1(\nu(f))$ implies $\nu(f) \leq g$;

(3) $S_1(h) = S_1(\nu(f)')$ implies $\nu(f)' \leq h$.

The set $\mathcal{E} := \{\nu(f) : f \in \mathcal{F}\}$ is the set of all *events* whose elements are denoted by $a, b, c, \ldots$ in the sequel.

Trivially, $\mathbb{O}$ and $\mathbb{I}$ are events. Moreover, for every effect f, the event $\nu(f)$ assured by Axiom CC is unique; hence the surjective mapping is well defined

$$\nu : \mathcal{F} \mapsto \mathcal{E}, \ f \to \nu(f).$$

In any complete SEFP, $f \sim g$ implies $\nu(f) = \nu(g)$; i.e., a proposition contains a unique event. This result allows the scheme

Effect $\mathcal{F}$	
f	
$\Theta \downarrow \qquad \searrow \nu$	
$[f]_\sim \quad \overset{\Omega}{\longleftrightarrow} \quad \nu(f)$	
Proposition $\mathcal{L}$	Event $\mathcal{E}$

Where the one-to-one and onto mapping $\Omega : \mathcal{L} \mapsto \mathcal{E}$ identifies propositions and events: $\nu(f)$ is the unique event belonging to the proposition $[f]_\sim$. Ω is also a partial ordering isomorphism

$$[f]_\sim \subseteq [g]_\sim \text{ iff } S_1(f) \subseteq S_1(g) \text{ iff } \nu(f) \leq \nu(g) \,.$$

We have seen that to every proposition $[f]_\sim$ it is associated a certain number of properties of the physical entity; we can say that any effect $\hat{f} \in [f]_\sim$ 'tests' all the properties associated to proposition $[f]_\sim$. The event $\nu(f) \in \mathcal{E}$ belongs to proposition $[f]_\sim$ and so it reveals those samples of the physical entity which posses all the properties associated to $[f]_\sim$; it is the more selective among all other measurements available in $[f]_\sim$ (i.e., $\hat{f} \in [f]_\sim$ implies $\nu(f) \leq \hat{f}$) and it does not reveal, as much as possible, the ones which do not posses these properties (i.e., $\nu(f)$ has the greatest certainly-no domain with respect to all the other effects in the same proposition: $S_0(\nu(f)) = \cup\{S_0(\hat{f}) : \hat{f} \in [\nu(f)]_\sim\}$); "it minimizes the randomness of the 'no' answer" (Mielnik 1976).

9.1. Brouwer–Zadeh Posets in Complete SEFP

In any complete SEFP the structure

$$\langle \mathcal{F}, \emptyset, \leq, ', \nu \rangle$$

where $<$ is the *phenomenological partial order relation on $\mathcal{F}$*

(po) $f \leq g$ iff $\forall w \in \mathcal{S}, \ p(w, f) \leq p(w, g)$;

the mapping $' : \mathcal{F} \mapsto \mathcal{F}$ is the *fuzzy orthocomplementation*

(oc) $f \mapsto f'$ with $\forall w \in \mathcal{S}, \ p(w, f') = 1 - p(w, f)$;

the mapping $\nu : \mathcal{F} \mapsto \mathcal{E}$ is the *necessity operator*

(ne) $f \mapsto \nu(f)$ which satisfies $CC(1) - (3)$;

is a *Brouwer–Zadeh (BZ)-poset* (Cattaneo and Nisticó 1989, Cattaneo, Garola and Nisticó, 1989) generated by the fuzzy orthocomplementation $'$ and the necessity operator ν. As usual, the intuitionistic orthocomplementation is defined as

$$f^\sim := \nu(f') \,.$$

Moreover, since $a \in \mathcal{E}$ iff $a = a^{\sim\sim}$, the set $\mathcal{E}$ of all events is the set of all *exact* or *sharp effects* and so $\mathcal{F} \setminus \mathcal{E}$ is the set of all *fuzzy* or *unsharp effects* (Cattaneo and Nisticó 1989); hence, the properties of a proposition $[f]_\sim$ are tested by a unique exact or sharp effect, the event $\nu(f)$, all the other effects from the same proposition represent *fuzzy* or *unsharp* realizations of this event.

With respect to the set of all events we have the following result.

THEOREM 9.1. *In any complete SEFP*, (see Cattaneo and Nisticó 1985, Cattaneo, Garola and Nisticó 1989)

(i) $\emptyset$ *and* $\mathbb{I}$ *are events.*

(ii) *Any proposition* $[f]_\sim$ *contains a unique event.*

(iii) *An effect* $a \in \mathcal{F}$ *is an event iff* a' *is an event (hence, the restriction to* $\mathcal{E}$ *of the mapping* $'$*, denoted by* $'$ *too, maps* $\mathcal{E}$ *onto* $\mathcal{E}$*).*

(iv) *If we denote by* $\leq$ *the restriction to* $\mathcal{E}$ *of the partial order relation defined in* $\mathcal{F}$*, then the set of all events* $\langle \mathcal{E}, \emptyset, \leq, ' \rangle$ *is a complete lattice (owing to Axiom JPc) with standard orthocomplementation in which all the phenomenological binary relations coalesce*

$$a \leq b \text{ iff } \left\{ \begin{array}{l} a < b \\ S_1(a) \subseteq S_1(B) \end{array} \right. \text{ iff } \left\{ \begin{array}{l} a <_0 b \\ S_0(b) \subseteq S_0(a) \end{array} \right.$$

(v) *The set* S *of all states strongly determines the ordering on* $\mathcal{E}$ *(Pool 1968a); i.e., for arbitrary* $a, b \in \mathcal{E}$,

$$S_1(a) \subseteq S_1(b) \text{ iff } a \leq b$$

(equivalently, $S_1(a) \subseteq S_0(b)$ *iff* $a \perp b$*). In particular Mielnik's* **Identity Axiom (ID)***:* $S_1(a) = S_1(b)$ *implies* $a = b$ *holds on the set of all events* $\mathcal{E}$.

(vi) *If Mackey's Axiom VIII (if* $f \neq \emptyset$ *there is a* $w \in S$ *such that* $p(w, f) = 1$*) is extended to SEFP, then the set of all states* S *is full (i.e., closed under mixtures, with identity and* VIII *axioms) (Gudder 1970).*

Of course, the *preclusivity* relation of *physical separation* of states by events, introduced in Section 7, can be extended to any SEFP as a preclusivity relation $\perp_\mathcal{F}$ of states by effects producing the *preclusivity space* $\mathcal{L}(S, \perp_\mathcal{F})$. One has the following

THEOREM 9.2. *In any complete SEFP*, (see Cattaneo and Nisticó 1985)

(i) $w_1 \perp_\mathcal{F} w_2$ *iff* $w_1 \perp w_2$*, since for every effect* $f \in \mathcal{F}$ *there exists the event* $\nu(f) \in \mathcal{E}$ *such that* $S_1(\nu(f)) = S_1(f)$ *and* $S_0(\nu(f)) \supseteq S_0(f)$.

(ii) *For any effect* $f \in \mathcal{F}$ *both the certainly-true domain* $S_1(f)$ *and the certainly-false domain* $S_0(f)$ *are exact with respect to the preclusivity relation* $\perp$*; i.e.,* $S_1(f) = S_1(f)^{\perp\perp}$ *and* $S_0(f) = S_0(f)^{\perp\perp}$.

(iii) *For any pair* (A_1, A_0) *of mutually preclusive (i.e.,* $A_1 \perp A_0$*) exact sets (i.e.,* $A_1 = A_1^{\perp\perp}$*,* $A_0 = A_0^{\perp\perp}$*) there exists an effect* $f \in \mathcal{F}$ *such that* $S_1(f) = A_1$ *and* $S_0(f) = A_0$ *(i.e., ext(f) = (A_1, A_0))*.

(iv) *Given an effect* $f \in \mathcal{F}$*,* $S_0(f) = S_1(f)^{\perp\perp}$ *iff* f *is an event.*

DEFINITION 9.2. In any preclusivity space $(S, \perp)$ a *proposition* is any pair $p = (A_T, A_F)$ of exact sets which is orthoconsistent (i.e., $A_T \perp A_F$). The set of all propositions over $(S, \perp)$ will be denoted by

$$\mathcal{L}_f(S, \perp) := \{(A_T, A_F) \ : \ A_T \perp A_F, A_T, A_F \in \mathcal{L}(S, \perp)\} \, .$$

The set A_T is the *certainly-true* domain and the set A_F is the *certainly-false* domain of p. Sometimes if p is a proposition, we denote by $A_T(p)$ and $A_F(p)$ the certainly-true domain and the certainly-false domain associated with p, respectively.

If $w \in A_T(p)$ then we say that the proposition is *true* in the state w, while if $w \in A_F(p)$ then the proposition is *false* in this state. If neither $w \in A_T(p)$ nor $w \in A_F(p)$, then in w proposition p is neither true nor false, that is its value is *indeterminacy*.

The two trivial propositions $\omega = (\emptyset, S)$ and $j = (S, \emptyset)$ are the *absurd* or *contradictory* proposition and the *certain* or *tautologous* proposition respectively. A proposition p is *self-contradictory* or, simply, a *contradiction*, iff $A_T(p) = \emptyset$.

Remark 1. If p is a proposition, then $A_T(p) \cap A_F(p) = \emptyset$ but in general $A_T(p) \nabla A_F(p) \neq S$.

THEOREM 9.3. *The set of all propositions has a natural structure of a* BZ^3-*complete lattice* (Cattaneo and Giuntini 1991)

$$\langle \mathcal{L}_f(S, \perp), \omega, j, \sqsubseteq, -, \Box \rangle$$

with respect to

(1) *The partial ordering*:

$\quad p \sqsubseteq q$ iff $A_T(p) \subseteq A_T(q)$ and $A_F(q) \subseteq A_F(p)$. $\hspace{2em}$ $(or - PC)$

(2) *The fuzzy orthocomplementation*:

$\quad -(A_T, A_F) = (A_F, A_T)$. $\hspace{6em}$ $(foc - PC)$

(3) *The necessity operator*:

$\quad \Box(A_T, A_F) = (A_T, A_T^{\perp})$. $\hspace{6em}$ $(nec - PC)$

The preclusivity *fuzzy-intuitionistic propositional logic* $\mathcal{L}_j(S, \perp)$ is bounded by the least element $\omega = (\emptyset, X)$ and the greatest element $j = (X, \emptyset)$; the proposition $O := (\emptyset, \emptyset)$ is the half element. The *possibility* and the *impossibility* proposition are, respectively

$$\Diamond(p) = -\Box - (p) = (A_F(p)^{\perp}, A_F(p)) \hspace{4em} \text{(possibility)}$$

$$\approx (p) = -\Diamond(p) = (A_F(p), A_F(p)^\perp) \, ; \qquad\qquad \text{(impossibility)}$$

the latter being the *intuitionistic-like* orthocomplementation of this BZ-lattice
(i.e., $\approx (p) = \Box - (p)$).

(PC1) The lattice g.l.b. and l.u.b. of any family of propositions $\{p_j = (A_T^{(j)}, A_T^{(j)}) : j \in J\}$ are given by

$$\sqcap(A_T^{(j)}, A_F^{(j)}) = (\cap A_T^{(j)}, \nabla A_F^{(j)})$$

$$\sqcup(A_T^{(j)}, A_F^{(j)}) = (\nabla A_T^{(j)}, \cap A_F^{(j)}) \, .$$

(PC2) The intuitionistic-like orthocomplementation satisfies both the de Morgan laws.

(PC3) For any family $\{p_j = (A_T^{(j)}, A_F^{(j)})\}$ of propositions their 'product' is the proposition

$$\Pi p_j := (\cap A_T^{(j)}, \cap A_F^{(j)}) \, .$$

The set of all *exact propositions* is

$$\mathcal{L}_e(S, \perp) \;=\; \{p \in \mathcal{L}_f(S, \perp) : p = \approx (\approx p)\} =$$
$$= \{(A, A^\perp) : A \in \mathcal{L}(S, \perp)\} \, .$$

The elements from $\mathcal{L}_j(S, \perp) \setminus \mathcal{L}_e(S, \perp)$ are *fuzzy propositions*. The restrictions to $\mathcal{L}_e(S, \perp)$ of the necessity and the possibility coalesce to the identity mapping; the restriction of the two orthocomplementations coalesce and define a unique standard orthocomplementation, moreover the following is a one-to-one correspondence

$$\mathcal{L}_e(S, \perp) \equiv \mathcal{L}(S, \perp)$$
$$(A, A^\perp) \longleftrightarrow A$$

which allows the identification of the orthocomplemented lattice structures

$$\langle \mathcal{L}_e(S, \perp), \omega, j, \sqcup, \sqcap, - \rangle \equiv \langle \mathcal{L}(S, \perp), \emptyset, S, \nabla, \Delta, {}^\perp \rangle$$

(hence, $\mathcal{L}_e(S, \perp)$ is an orthocomplemented complete lattice too). The following graph summarizes the above results

	EFFECT		PROPOSITION
	$f \in \mathcal{F}$	$\xrightarrow{\text{ext}}$	$\mathcal{L}(S, \perp) \ni (S_1(f), S_0(f))$
$\nearrow$	$\downarrow \nu$		$\downarrow \Box$
$[f]_\sim \in \mathcal{L}$	$\longleftrightarrow$ $\nu(f) \in \mathcal{E}$	$\longleftrightarrow$	$\mathcal{L}_e(S, \perp) \ni (S_1(f), S_1(f)^\perp)$
JP-PROP.	EVENT		EXACT PROP.

Wit respect to the set of all JP-propositions we quote the following

THEOREM 9.4. *In any complete SEFP, the following JP 'Axiom C' is vali-dated.*

Condition C. *Let $f \in \mathcal{F}$ be an effect, and let $\mathbf{a} = [f]_\sim$ be the corresponding JP-proposition. Then, $\mathbf{a}' := [\nu(f)']_\sim$ is a compatible complement of $\mathbf{a}$, i.e.,*

(CC–1) $\mathbf{a} \cap \mathbf{a}' = 0$ *and* $\mathbf{a} \cup \mathbf{a}' = 1$

(CC–2) $\nu(f) \in \mathbf{a}$ *and* $\nu(f)' \in \mathbf{a}'$.

10. QUANTUM SEFP AS A UNIFIED AXIOMATIC APPROACH TO UNSHARP QUANTUM THEORY

In the axiomatic approach to complete SEFP, the orthocomplemented complete lattice of all events in general is not orthomodular, the latter being an usually required condition. In this section we introduce a further axiom which assures this property for $\mathcal{E}$.

DEFINITION 10.1. A *quantum* SEFP is a complete SEFP in which the following further axiom holds.

Axiom OG. Let $\{a, b\}$ be orthogonal events $(a \leq b')$, i.e.,

$$\forall w \in \mathcal{S}, \ 0 \leq p(w, a) + p(w, b) \leq 1 \,,$$

then an effect $f_{a,b} \in \mathcal{F}$ exists, called the *sum* of a and b, such that,

$$\forall w \in \mathcal{S}, \ p(w, f_{a,b}) = p(w, a) + p(w, b) \,.$$

Axom OG assures the existence of the sum of every finite family of pairwise orthogonal events. Whenever an infinite countable family of pairwise orthogonal events is considered, the existence of the sum cannot be deduced from axiom OG.

THEOREM 10.1. *In any quantum SEFP* (Cattaneo, Garola and Nisticó 1989):

(i)   *The orthocomplemented complete lattice of all events $\langle \mathcal{E}, \emptyset, \leq, ' \rangle$ is orthomodular.*

(ii)   *For any $a, b \in \mathcal{E}$, with $a \perp b$, the l.u.b. $a \vee b$ (in $\mathcal{E}$) coincides with $f_{a,b}$.*

(iii) *Let $\mathbf{a} \in \mathcal{L}(\mathcal{F})$ be a proposition and $\mathbf{a}^*$ be any compatible complement of $\mathbf{a}$, then $\mathbf{a}^* = \mathbf{a}'$. (Uniqueness of the compatible complement of any proposition).*

(iv)   Let $\mathbf{a}, \mathbf{b} \in \mathcal{L}(\mathcal{F})$ and let $\mathbf{a} \subseteq \mathbf{b}$. Then, the sublattice of $\mathcal{L}(\mathcal{F})$ generated by $\{\mathbf{a}, \mathbf{b}, \mathbf{a}', \mathbf{b}'\}$ is boolean.

In conclusion, quantum SEFP structures furnish an axiomatic approach to quantum physics in which unsharp realizations of idealized yes–no devices are taken into account. Moreover, they give an unification of the following traditional approaches:

(1)   Ludwig's approach to effects, since in Cattaneo, Garola and Nisticó (1989) a Hilbert space model of quantum SEFP is given in which effects are represented by linear operators such that $\emptyset \leq F \leq \mathbb{I}$.

(2)   The Jauch–Piron approach to propositions outlined in Section 8, definitions (D1)–(D8); see also Cattaneo, Dalla Pozza, Garola and Nisticó (1988) and Cattaneo, Garola and Nisticó (1989).

(3)   The Mackey, Gudder and Pool approaches to events since our lattice of events $\mathcal{E}$ satisfies all their relevant axioms.

Finally, BZ structures (the BZ poset, which in general is not a lattice, of all effects and the complete BZ^3 lattice of all propositions) give a clear environment to correctly treat logical (and semantical) questions about sentential language of quantum theories. We notice that the sentential language of unsharp quantum physics, besides usual connectives 'or', 'and', 'not' involves the two *modal-like* connectives 'necessity' and 'possibility', and an *intuitionistic-like* negation, which is 'impossibility'; for an axiomatization of BZ quantum logic according to the semantic described in this paper see Cattaneo, Dalla Chiara and Giuntini (1991) (see also Dalla Chiara and Giuntini (1989) and Giuntini (1990)).

Dipartimento di Scienze dell'Informazione
Università di Milano

REFERENCES

Aerts, D.: 1983, 'Classical theories and non classical theories as special cases of a more general theory', *J. Math. Phys.* **24**, 2441.

Beltrametti, E. and Cassinelli, G.: 1976, 'Logical and mathematical structures of quantum mechanics', *Rivista Nuovo Cimento* **6**, 321.

Beltrametti, E. and Cassinelli, G.: 1981, *The logic of Quantum Mechanics*, Addison-Wesley, Reading, Mass.

Birkhoff, G. and Von Neumann, J.: 1936, 'The logic of quantum mechanics', *Ann. Math.* **37**, 823.

Bub, J.: 1973, 'On the completeness of quantum mechanics', in *Contemporary Research in the Foundations and Philosophy of Quantum Theory* (ed. by C.A. Hooker), Reidel, Dordrecht.

Bub, J.: 1974, *The Interpretation of Quantum Mechanics*, Reidel, Dordrecht.

Cattaneo, G., Dalla Chiara, M.L., and Giuntini, R.: 1991, 'Fuzzy-intuitionistic quantum logics', in print in Studia Logica.

Cattaneo, G., Dalla Pozza, C., Garola, C., and Nisticó, G.: 1988, 'On the logical foundations of the Jauch Piron approach to quantum physics', *Int. J. Theor. Phys.* **27**, 1313.

Cattaneo, G., Garola, C., and Nisticó, G.: 1989, 'Preparation–effect versus question–proposition structures', *Phys. Essays* **2**, 197.

Cattaneo, G. and Giuntini, R.: 1991, 'Solution of two open problems on Hilbert unsharp quantum physics', preprint.

Cattaneo, G., and Marino, G.: 1988, 'Non usual orthocomplementations on partially ordered sets and fuzziness', *Fuzzy Sets Syst.* **25**, 107.

Cattaneo, G. and Nisticó, G.: 1985, 'Complete effect-preparation structures: Attempt of an unification of two different approaches to axiomatic quantum mechanics', *Il Nuovo Cim.* **90B**, 161.

Cattaneo, G. and Nisticó, G.: 1989, 'Brouwer–Zadeh posets and three valued Łukasiewicz posets', *Fuzzy Sets Syst.* **33**, 165.

Dalla Chiara, M.L. and Giuntini, R.: 1989, 'Paraconsistent quantum logics', *Found. Phys.* **19**, 891.

Foulis, D.J., Piron, C., and Randall, C.H.: 1983, 'Realism, operationalism, and quantum mechanics', *Found. Phys.* **13**, 813.

Foulis, D.J. and Randall, C.H.: 1971, 'Lexicographic orthogonality', *J. Comb. Theo.* **11**, 152.

Foulis, D.J. and Randall, C.H.: 1972, 'Operational statistics, I, Basic concepts', *J. Math. Phys.* **13**, 1667.

Foulis, D.J. and Randall, C.H.: 1974a, 'The empirical logic approach to the physical sciences', in *Foundations of Quantum Mechanics and Ordered Linear spaces* (ed. by A. Harkämper and H. Neumann), Lecture Notes in Physics, Vol. 29, Springer-Verlag, Berlin.

Foulis, D.J. and Randall, C.H.: 1974b, 'Empirical logic and quantum mechanics', *Synthese* **29**, 84.

Foulis, D.J. and Randall, C.H.: 1984, 'A note on misunderstanding of Piron's Axioms of Quantum Mechanics', *Found. Phys.* **14**, 65.

Garola, C.: 1980, 'Propositions and orthocomplementation in quantum logic', *Int. J. Theor. Phys.* **19**, 369.

Garola, C.: 1985, 'Embedding of posets into lattices in quantum logic', *Int. J. Theor. Phys.* **24**, 423.

Garola, C. and Solombrino, L.: 1983, 'Yes–no experiments and ordered structures in quantum physics', *Il Nuovo Cim.* **77B**, 87.

Giles, R.: 1970, 'Foundations for quantum mechanics', *J. Math. Phys.* **11**, 2139.

Giuntini, R.: 1990, 'Brouwer–Zadeh logic and the operational approach to quantum mechanics', *Found. Phys.* **20**, 701.

Giuntini, R.: 1991, 'A semantical investigation on Brouwer–Zadeh logic', *J. Phil. Logic.* **20**, 411.

Gudder, S.P.: 1965, 'Spectral methods for a generalized probability theory', *Trans. Amer. Math. Soc.* **119**, 428.

Gudder, S.P.: 1966, 'Uniqueness and existence properties of bounded observables', *Pac. J. Math.* **19**, 81.

Gudder, S.P.: 1967, 'Systems of observables in axiomatic quantum mechanics', *J. Math. Phys.* **8**, 2109.

Gudder, S.P.: 1968, 'Joint distributions of observables', *J. Math. Mech.* **18**, 335.

Gudder, S.P.: 1969, 'On the quantum logic approach to quantum mechanics', *Comm. Math. Phys.* **12**, 1.

Gudder, S.P.: 1970, 'Axiomatic quantum mechanics and generalized probability theory', in *Probabilistic Methods in Applied Mathematics*, Vol. 2, (ed. by A. Bharucha-Reid), Academic Press, NY.

Hadjisavvas, N., Thieffine, F., and Mugur-Schächter, M.: 1980, 'Study of Piron's system of questions and propositions', *Found. Phys.* **10**, 751.

Hadjisavvas, N. and Thieffine, F.: 1984, 'Piron's axioms for quantum mechanics, a reply to Foulis and Randall', *Found. Phys.* **14**, 83.

Jammer, M.: 1974, *The Philosophy of Quantum Mechanics*, J. Wiley & Sons, NY.

Jauch, J.M.: 1968, *Foundations of Quantum Mechanics*, Addison-Wesley, Reading, Mass.

Jauch, J.M.: 1971, 'Foundations of quantum mechanics', in *Foundations of Quantum Mechanics* (ed. by B. D'Espagnat), Academic Press, NY.

Jauch, J.M. and Piron, C.: 1969, 'On the structure of quantal proposition systems', *Helv. Phys. Acta* **42**, 842.

Jauch, J.M. and Piron, C.: 1970, 'What is quantum logic?', in *Quanta* (ed. by P.G.O. Freund, C.J. Goebel and Y. Nambu), Chicago University Press, Chicago.

Kraus, K.: 1983, *States, Effects, and Operations*, Lecture Notes in Physics, Vol. 190, Springer-Verlag, Berlin.

Kraus, K.: 1986, 'The classical behaviour of measuring instruments', in *Fundamental Aspects of Quantum Theory* (ed. by V. Gorini and A. Frigerio), Plenum, NY.

Ludwig, G.: 1971, 'The measuring process and an axiomatic foundation of quantum mechanics', in *Foundations of Quantum Mechanics* (ed. by B. D'Espagnat), Academic Press, NY.

Ludwig, G.: 1974, 'Measuring and preparing processes', in *Foundations of Quantum Mechanics and Ordered Linear Spaces* (ed. by A. Hartkämper and H. Neumann), Lecture Notes in Physics, Vol. 29, Springer-Verlag, Berlin.

Ludwig, G.: 1977, 'A theoretical description of single microsystems', in *The Uncertainty Principle and Foundations of Quantum Physics* (ed. by W.C. Price and S.S. Chissick), J. Wiley & Sons, NY.

Ludwig, G.: 1981, 'Quantum theory as a theory of interactions between macroscopic systems which can be described objectively', *Erkenntnis* **16**, 359.

Ludwig, G.: 1983, *Foundations of Quantum Mechanics*, Vol. I, Springer-Verlag, NY.

Mackey, G.W.: 1963, *The Mathematical Foundations of Quantum Mechanics*, Benjamin, NY.

Maczynski, M.J.: 1973, 'The orthogonality pustulate in axiomatic quantum mechanics', *Int. J. Theor. Phys.* **8**, 353.

Maczynski, M.J.: 1974, 'Functional properties of quantum logics', *Int. J. Theor. Phys.* **11**, 149.

Mielnik, B.: 1976, 'Quantum logic: is it necessarily orthocomplemented?, in *Quantum Mechanics, Determinism, Causality and Particle* (ed. by M. Flato *et al.*), Reidel, Dordrecht.

Piron, C.: 1964, 'Axiomatique quantique', *Helv. Phys. Acta* **37**, 439.

Piron, C.: 1972, 'Survey of general quantum physics', *Found. Phys.* **2**, 287.

Piron, C.: 1976a, *Foundations of Quantum Physics*, Benjamin, Reading, Mass.

Piron, C.: 1976b, 'On the foundations of quantum physics', in *Quantum Mechanics, Determinism, Causality and Particles* (ed. by M. Flato *et al.*), Reidel, Dordrecht.

Piron, C.: 1977, 'A first lecture in quantum mechanics', in *Quantum Mechanics, a Half Century After* (ed. by J. Leite Lopes and M. Paty), Reidel, Dordrecht.

Piron, C.: 1978, 'La description d'un systeme physique et le presuppose de la theorie classique', *Ann. de la Fondation L. de Broglie* **3**, 131.

Piron, C.: 1981, 'Ideal measurement and probability in quantum mechanics', *Erkenntnis* **16**, 397.

Pool, J.C.T.: 1968a, 'Baer*-semigroups and the logic of quantum mechanics', *Comm. Math. Phys.* **9**, 118.

Pool, J.C.T.: 1968b, 'Semimodularity and the logic of quantum mechanics', *Comm. Math. Phys.*

9, 212.

Randall, C.H. and Foulis, D.J.: 1973, 'Operational statistics, II, Manual of operations and their logic', *J. Math. Phys.* **14**, 1472.

Randall, C.H. and Foulis, D.J.: 1977, 'The operational approach to quantum mechanics', in *The Logico-Algebraic Approach to Quantum Mechanics*, Vol. III (ed. by C.A. Hooker), Reidel, Dordrecht.

Randall, C.H. and Foulis, D.J.: 1983, 'Properties and operational propositions in quantum mechanics', *Found. Phys.* **13**, 843.

Thieffine, F., Hadjivsavvas, N., and Mugur-Schächter, M.: 1981, 'Supplement to a critique of Piron's system of questions and propositions', *Found. Phys.* **11**, 645.

Thieffine, F.: 1983, 'Compatible complement in Piron's system and ordinary modal logic', *Lett. Nuovo Cim.* **36**, 377.

van Fraassen, B.: 1973, 'Semantical analysis of quantum logic', in *Contemporary Research in the Foundations of Quantum Theory* (ed. by C.A. Hooker), Reidel, Dordrecht.

van Fraassen, B.: 1974, 'The labyrinth of quantum logics', in *Boston Studies in the Philosophy of Science*, Vol. XIII (ed. by R.S. Cohen and M.W. Wartofsky), Reidel, Dordrecht.

Varadarajan, V.S.: 1962, 'Probability in physics and a theorem on simultaneous observability', *Comm. Pure Appl. Math.* **15**, 189.

Varadarajan, V.S.: 1968, *Geometry of Quantum Theory*, Vol. I, Van Nostrand, Princeton, 1968.

Von Neumann, J.: 1932, *Mathematical Foundations of Quantum Mechanics*, Princeton Univ. Press, Princeton (Translated from 1932).

M.L. DALLA CHIARA AND G. TORALDO DI FRANCIA

INDIVIDUALS, KINDS AND NAMES IN PHYSICS

PART I. SEMANTIC CONSIDERATIONS ON THE LANGUAGE OF PHYSICS

1. INTRODUCTION

It has been recognised for a long time that language plays a crucial role in science. The neopositivist philosophers were right when they put special emphasis on the analysis of the language of science. But is this analysis sufficient to clarify all the problems of the philosphy of science? Can science be reduced to a pure *linguistic game* – as Wittgenstein would put it – or is there something else? Surely science *talks of something*. Accordingly, the scientist must be aware of the relation existing between the language he uses and the things he is talking about. But here ends the 'game', if ever there was one, and formidable problems arise.

Whether semantics is at all possible or under what conditions it is possible has been the subject of profound speculation by many contemporary philsophers. Opinions may vary from Quine's pessimism[1] to the less negative attitude of those who, like Putnam[2] and Kripke[3], think of semantics as a more or less social enterprise. But all seem to agree that the analysis is difficult.

It is customary in this kind of discussions to refer either to the terms used in everyday life or to the terms of classical science. Now it has occurred to us that the task, which is already involved in classical science, may become a terrible challenge in modern physics, especially in particle theory.

To be more precise, let us recall that a science can talk of different things, like *numbers, operations, phenomena*, and so on. But our analysis will not be so general. We shall concentrate our attention solely on the problems that arise when we talk of *physical objects*, whatever this term may mean. It may seem obvious that in order to talk of physical objects one should be able to *name* them and to divide them into different *species*. Let us list some of the questions – all well known in modern philosophy – that can arise when we do so:

(1)   Is a proper name a rigid designator?

G. Corsi et al. (eds), Bridging the Gap: Philosophy, Mathematics, and Physics, 261–283.

(2)   Can a description be substituted for a proper name?
(3)   Is a proper name the abbreviation of a description?
(4)   Is a description a necessary and sufficient condition for fixing a referent?
(5)   Is identity a relation between names or between individuals?
(6)   Does the intension of a natural-kind term determine its extension?
(7)   Can different extensions have the same intension or viceversa?
(8)   Is the referent of a natural-kind term fixed by a cluster description?

The list is by no means exhaustive and is intended only to give a typical exemplification. Now, it turns out that not only are questions of this kind difficult to answer, but some of them may even make no sense when we are dealing with theories in modern physics. The reason does not reside so much in the way we are talking about things as in the very nature of things, that do not allow to be talked about in that way! If this is true, the situation is really serious. For we are so far unable to dispense with naming objects and species.

2. THE CASE OF THE HEAVENLY BODIES

The most intriguing problems of physical semantics are met in the realm of particle theory. But frontier research in physics does not deal only with particles. Also the objects of astrophysics and cosmology, although they are *classical* and *macroscopic* bodies, may present some interesting case.

Astrophysics is perhaps the only domain of physical science where proper names are largely used. The older ones, like Jupiter, Sirius, Andromeda, are of the traditional type. But also such names as SS 433 that can be found in a stellar catalogue may appear to present no new problem with respect to those already debated in modern semantics. Even when one and the same star is given different names in different catalogues, the case is perfectly analogous to that of *Cicero* being *Tully*[4].

However, we may ask in what sense are names used in astrophysics; are they meant to be abbreviated descriptions or rigid designators? Is there any fundamental difference between proper names and descriptions?

To a certain extent descriptions in astrophysics can be much more 'rigid' than in ordinary language. For instance, one can describe the star whose name is 'M 831'[5] by specifying its celestial coordinates. Suppose that – at a given epoch – the star M 831 has the right ascension $\alpha = 12^h47^m$ and the declination $\delta = 25°.4$. One might be induced to think that because $\alpha = 12^h47^m$, $\delta = 25°.4$ uniquely and exactly determines M 831, the description 'the star having $\alpha = 12^h47^m$, $\delta = 25°.4$' can be substituted for the name 'M 831'. But this is not so. In fact it is perfectly conceivable for an astronomer to say: "It has turned out that M 831 does not have the coordinates $\alpha = 12^h47^m$, $\delta = 25°.4$, but $\alpha = 12^h43^m$, $\delta = 25°.8$"; whereas it is inconceivable for him to say: "It has turned out that the star having $\alpha = 12^h47^m$, $\delta = 25°.4$ does not have the coordinates $\alpha = 12^h47^m$, $\delta = 25°.4...$"

Let us then accept 'M 831' as a rigid designator *à la* Kripke, one that has the same referent in all possible worlds. But what kind of referent? A star? Note that in this case a causal chain having the origin in a baptismal ceremony by ostension can refer solely to a light spot in the sky. One cannot even be certain that it is a star! Then the idea suggests itself that the referent should be the *light spot*. After all, stellar catalogues are catalogues of light spots. However, one might try to dispute this idea with several arguments. But, instead of entering into these, we want to illustrate a case in modern astrophysics that can be illuminating.

A few years ago two quasars were discovered in the sky – say Q 33 and Q 34 – very close to one another and showing the same spectra and the same red-shift. Their appearance was in many ways a puzzle, until it was suggested that they are one and the same object! What happens is that the light rays are bent by an intervening galaxy that acts – by general relativity – as a huge lens; rays coming from one and the same quasar but arriving to us from two different directions in the sky give rise to two different light spots. Now, what do we mean when we say: "Q 33 is equal to Q 34"? Surely we do not intend to say that the light spot is the same, because that is contradicted by experience. We intend instead to say that one and the same object – of whatever nature it may be – is the source giving rise to both light spots. Note that the situation is not identical with that obtaining in the venerable case of 'Hesperus' and 'Phosphorus'; for Q 33 and Q 34 are seen *at the same time*; you can turn your telescope either to Q 33 or Q 34 and the operations you do in both cases are clearly different. You could eliminate one of the two names 'Hesperus' or 'Phosphorus' – or both, and just talk of 'Venus' – but you cannot dispense with either of the two names 'Q 33' or 'Q 34'.

A way out that seems reasonable in the present case can have profound consequences. Let us stipulate that the referents of stellar catalogues are *light spots*. Then 'Q 33' and 'Q 34' have different referents. If we say that "Q 33 is equal to Q 34" we make a false statement. What we want to say is: "One and the same object gives rise to both Q 33 and Q 34".

But then, when we use names to designate *stellar objects* instead of light spots, we use them as abbreviated descriptions! In the foregoing case the description terms are: "the unique X giving rise to Q 33", and "the unique X giving rise to Q 34", having stipulated in our metalanguage that 'Q 33' and 'Q 34' are really names for light spots.

In our opinion, this discussion shows that even in macrophysics, the legitimacy of a sharp distinction between *true proper names* (in the sense advocated by Mill[6], Russell[7], and Kripke[8]) and descriptions may be highly questionable[9].

A case of special interest is offered by those kinds of objects that have first been predicted theoretically and later have been found to exist – with a more or less stringent proof – like neutron star or black holes. One starts by building a well-determined intension that – so far as is known at the time – can have a non-null extension in some possible world. Then it is found that the extension is not empty in the actual world, and one identifies the individuals so detected with those of a previous possible world. Thus the species-names like 'neutron star' or 'black hole' are preserved after the discovery. This interpretation seems to present no difficulty. However, we may recall with some worry the opinion expressed by Kripke about *unicorns*: "I think that even if archaeologists or geologists were to discover tomorrow some fossils conclusively showing the existence of animals in the past satisfying everything we know about the unicorns from the myth of the unicorn, that would not show that there were unicorns."[10] If we were to accept this opinion also for such objects as neutron stars or black holes, the entire scientific procedure would be shaken from its foundations. But we cannot help asking: where is the difference – if there is any – between unicorns and neutron stars? Of course, if one argues that unicorns were *meant* from the start to be mythical and that property is essential, the case becomes trivial. But we are not sure that everything resides there. Anyway, we believe that it is correct to say that neutron stars, first proposed theoretically, were later found to exist. And the same can be said for a number of other natural-kind names occurring in physics.

3. THE LAND OF ANONYMITY

We come now to the realm of microphysics, whose most striking characteristic from the semantical point of view is that there are *no proper names*. No one has ever attempted to label single atoms or electrons or protons with mythological names (or with conventional numbers or letters, for that matter). Particles – perhaps with one exception of which we shall talk shortly – are *anonymous*.

This raises two main questions. The first one is: why are there no proper names in that country? And the second is: how can one talk about what happens in a country, where there are no proper names? Such questions, in turn, may lead to profound problems which in our opinion have not yet been fully solved.

The lack of proper names is due fundamentally to the fact that the objects of microphysics are *nomological*.[11] All their characteristics are fixed by physical law and are identical for objects of one and the same kind. As far as we know today, one electron is identical to another electron, one proton to another proton, and so on. Think of a country where all men were monoovular twins and looked all alike. How can you talk of Peter or Paul if you cannot distinguish the one from the other and from all the other men? Of course, you can ask: are you Peter or Paul? But two electrons would give you no answer! Nor can two electrons bear a red or a blue ribbon respectively on their arms.

It is a fundamental principle of quantum mechanics that two particles of the same kind are absolutely *indistinguishable*. Are they therefore *identical* or the same object? A hasty application of Leibniz' principle of *indiscernibels* might seem to lead to this absurd conclusion. For Leibniz says: "There are never in nature two exactly similar entities in which one cannot find an internal difference". But, of course, the difference can merely be represented by a different location in space and time.

Let us then stipulate for the moment that two equal particles are *two* different objects, because at a certain time they occupy two different space regions. We shall now recall – in a simplified version that is sufficient for our purpose[12] – how the particle system is dealt with in quantum mechanics. Let $\psi_\alpha(x_1)$, $\psi_\beta(x_2)$ represent the wave functions of both particles, having denoted their coordinates by x_1, x_2 respectively. The joint probability amplitude of finding the first particle at x_1 and the second at x_2 might seem to be simply the product $\psi_\alpha(x_1)\,\psi_\beta(x_2)$. However, this expression cannot be accepted, because in general its value would change by interchanging the particles, becoming $\psi_\alpha(x_2)\,\psi_\beta(x_1)$. And this would go against the indistinguishability principle. It can be shown that the correct wave function ψ to use is the following combination of both products[13]

(1) $$\psi(x_1, x_2) = \psi_\alpha(x_1)\,\psi_\beta(x_2) \pm \psi_\alpha(x_2)\,\psi_\beta(x_1)$$

where the plus sign holds in the case of *bosons* (i.e. particles that can have all quantum numbers equal) and the minus sign for *fermions* (i.e. particles that cannot have all quantum numbers equal). Hence, the joint probability density is

(2) $$|\psi(x_1, x_2)|^2 = |\psi_\alpha(x_1)|^2\, |\psi_\beta(x_2)|^2 + |\psi_\alpha(x_2)|^2\, |\psi_\beta(x_1)|^2 \pm$$
$$\pm 2\Re\left[\psi_\alpha(x_1)\,\psi_\beta(x_2)\,\psi_\alpha(x_2)\,\psi_\beta(x_1)\right].$$

Let us now suppose that ψ_α and ψ_β vanish everywhere except in two small regions A, B around two points x_A, x_B respectively, located a large distance apart. Then the last term in (2) is identically zero and we have

(3) $|\psi(x_1 x_2)|^2 = |\psi_\alpha(x_1)|^2 \, |\psi_\beta(x_2)|^2 + |\psi_\alpha(x_2)|^2 \, |\psi_\beta(x_1)|^2 \, .$

This probability does not vanish only if either $x_1 \simeq x_A$, $x_2 \simeq x_B$, or $x_1 \simeq x_B$, $x_2 \simeq x_A$. Hence we are certain to find *one* particle in A and *one* particle in B. We are then tempted to forget the somewhat metaphysical question of which is which and call Peter whichever particle we find in A and Paul whichever particle we find in B. This is particularly convenient because x_A and x_B are by hypothesis two distant points. Then, by continuity[14] we can be sure that – at least for some time – the regions A and B where ψ_α and ψ_β do not vanish will evolve without overlapping: so we can stipulate that Peter follows the path of A, and Paul follows the path of B. Thus Peter and Paul behave exactly as two *distinguishable* particles would do. They can legitimately bear their own names, though strictly speaking they are not *two particles*, but *two regions of space* in each of which one and only one particle can be detected.

But there is a catch! First of all, due to mathematical reasons, it is virtually impossible for ψ_α and ψ_β to be *exactly* confined to two finite regions. There will always be some degree of overlapping between regions A and B. Of course, it is easy to guess that when one electron is on Sirius and the other is on the Sun, the overlap will be absolutely negligible. But it is not necessary to go to such extreme conditions. The overlap is fairly negligible – at least most of the time – even when the particles – say electrons – travel inside a cathode ray tube or a high energy accelerator. The overlap becomes appreciable only when A and B approach a distance apart not much larger than the de Broglie wavelength $\lambda = h/p$ (where h is Planck's constant and p the electron momentum). In a typical TV tube one may have $\lambda = 10^{-9}$ cm. Hence, during the flight, each electron is sufficiently far apart from all the others for their significant regions not to overlap. Incidentally, in this situation, the path of one single electron could – in principle – be tracked without appreciable disturbance by illuminating it with visual photons ($\lambda \simeq 10^{-5}$ cm), whose momenta are negligible with respect to the electron's momentum.

This is the reason why an engineer, when discussing a drawing, can *temporarily* make an exception to the anonymity principle and say for instance: "Electron a issued from point S will hit the screen at P while electron b issued from T hits it at Q". But this mock individuality of the particles has very brief duration. When the electron hits the screen (or when – going back in time – it is still inside the source filament) it meets with other electrons with substantial overlapping, and the individuality is lost. In fact the de Broglie wavelength of an electron inside an atom is on the same order of magnitude as the atomic diameter.

4. PARTICLE QUASETS

We have thus established that under certain conditions a quantum particle can mimic the behaviour of a classical particle. Precisely:

(1) There can exist a small region of space A which travels with continuity and within which one and only one particle can be detected throughout. This holds only *approximately*, but the approximation can be extremely good.

(2) Such situation has generally a very short duration on earth. But it can last for a very long time in a rarefield environment (e.g. outer space).

Region A can conventionally be called *particle* and even be given a proper name 'a'. But is this name a rigid designator? We strongly doubt it.

Let us skip the *metaphysical* question, to which we are not particularly sensitive. We can even tolerate without much worry the *approximation* mentioned in (1), for its meaning is that the probability of finding no particle or more than one particle in A is extremely small. In this connection, we note that *all* assertions of physics have that kind of approximation. When we state that heat passes spontaneously only from a hotter to a cooler body in contact, we really mean that in a real case it is extremely *probable* that it should do so.

However, the limited duration mentioned in (2) raises more serious concern. *Prima facie* one may be tempted to think that the case of particle a is not different from that of Aristotle. After all, there was no Aristotle before 384 b.C. or after 322 b.C. But, in general, the particle does not die! An electron may very well survive a close encounter with another electron. Suppose that we follow with continuity an electron – say Peter – going from point P to point Q in a vacuum. We would like to be able to say that in a possible world Peter might encounter other electrons on its path and finally be scattered to Q. But then no one could tell that that electron is still Peter. There is no *trans-world* identity. In this situation the meaning of 'rigid designator' becomes very fuzzy. Anyway, the term seems useless.

The more massive is a particle and the smaller becomes its de Broglie wavelength for a given energy.[15] Therefore, heavy particles can preserve their mock individuality at closer distances than electrons. In the case of a macroscopic body one can absolutely forget the foregoing quantum considerations and come to the domain of ordinary semantics. As already stated, we do not believe that everything is settled in that domain. But at least the ground has been explored by many authors.

Let us therefore come back to microphysics. What can we say when the particles *are* at close quarters, like the electrons inside an atom or the nucleons inside a nucleus?

Consider the electrons of an atom. We generally know perfectly well how many electrons there are, but cannot tell which is which. It is customary to talk of the 'set' of the electrons of that atom. But do they really constitute a *set*? Certainly not in a classical sense. How can we verify that the collections of electrons that are found around the atoms satisfy, say, the Zermelo–Fraenkel axioms, without being able to distinguish one element from another?

Yet the electrons of an atom, taken as a whole, possess some properties that are characteristic of a set. For instance, they have a *cardinality*, even if we cannot *count* (or well-order) them, hence cannot make an *ordinal* number to correspond to each electron. How can we establish the cardinality? We can follow two different ways: one is theoretical, the other one is experimental.

Take the case of a helium atom. *Theoretically*, we can establish that the electrons are *two*, because the wave function (1) depends on the *six* coordinates $x_1, y_1, z_1, x_2, y_2, z_2$; we can therefore say that the wave function has the same degrees of freedom as a system of two classical particles. *Experimentally*, we can ionise the atom (by bombardment or other means) and extract *two* separate electrons in the sense of (1) and (2) of this section.

It is intriguing to note that there are even '*subsets*' inside the 'set', each one with its own cardinality. For example, you can say that inside a sodium atom there are *two* electrons in the shell $1s$, *two* electrons in $2s$, *six* electrons in $2p$, *one* electron in $3s$. Thus there is a sort of *isolation* procedure. You can state a property (e.g. having the quantum number $\ell = 1$) and you can tell how many electrons form the 'subset' having that property (six), even if you cannot distinguish those electrons from all the others!

At this point we suggest to introduce the concept of *quaset* (abbreviation for *quasi-set*) for a collection of elements which may be *indistinguishable* from one another. It is an entity having a cardinal, but not necessarily an ordinal number.[16] What is ordinarily termed a 'set of equal particles' is a quaset. The electrons of an atom form a quaset. The photons of a given mode inside a resonant cavity form a quaset. When a property is shared by a number of elements less than the cardinal of the quaset you can *isolate* a *subquaset*. For instance, the electrons in the shell $2p$ inside an atom form a subquaset of the complete quaset of all the electrons. The quaset formed by the photons of a given mode inside a resonant cavity does not yield any subquaset by this procedure, because there is *no* property that is not shared by all the photons of the mode.

When a particle quaset is sufficiently isolated in *space* it can mimic some properties of a classical body. An isolated particle in the sense of (1) corresponds to a *singleton quaset* sufficiently removed from all other quasets of equal particles. Quasets of particles – as long as they last – can be *named*. Thus you can talk of the 'free particle a' or of the 'electrons $2s$' inside a given atom. But, as already explained, the locution is conventional and has nothing to do with *substantiality*.

The peculiar behavior of subatomic particles can force us to revise the theory of *natural kinds*. Take for instance the electrons. There can be little doubt that in nature there are a kind of particles that we call 'electrons'. How is this concept to be determined? Certainly not by giving its *extension* in an ordinary way. For an extension can be specified either by stating that it comprises Peter, Paul, Edward, and so on; or by pointing out this, that, that other, and so on; or, finally, by determining a set through the specification of a convenient property, and assuming at the same time a kind of *comprehension* and *extensionality* principle. But none of these procedures makes sense in our case.

Note now that what we actually find in nature are not so much *electrons*, as *electron quasets*. Thus the *extension* of the natural kind 'electron' can be defined as a *quaset* whose subquasets are all the electron systems that can – at least in principle – be detected experimentally. And the same definition would be assumed for all other fermion or boson terms.

Note that this definition of the extension does not clash with the older one that applies to macroscopic bodies. For in the case of macroscopic bodies we are usually dealing with a set in the standard sense, which is a special case of a quaset.

What about *intension*? Here the situation seems – at least to a first approach – definitely simpler than in the macroscopic domain. The notion that the intension is represented by the *conjunction of a number of properties* has been criticised in the case of macroscopic natural-kind names; but for particle names it seems to be all right. For instance, the 'electron' can today be defined by mass 9.1×10^{-28} g, electron charge $= 4.8 \times 10^{-10}$ e.s.u., spin $= 1/2$. The well-known case of the 'three-legged tiger'[17] cannot be encountered in this domain, for all the properties mentioned are taken to be *essential*. For example, when a particle was discovered having all the properties of the electron, except for the mass $= 1.9 \times 10^{-25}$ g, physicists did not say that it was a peculiar electron, but concluded that a *new* particle had been found. It was later baptised the 'muon'. Similar cases have occurred repeatedly.

It is interesting to recall that at first the muon was thought to be the Youkawa meson mediating nuclear forces. When it was proved, that the muon is insensitive to the strong force, it was not concluded that "muons do not exist" but that "muons are not mesons". This reminds us of the imaginary case treated by Kripke[18] where cats are found to be demons. One should not conclude "that there turned out to be no cats, but that cats have turned out not to be animals".

Note that the *intension* of a particle term determines – at least in principle – an *extension* of the term. For instance, once the intension of the term 'electron' has been stipulated, we have the possibility of recognising – by theoretical or experimental means – whether a given physical system is an electron quaset or not; if yes, we can also enumerate all the quantum states available within it. But we can do so in a number of different ways! For example, take the spin. We can choose a z-axis and state how many electrons have $s_z = +1/2$ and how many have $s_z = -1/2$. But we could instead refer to the x-axis, or the y-axis, or any other direction, obtaining different sets of quantum states, all having the same cardinality. We thus arrive at a situation, which is usually believed to be impossible in classical semantics: different extensions can correspond to one and the same intension. Of course, the reverse situation of one and the same extension corresponding to different intensions is trivially possible, as in classical semantics (for instance, instead of giving the mass of a particle, one could give its rest energy).

What kind of entities are the properties that constitute a particular intension? Are they linguistic or extralinguistic entities? As is well known, this is a temendous problem of classical semantics, but it seems to us that it does not represent a peculiar difficulty of microphysics.

5. INTERACTION WITH POSSIBLE WORLDS

It is today a common notion that particles can appear or disappear via the *creation* and *annihilation* processes. Thus a photon, under certain conditions, can disappear and give rise to an electron–positron pair. A necessary condition for this to happen is that the photon should have an energy equal to at least the rest energy $2mc^2$ of the two new particles. In the reverse process an electron and a positron may meet and be annihilated, giving rise to two photons,[19] which share the total energy available. Similar processes can occur for all kinds of particles.

Creation and annihilation processes do not seem to pose new semantic problems, with respect to what we have been discussing so far. Indeed, we have already realised that, even without such processes, an isolated particle has only a limited lifetime. If the disappearing or appearing particle is part of a quaset, the quaset changes its cardinality; but no new difficulty seems to arise. We have only to stipulate that the extension of any neutral-kind term is a function of time, which, of course, is true also in the ordinary macroscopic world. However, we get into trouble when we come to considering *virtual* particles.

We have stated that a necessary condition for a particle to be created is that its rest energy may be provided by other particles (either disappearing or yielding a part of their kinetic energy), as is required by the law of energy conservation. However, this condition may not be strictly respected. A certain degree of violation is allowed because of the uncertainty principle.

To measure an energy E with a precision ΔE, we need at least a time interval Δt, such that

(4) $\Delta E \cdot \Delta t \simeq h$.

But this is not only an *epistemic* limitation. A much more substantial interpretation is possible and is warranted by a host of experimental results. The energy of the system *is* intrinsecally indefinite by $\Delta E \simeq h/\Delta t$ during the time interval Δt, and by the same amount can energy conservation be violated. If Δt is sufficiently small for ΔE to be equal to or greater than the rest energy of a new particle – all other selection rules being respected – that particle *can* be created. Of course, it has only a fleeting existence and must be annihilated before the expiration of Δt. Yet during its very short lifetime such a *virtual* particle can have very important effects.

Indeed, all elementary interactions between particles are mediated by virtual particles. For example, the interaction between a proton and an electron is mediated by virtual photons that are continually emitted by one particle and soon reabsorbed by the other particle, and viceversa. But even an 'isolated' particle steadily emits into virtual states and soon reabsorbs all kinds of particles with which it can interact. It is customary to say that a particle is never *naked*, but always travels accompanied by the whole cloud of its virtual particles.

Now the *intension* of the *virtual* particles of a given kind can perhaps be specified by assigning all the properties that characterise the *real* particles of that kind plus a total energy less than the rest energy (and correspondingly a very short lifetime Δt). But the specification of an *extension* poses formidable problems that we are not sure to be able to solve. Of course, the extension should be infinite, even within a finite volume, and have the power of the continuum, because E can have any value, from zero to infinity.

We limit ourselves to making an intriguing remark. The effect of the *virtual* particles on the behaviour of *real* particles could perhaps be interpreted as an interaction of *possible* worlds with the actual world! Indeed each virtual particle must correspond to a possible real particle – i.e. to a particle whose creation would respect charge, spin, baryon number etc. conservation – in a world where the necessary energy were available. What *might* happen under suitable conditions interacts with what *does* happen. We are aware that this is a daring speculation. However, it is in the gist of quantum mechanics.

On the other hand, it is possible to interpret virtual particles in a different way, perhaps by saying that they are completely new kinds of particles or entities? Certainly not. A virtual electron has all the properties that characterise the intension 'electron' except for a lack of energy than can be arbitrarily small. Its interpretation as an electron in a possible world may perfectly agree with everything we know about virtual particles.

That the actual world should not be so independent of or indifferent to what happens in possible worlds has perhaps been suspected by several workers. After all, what would be the purpose of discussing possible worlds, if they did not in the least conern our own world? But such a concrete and heavy interaction as the one we now suggest, may create some amazement.

Be that as it may, the task of building a reasonable semantics of science in such a complicated world becomes extremely difficult and will probably require much further work.

PART II. A FORMAL APPROACH TO THE PROBLEM OF INDIVIDUALS IN MICROPHYSICS

6. DENOTABILITY AND DISTINGUISHABILITY

In the previous part we have often used the expressions "one can (or one cannot) give a name to ...", "one can (or one cannot) distinguish which is which". Let us now try and explicate with a somewhat more formal analysis these intuitive semantical ideas. How can we characterize a logical situation where "any individual can be denoted by a name and one can always distinguish which individual is which"?

Let L be a generic (at least first-order) language and let M be a possible *interpretation* of L. Since we are not referring to a particular logic, it seems reasonable to suppose that M satisfies the following general requirements: for any semantic *situation* (or *world*) which is considered (and which is labeled by an index i), M determines a *domain of individuals* D_i and an *interpretation-function* ρ_i which associates appropriate meanings to the non logical constants of the language. As a particular case (as happens for instance in classical model theory) the set of situations may be a singleton. We will consider the following conditions:

(a) the main relation which may hold between the individuals and the domain D_i is the set-theoretical $\in$ (supposed governed by a standard formal system like for instance the Zermelo–Fraenkel theory (ZF)); any *property* of M univocally determines a subset of D_i (and similarly for the *relations*).

(b) for any individual d of D_i the object-language L can be extended to an L' which contains a *name* a and the interpretation-function ρ_i can be extended to a ρ_i' such that $\rho_i'(a) = d$.

(c) If L is at least a second-order language, the Leibniz principle holds:

$$\forall xy[\neg\, x = y \to \exists\, P(Px \wedge \neg Py)]\ .$$

In other words, individuals which are not one and the same are distinguished by at least one property.

(d) Let us suppose that M refers to a set of *situations* (or *worlds*) I that is not a singleton. The world may be correlated (as happens in the usual Kripke semantics) by a number of *world-relations*. A particularly interesting world-relation may correspond to a time-order relation : $i < j$ if and only if the situation i temporally precedes the situation j. We suppose that a binary relation $\approx$, which we will call *transworld-identity*, is defined on the set $U = \cup_{i \in I}\{D_i\}$, such that at least the following conditions are satisfied:

(I) for any $d \in D_i$ and any D_j there exists at most one $d' \in D_j$ such that $d \approx d'$;

(II) if a name a has a *denotatum* in two different domains D_i and $D_j (\rho_i(a) = d, \rho_j(a) = d')$ then $d \approx d'$.
In the particular case where $i < j$ (and $<$ represents the temporal order), the transworld-identity between individuals of D_i and of D_j is usually called *genidentity-relation*.

It is easy to imagine abstract logical situations where the conditions (a)–(d) are satisfied. Our claim is that in microphysics all these conditions may be simultaneously violated.

7. LANGUAGES AND METALANGUAGES OF PHYSICAL THEORIES

Let us first ask: what is the appropriate linguistic level where physical theories may refer to individuals? The object-language of a physical theory T is usually built up in order to express quantitative physical laws which have the form:

$$\alpha(q_{i_1}, \ldots, q_{j_n})$$

where $\alpha(x_1, \ldots, x_n)$ is a well-formed formula of the mathematical subtheory of T, and $q_{i_1}, \ldots, q_{j_n}$ are physical variables ranging over the possible values of the physical quantities considered by the theory (as an example think of the second law of dynamics: $f = ma$). Adequate *physical models* of T may be described[20] as consisting of:

(a)   a *mathematical part* M_o (a *model* – in the usual model-theoretic sense – of the mathematical subtheory of T);
(b)   an *operational part* O determined by a set S of *physical situations* and by a sequence of *operationally defined physical quantities*;
(c)   a *translation function* τ which associates a mathematical interpretation in M_o to the elements of O.

Since any physical situation of S consists of physical systems in well determined states, it seems natural – at least *prima facie* – to admit that the semantical metalanguage of T may "speak" about singular physical systems regarded as a kind of "individuals evolving in time".

When the mathematical interpretation $\tau(\sigma_t)$ of a physical system σ (at a time t) corresponds to a *maximum of information* about the system, one usually says that it represents an *(ideal) pure state* of σ. For instance, in classical particle mechanics (CPM) pure states can be identified with points in an appropriate phase-space; in quantum mechanics (QM), instead, pure states can be represented as wave-functions (or, equivalently, as unitary vectors in an appropriate Hilbert space). Being a maximum of information, any pure state p may be regarded as a *metatheoretical description* of a particular physical object. As is well known – and as has been stressed in Part I – differently from CPM, in QM a pure state associated to a system does not determine exact values for all the relevant physical quantities concerning the system. In particular, it may happen that p does not exactly localize σ_t in a precise space-region.

Let us now consider, instead of a single system, a *compound* σ, consisting of n subsystems: in what sense can (or cannot) the n parts of σ be semantically dealt with as 'individuals'?

We will first discuss a simple case, arising in CPM. Let σ represent an n particle system considered in the time interval $[t_0, t_1]$, and let $p(t_0)$ be a pure state which describes the state of σ at the initial time t_0. The equations of motion of CPM predict the time-evolution of the system in such a way that for any $t \in [t_0, t_1]$ the pure state $p(t)$ is determined. What about the component systems? For any t, $p(t)$ univocally determines n pure states $p_1(t), \ldots, p_n(t)$, which represent a *maximal and complete information* about each one of the n parts of σ_t.

This physical situation suggests in a natural way a kind of second-level Krip-kean semantical construction for a particular sublanguage L^σ of the metatheory of T. This is a language where we want to use names $a_1, \ldots, a_n$ for the n subsystems of σ and monadic predicates Q_i^E expressing the idea that "the value of the quantity Q_i lies in the interval E"; in other words, we want to be allowed to formulate statements like for instance "the velocity of the first component of σ is less than c".

Adequate models for such a language may have the form:

$$< I, <, D, \rho >$$

where:

(1) the set of worlds I is identified with the time-interval $[t_0, t_1]$;
(2) the world-relation $<$ is the time-order relation;
(3) D is the domain-function such that for any time t, D_t is the set of the n physical subsystems of σ, which are completely described by $p(t)$.
(4) ρ is the interpretation-function such that for any time t and any name a_i, $\rho_t(a_i)$ is univocally determined as an element – in standard set-theoretical sense – of D_t; further, for any monadic predicate Q_i^E, $\rho_t(Q_i^E)$ is determined as a particular subset of D_{t_1}, on the basis of the information given by $p(t)$.

One can easily recognize that, in this semantical situation, the n parts of σ behave like 'good individuals' in the sense of our conditions (a)–(d). Indeed the relations between elements, properties and domains are the set-theoretical relations of standard extensional semantics. Any individual is denoted by a name and the denotation-function is strongly determined not only in a logical sense, but also in a physical sense, since $p(t)$ fixes, among others, an exact space-localization of each one of the subsystems of σ. If the language L is built up as a second-order language, one can easily take as models only structures where the set of properties coincides with the *full* power-set of the total universe $U = \cup_{t \in I}\{D_i\}$, in such a way that the Leibniz principle holds.

Finally, a transworld-identity relation can be easily defined, using the equations of motion of the theory.

8. A QUASET-THEORETICAL SEMANTICS FOR MICROPHYSICS

We will now consider an analogous semantical situation which may arise in QM. As an example, let us refer to the case of an n-electron system (which has been discussed in Part I). Let σ represent our compound system considered in a time interval $[t_0, t_1]$; as in CPM, for any $t \in [t_0, t_1]$ the whole system σ_t may be represented as a pure state $\psi(t)$. The time-evolution of the system is now governed by the Schrödinger equation.

Can we construct, as in the previous case, models of the form

$$< I, <, D, \rho >$$

for a language L^σ with monadic predicates Q_i^E, which represent 'meaningful properties' of single particles, and names $a_1, \ldots, a_n$, which should denote the n subsystems of σ at different times? This problem gives rise to two questions:

(I)   Does $\psi(t)$ determine a set of n elements in the standard set-theoretical sense?

(II) Is the denotation-function definable in such a way that $\rho_t(a_i)$ univocally determines an element of D_t?

The arguments set forth in Part I show clearly that both these questions have a negative answer. Naturally, the critical point of the problem is not "whether or not *we are allowed* to introduce names $a_1, \ldots, a_n$ for the n subsystems of σ", but rather "whether or not *we are able* to determine a reasonable denotation-function ρ_t for such names".

One might object that the possibility of exactly localizing in space a physical object $\rho_t(a_i)$ is not a necessary condition in order that the denotation-function ρ_t be logically defined. Such an observation may be correct in the case of a single isolated physical system. However, in our example, the situation seems to be more serious, since we are dealing with a domain of n individuals and there are no *theoretical means* to 'capture' any one of them, in order to 'attach' a name to it. It will be useless to recall, that if we really trust orthodox QM (and do not accept any form of *hidden variables hypothesis*) we cannot expect to escape these logical difficulties simply by maintaining that *in mente Dei* a kind of denotation-function must be defined.

This critical situation can be – at least partially – overcome by means of a 'quaset-theoretical semantics'. As explained in Part I, from an intuitive point of view a quaset may be imagined as a kind of non-standard set, whose cardinal number is always determined, even if it is not generally determined which elements belong to it. As a consequence, a quaset cannot be identified with the set of its elements (its extension).

Let us sketch a minimal nucleus of a possible axiomatization of a 'quaset theory' (QST), which will be presented as a theory in which ZF is relatively interpretable. Readers who are not familiar with classical set theory will certainly understand the semantical applications of QST even if they will not follow all the details of the axiomatic presentation of the theory.

The language of QST is supposed to contain (besides the variables $x, y, z, x_1,$ $x_2, \ldots$), the following primitive constants: the identity predicate (dealt with as a logical constant), the inclusion predicate $\subseteq$; the predicate $\in$ (which is read ". . . certainly belongs to . . .") and the predicate $\notin$ (which is read ". . . certainly does not belong to . . ."); the 1-ary functional symbol card* (the cardinal number of the quaset . . .), the binary functional symbol $\cap_*$ (the quaset-theoretical intersection).

The axioms are the following:

(A1) $\subseteq$ is a partial order:

$$\forall x(x \subseteq x)$$
$$\forall xy(x \subseteq y \wedge y \subseteq x \rightarrow x = y)$$
$$\forall xyz(x \subseteq y \wedge y \subseteq z \rightarrow x \subseteq z)$$

(A2) $\forall xy(x \notin y \rightarrow \neg x \in y)$
In other words: 'certainly does not belong to . . .' is stronger than 'does not belong to . . .'.

(A3) $\forall xy(x \subseteq y \rightarrow (\forall z(z \in x \rightarrow z \in y) \wedge (z \notin y \rightarrow z \notin x)))$
(A4) $\forall x \exists!!y \forall z((z \in y \leftrightarrow z \in x) \wedge (z \notin y \leftrightarrow \neg z \in x))$
where $\exists!!$ is the quantifier 'there exists exactly one'.

(A4) justifies the definition of a 1-ary functional symbol Ext(x) (the extension of x):

DEFINITION 1.

$$\forall xy(y = \text{Ext}(x) \leftrightarrow \forall x(z \in y \leftrightarrow z \in x) \wedge (z \notin y \leftrightarrow \neg z \in x))).$$

In other words, the extension of a quaset x is the unique quaset that certainly contains all the certain elements of x and certainly does not contain all the other elements. On this basis one can define the predicate 'to be a set' (Sx); sets are quasisets which are identical with their extensions:

DEFINITION 2. $\forall x(Sx \leftrightarrow x = \text{Ext}(x))$.

One can easily show that sets satisfy the extensionality principle:

$$\forall xy(Sx \wedge Sy \wedge \forall z(z \in x \leftrightarrow z \in y) \rightarrow x = y).$$

(A5)

$$\exists y \forall x(x \notin y)$$

One can show that the quaset to which anything certainly does not belong is a set and is unique. This justifies the definition of the individual constant $\emptyset$ (the empty set). For any formula α of ZF, we will indicate by α^s the corresponding formula of QST relativized to sets.

(A6) If α is any instance for an axiom of ZF then α^s is an axiom of QST.

The definitions of the notions of *cardinal number* (Cx), *cardinal numbers of a set* (card(x)), order-relation between cardinal numbers (<) are the standard definitions of ZF (relativized to sets).

(A7) $\forall x \exists !! y(Cy \wedge \text{card}^*(x) = y \wedge (Sx \rightarrow \text{card}^*(x) = \text{card}(x)))$

(A8) $\forall x(\text{card}^*(x) \geq \text{card}^*(\text{Ext}(x)))$

(A9) $\forall xy(x \subseteq y \rightarrow \text{card}^*(x) \leq \text{card}^*(y))$

(A10) $\forall xy(x \cap_* y \subseteq x \wedge x \cap_* y \subseteq y) \wedge (Sx \wedge Sy \rightarrow x \cap_* y = x \cap y))$

where $\cap$ is the standard set-theoretial intersection.

Notice that our axiomatization does not require that *proper* quasets (which are not sets) exist. Any application of QST may specify a certain collection of proper quasets. Being non-extensional, proper quasets can be construed as a kind of 'semiintensions'; and the last axiom (A10) warrants the possibility of applying and 'isolation procedure' to such 'intensions'.[21]

We now try and apply QST to our semantical problems. Let us go back to our example of an n-electron system σ considered in the time-interval $[t_0, t_1]$. Can we reasonably construct models of the form

$$< I <, D, \rho >$$

for the language L^σ (with names $a_1, \ldots, a_n$ and predicates Q_i^E)? One might be tempted to require: for any $t \in [t_0, t_1]$, D_t is a quaset determined by $\psi(t)$; for any t, $\rho_t(Q_i^E)$ is a quaset (determined by $\psi(t)$) such that $\rho_t(Q_i^E) \subseteq D_t$. But what about $\rho_t(a_i)$?

Intuitively, it seems very reasonable to accept that $\psi(t)$ determines a quaset D_t. Indeed, $\psi(t)$ describes a kind of 'collection' whose cardinal number is definite (equal to n). However, unlike what would be the case in a classical set-theoretical situation, each element of this collection is not precisely determined: for instance anyone does not generally belong with certainty to a corresponding singleton. As a consequence, each singleton is not identical with its extension (and turns out to be a proper quaset). In this situation, $\rho_t(a_i)$ will be generally undefined. In spite of this, using the information provided by $\psi(t)$, one can determine reasonable truth-conditions for sentences like $Q_j^E\, a_i$. Although, in this framework, we cannot go into technical details, we will only recall that such truth-conditions can be described as governed by a *quantum-logical semantics*.[22]

9. LEIBNIZIAN AND ANTILEIBNIZIAN PARTICLES

As a final question let us briefly discuss in what sense the Leibniz principle

(LP) $\forall xy(\neg x = y \rightarrow \exists P(Px \wedge \neg Py))$

can be violated in microphysics.[23]

Clearly, a strong 'quaset-theoretical version' of LP

$\forall xy(\neg x = y \rightarrow \exists z(x \in z \wedge y \notin z))$

is generally not valid. Indeed, the classical argument, founded on the implication $\neg x = y \rightarrow x \in \{x\} \wedge y \notin \{x\}$, cannot be repeated ni QST.

In spite of this, in the case of fermions, one can extend the semantics sketched at the end of the previous section to a second-order semantics – by specifying for each model a convenient set Π of quasets which correspond to 'meaningful' physical properties – in such a way that LP turns out to be valid. Models for boson systems, instead, will have a less rich set of properties Π: for instance, singleton quasets will not generally correspond to meaningful properties; and this gives easily rise to violations of LP.

But how can fermions be at the same time *indistinguishable* and *leibnizian* particles? Intuitively, one might observe that: if two electrons are (according to LP) distinguished by at least one property P, then they are *distinguished* and hence cannot be *indistinguishable*.

This observation would be correct in classical logic. Nevertheless, in quantum logic it may happen that a sentence like $\exists P(Pa \wedge \neg Pb)$ is true, even if any possible choice of P does not satisfy the formula $Pa \wedge \neg Pb$. As an example, let us consider the two electrons of a helium atom in the fundamental state. Let a, b be names for our two different electrons and let us suppose that both the extension $\rho_t(a)$ and $\rho_t(b)$ are indetermined (in the sense discussed in the previous section): let P^+ represent the property 'having spin up' and P^- the property 'having spin down'. Now the sentence $(P^+a \wedge \neg P^+b) \vee (P^-a \wedge \neg P^-b)$ is physically true; hence is true also the following instance of LP: $\neg a = b \rightarrow \exists P(Pa \wedge \neg Pb)$. Nevertheless the truth-state of each member of the disjunction $(P^+a \wedge \neg P^+b) \vee (P^-a \wedge \neg P^-b)$ is indetermined, and this holds for any other possible choice of P.

The problems discussed in this Part II regard only the case of *real particles*. As has been observed at the end of Part I, virtual particles give rise to a completely new logical situation, where the eccentric interaction between actual and possible worlds seems to be hardly analyzable by means of the usual tools of 'normal' Kripke semantics. A main difficulty is represented by the fact that in order to describe the peculiar behaviour of virtual particles, one cannot simply label the possible worlds by indices representing time-instants, ordered according to the natural order.

The possibility of a logical analysis of the strange universe of virtual particles appears to be a challenge for the future semantical investigations about physics.

10. GENERAL CONCLUSIONS

It is hard to derive clear-cut conclusions from what we have been discussing above. However, some remarks may be justified.

Ordinary semantics seems to be modeled after the macroscopic world of everyday life, so as to meet the needs of verbal intercourse in that world. But the domains of the infinitely large and of the infinitely small – two worlds very far removed from our own and, in a sense, 'inaccessible' to man – pose a number of problems that may even lead to reverse some of the more largely received views.

The concept of *extension* turns out to become more and more blurred as we go deep into the intricacies of modern physics, until it breaks down altogether. The case for taking a name as an abbreviation for a description becomes stronger. At the same time, the existence of rigid designators appears doubtful. And anyway, when we deal with physical objects, what is the referent of a name?

Is it the physical *object* we want to name, or a *phenomenon* caused by that object, as in the case of the two quasars Q 33, Q 34 (discussed in Section 2 of Part I)? Note that the statement that a given phenomenon is produced by a given object is the *result of a theory*. And this is not true only of quasars. Suppose you baptise Aristotle by ostension. What you really do is to say that you call 'Aristotle' that *image* (dark hair, brown eyes, straight nose, short legs ...). That that image is caused by a special object (a man) is the result of a theory. And you cannot transmit the stipulation of that baptismal cermony to a chain of other people, unless they share the same theory. Stating the identity between two individuals amounts to affirming a relation between two *appearances*, that are deemed to have their origin in the same individual. For instance, you may point out Aristotle's image reflected in a mirror and say: "That is Aristotle".

Mircophysics is a world of *intensions*, where individual objects and names appear, so to speak, to be *unnatural* notions. Equal objects are fundamentally inseparable[24] and only in special cases and with *approximation* can be treated as separate.

The *actual* world is constantly accompanied by an infinity of *possible* worlds, whose role is not so abstract as was previously thought. There is a substantial interaction between the actual world and a large class of possible worlds, that gives rise to tangible effects.

Natural-kind terms in microphysics have strictly rigid descriptions, consisting in a conjunction of properties. Such properties must be verifiable or measurable. If one of the stipulated properties turns out not to be valid in an actual observation, we simply conclude that the object is of a different kind.

In the macroscopic world one can still accept the notion that a natural kind is defined by a *cluster* of properties, only a percentage of which may be satisfied in an actual case. But the description must become increasingly rigid when we go towards the microscopic world.

Finally, we can also accept the notion of the *division of labor*[25] in semantics. Most people who today may talk of electrons or quarks do not have a clear idea of how such particles should be recognised, and must rely on the authority of experts, who affirm that such particles exist and can be recognised. In common talk, people make use of more or less vague *sterotypes* of scientific objects.

MARIA LUISA DALLA CHIARA
Dipartimento di Filosofia,
Università degli Studi di Firenze.

GIULIANO TORALDO DI FRANCIA
Dipartimento di Fisica,
Università degli Studi di Firenze

NOTES

[1] See Quine (1960).
[2] See Putnam (1975).
[3] See Kripke (1980).
[4] See Kripke (1980), p. 20.
[5] These names will be imaginary throughout.
[6] See Mill (1950).

[7] See Russell (1919).

[8] See Kripke (1980).

[9] This position has been maintained mainly by Linski (1977).

[10] Kripke (1980), p. 24.

[11] Toraldo di Francia (1981), p. 222.

[12] We ignore the direction of the spin.

[13] Apart from a normalisation coefficient that does not interest us here.

[14] The solutions of Schrödinger's equation are continuous.

[15] The wavelength is inversely proportional to the square root of the mass.

[16] The notion of *quaset* has some aspects in common with the concepts of *fuzzy set*, of *multiset*, of *semiset* (see Vopenka 1977), of *quasiset* (see Da Costa and Krause (to appear), Krause (to appear).

[17] See Ziff (1960).

[18] Kripke (1980), p. 122.

[19] The one-photon process is impossible because of the energy and momentum conservation laws.

[20] See Dalla Chiara and Toraldo di Francia (1979).

[21] In many application – including the physical ones – it might be useful also to have *urelements* (primitive individuals). In such a case collections of urelements will be represented by quasets, that are not generally sets. As a consequence, an urelement will not necessarily belong to its singleton.

[22] See, for instance, Dalla Chiara (1986).

[23] The question is highly controversial. One may propose different solutions, that strictly depend on the technical aspects of the kind of logical analysis that is assumed. See, for instance, Van Fraassen (1985) and (1991), Margenau (1950), Mittelstaedt (1985), Da Costa and Krause (to appear), Krause (to appear), Costantini and Garibaldi (1990), Stoeckler (1988).

[24] See D'Espagnat (1981).

[25] See Putnam (1975).

REFERENCES

Beltrametti, E. and Cassinelli, G.: 1981, *The Logic of Quantum Mechanics*, London, Addison-Wesley.

Costantini, D. and Garibaldi, U.: 1990, 'The non frequency approach to elementary particle statistics', in *Statistics in Science* (ed. by R. Cooke and D. Costantini), Dordrecht, Kluwer.

Da Costa, N. and Krause, D.: to appear, 'Schrödinger logics'.

Dalla Chiara, M.L.: 1985, 'Names and descriptions in quantum logic', in *Recent Developments in Quantum Logic* (ed. by P. Mittelstaedt and E.W. Stachow), Mannheim, Bibliographisches Institut.

Dalla Chiara, M.L.: 1986, 'Quantum Logic', in *Handbook of Philosophical Logic III* (ed. by D. Gabbay and F. Guenthner), Dordrecht, Reidel.

Dalla Chiara, M.L. and Toraldo di Francia, G.: 1979, 'Formal analysis of physical theories', in *Problems in the Foundations of Physics* (ed. by G. Toraldo di Francia), Amsterdam, North Holland.

D'Espagnat, B.: 1981, *A la recherche du réel*, Paris, Bordas.

Jammer, M.: 1974, *The Philosophy of Quantum Mechanics*, New York, Wiley.

Jauch, J.M.: 1968, *Foundations of Quantum Mechanics*, London, Addison-Wesley.

Krause, D.: to appear, 'On a quasi-set theory'.

Kripke, S.: 1980, *Naming and Necessity*, Oxford, Blackwell.

Linski, L.: 1977, *Names and Descriptions*, Chicago, University of Chicago Press.

Margenau, H.: 1950, *The Nature of Physical Reality*, New York, McGraw-Hill.

Mill, J.S.: 1950, 'A system of Logic', in *John Stuart Mill's Philosophy of Scientific Method* (ed. by E. Nagel), New York, Hafner.

Mittelstaedt, P.: 1985, 'Constituting, Naming and Identity in Quantum Logic', in *Recent Developments in Quantum Logic* (ed. by Mittelstaedt and E.W. Stachow), Mannheim, Bibliographisches Institut.

Putnam, H.: 1975, 'The meaning of meaning', in *Language, Mind and Knowledge* (ed. by K. Gunderson), Minnesota Studies in the Philosophy of Science.

Quine, W.V.O.: 1960, *World and Object*, New York, Wiley.

Russell, B.: 1919, *Introduction to Mathematical Philosophy*, London, Allen and Unwin.

Stoeckler, M.: 1988, 'Individualität, Identität, Ununterscheidbarkeit', *Conceptus* 12, 5.

Toraldo di Francia, G.: 1985, 'Connotation and Denotation in Microphysics', in *Recent Developments in Quantum Logic* (ed. by P. Mittelstaedt and E.W. Stachow), Mannheim, Bibliographisches Institut.

Toraldo di Francia, G.: 1986, *Le cose e i loro nomi*, Bari, Laterza.

van Fraassen, B.: 1985, 'Statistical behaviour of indistinguishable particles: problems of interpretation', in *Recent Developments of Quantum Logic* (ed. by P. Mittelstaedt and E.W. Stachow), Mannheim, Bibliographisches Institut.

van Fraassen, B.: 1991, *Quantum Mechanics: an Empiricist View*, Oxford, Oxford University Press.

Vopenka, P.: 1979, *Mathematics in the Alternative Set Theory*, Leipzig, Teubner.

Ziff, P.: 1960, *Semantic Analysis*, Cornell University Press.

BAS VAN FRAASSEN

SYMMETRIES OF PROBABILITY
KINEMATICS[1]

Most of probability theory is resilient under shifts of interpretation: the same theorems are proved for personal probability as for objective chance. The question of how probabilities change with time, however, is approached quite differently. In quantum theory for example, one can see a veritable dynamics of objective probabilities, which evolve in time, constrained by symmetries which induce conservation laws. This topic I shall leave aside here, to concentrate on rational change of opinion, probabilistically conceived.

Recent literature has often given the appearance of strong opposition between those who do, and those who do not, look to Simple Conditionalization as the alpha and omega of this subject. I shall argue that this is only superficial appearance. It is indeed true that no admissible rule can rival Conditionalization on its own ground, and also (trivially) true that every rational opinion changer can be simulated by a pure Conditionalizer – but those truths place no severe limit on general probability kinematics, nor answer many of its questions.

1. A GENERAL APPROACH TO OPINION CHANGE

Epistemology is the theory of knowledge and opinion. For empiricism, fascinated with the erosion of certainty in our view of nature, opinion is the more important topic.

Personal probability is one model of opinion. Or rather, to be more accurate, it is the main ingredient in several competing, but related models. Opinion is expressed in judgements. All such judgements can be expressed as judgements of personal probability, assigned to the factual propositions involved. We can represent those propositions by areas on a Venn diagram, and their proportional probability by mud heaped on them. That is, we imagine that a quantity of mud is heaped on each area, in proportion to the probability assigned to the proposition it represents. This Muddy Venn Diagram model is a better guide than any axioms or rules for calculation, and it does equally well. (There are in fact deep theorems to show that we have here the most general model of probability theory, provided the mud is so fine as to be continuous; see Section 3

G. Corsi et al. (eds), Bridging the Gap: Philosophy, Mathematics, and Physics, 285–314.
© 1993 Kluwer Academic Publishers.

below.)

Let us now turn to change of opinion. Here I want to concentrate solely on change in response to experience, and its rationality. (In other words, I leave out at this point deliberation, theoretical innovation, conjecture, conceptual change, or whatever else here be.) *There is a general obstacle.* Suppose I were to write a recipe, or book of recipes, that would tell you how to amend your opinion rationally in view of your experience. The recipe would begin with a description of a state of opinion, and then a description of experience, or of the deliverances of experience, and then prescribe new opinions. You could use this recipe to evaluate how well someone else was doing, if you thought you knew what his or her experience and opinion were. But if you tried to use it on yourself, you would first have to describe your experience. And that is already a response to experience, and would be expressed in term of a whole new set of judgements that are yours. So you would already have done a great deal of the job that the recipe is meant to guide. You cannot really step out of yourself and compare the representation with what it represents.

I didn't describe this problem because I think there could be a serious use for recipes here, nor because I want to discuss foundationalism of any sort. I wanted to bring out instead the need for a model of experience, if we want to continue the epistemological story. And I do not mean a physiological or psychological model, because the focus of interest is not so much on how things happen, as on the rationality of the response that it must somehow be possible to evaluate. The model must pertain to phenomena presumably reported in such utterances as: "I saw a flying saucer last year; and ever since I have been a firm believer in reincarnation".

There are three models. The first was inherited I think from the main tradition. It says that in experience, some proposition E is received as evidence (or, it is taken as evidence). The subject becomes immediately totally certain that E, and adjusts his opinions accordingly. The rule for adjustment is *Simple Conditionalization*, to which I will return below. Perhaps also this model is perfectly suitable to another source for the Bayesian tradition: the working statistician, after all, is paid to accept certain propositions as data, not to question them but to use them as input for his calculations. But as a model for experience it is a bit simple-minded. I call it the *Revelation Model* because in it, experience speaks with the voice of an angel and gives you new total certainties.

A second model was described by Hartry Field in his article about Jeffrey.[2] He takes it as an article of faith that any epistemology must be compatible with materialism. So the input is a physical stimulus; the opinion is in physical storage, and is physically modified. I call this the *Robot Model*. Here the input is not a proposition, obviously, and need have nothing to do with propositions directly. Nor is there any question of an evaluation of the rationality of the response. It may be possible to speculate about general features of the mechanism, perhaps with an eye to survival value.

The third model, which I propose and endorse, also does not take the deliverances of experiences to be propositions, though still intimately connected with them. There is undoubtedly some level of response which we cannot criticize effectively in ourselves except retrospectively, at some later point. Let me call this the primary response. But I do not think that this primary response must already be a judgement, let alone a new total certainty. It is instead, I think, *the acceptance of some constraint* on what your opinion (henceforth) should be. Thus the deliverances of experience are not propositions, but commands (to oneself). A limiting case is possible: the command to become totally certain that E. If you accept that as a constraint, what happens next must be the same as what happens next in the revelation model. But many different sorts of constraints are possible – for example, to raise a subjective probability a little, or to accept new odds, or a new conditional probability, or indeed anything that could constrain opinion. Because of the crucial roles of the terms 'accept', 'constraint', 'command', it seems natural to call this model a *Voluntarist* one.[3]

Here questions of evaluation can and do arise. Suppose you accept a constraint on your actions by saying you will post a letter for me. I can later evaluate (a) whether your actions satisfied the accepted constraint, and (b) if so, how well you did. If for example you dropped the letter in the mud and got it to the mail box several days later, you satisfied the constraint, but not optimally so, in some interesting respects. Similarly for adjustments to opinion, made as secondary response, when certain constraints have been accepted.

In the remainder of this essay I will discuss examples of primary response, and putative rules to govern the corresponding secondary response. Let us begin with the very simplest sort: the limiting case in which the constraint is simply a new certainty. In this case it is indeed as if experience has simply handed us some proposition E on a platter – our 'total new evidence' – and has spoken as if with the voice of an angel.

Suppose I go for a walk in the garden and come away absolutely convinced that a flying saucer has landed there. I reconstruct this as follows: I had originally a certain state of opinion, but accepted the constraint to become certain of this new proposition E, and adjusted my opinion accordingly. There is, as I mentioned, a rule for this adjustment, *Simple Conditionalization*. It is easily explained in terms of the muddy Venn diagram. You simply wipe away all the mud on the area representing not-E. This has two effects: it raises the probability of E to 1 (for all the mud remaining is on E) and keeps the odds between propositions that entail E the same (for the mud 'inside' E was not disturbed). This is a complete description of the rule. If probability function P is changed in this way we say it is *conditionalized* on E and we call the result P_E or $P(-|E)$.

It is also possible to state this rule in algebraic form (which follows logically from its description in terms of the mud model):

$$P(X|E) = P(X \text{ and } E) / P(E) .$$

In either form it is obvious that the rule cannot apply if the prior probability $P(E)$ equals zero. In that case wiping away the mud on (not-E) would remove all mud, thus destroying the model.

So much for the geometric and algebraic descriptions of the rule – *but is it right*? This question of justification is a very fair one. Ian Hacking, writing in 1967, noted that Bayesians took this rule for granted and he called it the *Bayesian Dynamic Assumption*.[4] Textbook presentations tend to darken counsel, as usual, by suggesting that the meaning of 'conditional probability' is 'the probability you would have if you had learned that'. This cannot help because if it were the meaning (which it is not), that meaning has logically nothing to do with my present odds for (X and E) to (not X and E) – which is the information conveyed by whatever number $P(X \text{ and } E) / P(E)$ is.[5] But the rule has an all-but-completely *a priori* justification, namely, by means of a symmetry argument.

2. A SYMMETRY ARGUMENT FOR CONDITIONALIZATION

In this section I will outline a simple symmetry argument to support the following view: if we want a *rule* for changing opinion when we simply become certain of new evidence, then Simple Conditionalization is the only candidate. The proof given in Section 4 will be more rigorous, more general and have the present result as corollary. That will require the technical precision to be introduced in Section 3, while here the reasoning will be intuitive only.

Three preliminary remarks first. Each rule has a domain of application, which may be more or less wide. Simple Conditionalization is the only admissible rule, *when it is applicable*. This leaves room for other, or more general rules if it is not always applicable – a question we shall take up later. Secondly, this discussion is about *rules*. Whether rationality requires rule-following is a separate question. Thirdly, there is a trivial sense in which Conditionalization is supreme, and this triviality tends to crop up in rhetoric. We had better discuss it first.[6]

Suppose some one never violates the probability calculus in describing his opinions at any one time. Imagine however that he changes his opinion *apparently* by leaps and bounds. Then we can still always claim consistently that his opinion never changes except by Conditionalization. This sounds spectacular, but it is a pure triviality. For we can simply postulate a hidden (unconscious) event $f(t)$ for each time t, which we call this person's (unconscious) insight. Then we can embed his opinion, represented as a function $P(t)$ – which is a probability function for each time t – in one with larger domain, call it $P'(t)$ such that $P'(t + 1)$ is the conditionalization of $P'(t)$ on a proposition $F(t)$. The latter proposition is the conjunction of the evidence $E(t)$ which this person consciously acknowledges and the 'insight' proposition $f(t)$ of which he is not aware.[7]

There is a nice construction of $f(t)$ in higher-order probability theory (where propositions may describe the person's own states of opinion). That is interesting, but does not remove the fact that the above claim – 'every rational person can be regarded as a Conditionalizer' – is already true on trivial grounds.

What substantive questions remain, after this trivial point is made? We should avert our eyes from the useless *post hoc* question of how all behavior can be *retrospectively* simulated by machine. That is trivially so. No such trivial result can give us real insight. We should look instead to possible answers to the question posed by the conscious person: how *shall* I conduct myself, what *shall* I take to guide me, as I change my opinion in response to my experience?

Let us first of all narrow this problem so that Conditionalization becomes clearly applicable. This person's opinion is well-defined for a certain family of propositions, call it F, which is closed under all logical operations. His opinion at a given time is represented by a probability function P, and the new deliverances of experience are summed up entirely in a proposition E which belong to that family F. Call P his *prior* opinion and E his new *evidence*. What should be his new, *posterior* probability P', which accords certainty to the evidence E? Abstractly speaking there are many possibilities, that $P'(E) = 1$ can be satisfied in many ways.

So we narrow the problem to: what *rule* could define P', in terms of P and E? Now we have a problem of the typical sort for which a symmetry argument can be constructed.

Is anything tacitly assumed about P and E? We were narrowing the problem so that Conditionalization would become at least applicable – hence we require that $P(E)$ be positive. But apart from that the names F, P, E can stand for *any* probability function with domain F, and E in F. This is a very large class; what are its symmetries?

Structure in this class appears under two headings: logic and probability. There are the relations of logical implication among the propositions; and there are the numbers assigned as probabilities. So a transformation g which is applied to domain F_1 and probability function P_1, and sends them into a new domain F_2 and probability function P_2, is a symmetry if it preserves *that* structure.

The overall argument for Conditionalization must therefore be that this is the only rule which does not violate symmetry. Assume we have some rule, call it R, which yields a posterior probability function $R(P)$ when we specify prior P, its domain F, and the evidence E. An essentially similar problem will be one which has another prior P', domain F', and evidence E, which are connected to the former set-up by some symmetry transformation g. There the rule R will specify posterior $R(P')$. The symmetry requirement is then that also $R(P')$ must be connected to $R(P)$ by the same symmetry transformation.

To see intuitively how this symmetry argument can be carried through, three simple points need to be appreciated. The first is that if the rule R leaves ratios of probabilities invariant, for propositions which have no mud wiped off (i.e. propositions that imply E), then clearly R is just Conditionalization. For after all the posterior $R(P)$ must give 1 to E, so if A is any other proposition $R(P)$ assigns to A the same as to $A \cap E$. And secondly, the ratio between that number and what $R(P)$ assigns to E – i.e. 1 – then equals the ratio $P(A \cap E) / P(E)$.

The second point is that a symmetry can be an embedding, that is, it can relate a probability function defined on a small domain of propositions to one defined on a large domain. If we think about this momentarily in terms of language, suppose that one person's language does not include the sentence "Snow is white" while another does. Then the second person's probabilities can still be the same as the first with respect to all propositions that both understand. This point becomes mathematically important when we think of how much mathematics deals with continua. That a rule should have essentially the same effect when we embed a problem set-up into a similar but continuous one, is very informative.

The third point is that the symmetry requirement explained entails that, in intuitive terms, the rule will depend solely on two parameters: the prior probability function and the evidence proposition. From this we can deduce that there is a functional relationship between the prior and posterior probabilities assigned to propositions which imply the evidence. The argument for this, as well as the argument that ties all these threads together will be given in Section 4. But for now I'll just add this: what ties them together is a lemma that in the continuous cases (see point 2), the functionality (point 3) implies the invariance of probability ratios (point 1). This lemma is due to Paul Teller and Arthur Fine (see reference below).

3. PROBABILITY AS MEASURE: THE HISTORY

It is time to become very precise. Before we go on, we must examine the mathematical foundations of probability theory. To have a well-defined probability, you must specify to what object the probability is assigned. This may be a family of events that may or may not occur, of propositions that may or may not be true, or of (purported) facts which may or may not be the case. The family must have at least the simple sort of structure that allows representation by means of sets.

The Muddy Venn Diagram described above is the perfect intuitive guide to probability. From that model we can immediately derive (where Λ is the empty set):

I. $P(\Lambda) = 0 \ P(K) = 1 \ \ 0 \le P(A) \le 1$

II. $P(A \cap B) + P(A \cup B) = P(A) + P(B)$.

To understand probability properly, though, one needs to know why this model is such a good guide. I will first state the formal theory in its current form, and then describe how it came to have this form. After that I will state the results that convey such a privileged status on the muddy diagram models. It will be seen that the geometric probabilities (which Buffon is rightly claimed to have introduced in the 18th century) are still, also in the abstract expressionism of our day, the core of the subject.

Let us distinguish between a *probability function* and a *probability measure*. The first is the subject of probability theory when we do not impose any continuity requirements or other concerns about infinity. (Bruno De Finetti insisted this should remain the complete subject). In any case, a probability measure is a special kind of probability function.

A *field* of sets on a set K is a class F of subsets of K such that:

  K and Λ are in F

  If A, B are in F, so are $A \cap B$, $A \cup B$, $A - B$.

A field F on K is a *Borel field* or *sigma-field* on K if in addition F contains the union of any countable class $A_1, \ldots, A_n, \ldots$ of its members.

It follows automatically that if $A_1, \ldots, A_n, \ldots$ are in a Borel field, so is their intersection.

P is a *probability function* on set K, defined in field F, exactly if

(1) $P(A) \in [0, 1]$ for each member A of F

(2) $P(\Lambda) = 0$, $P(K) = 1$

(3) $P(A \cap B) + P(A \cup B) = P(A) + P(B)$
       for any members A, B of F.

P is a *probability measure* on K, defined on F, exactly if F is a Borel field and

4. $P(A) = \lim_{n \to \infty} P(A_n)$ if A is the union of the series $A_1 \subseteq \ldots \subseteq A_n \ldots$ of members of F.

Here 3 and 4 have equivalents some of which may look more familiar:

(3a) $P(A \cup B) = P(A) + P(B)$ if $A \cap B = \Lambda$

(4a) $P(\cup A_i) = \sum P(A_i)$ if $A_i \cap A_j = \Lambda$ for all $i \neq j$

(4b) $P(\cap A_i) = \lim_{n \to \infty} P(A_n)$ if $A_1 \supseteq \ldots \supseteq A_n \ldots$.

The property described by 3a is *finite additivity* and that described by 4a is *sigma-additivity* or *countable additivity*. It is clear from formulations 4 and 4b that the additional property that makes a probability function a measure is that it satisfies a continuity requirement.

So defined, probability theory is a part of measure theory. For in mathematics, a *measure* is a function defined exactly like a probability measure except that it need not have 1 as upper bound; indeed some sets may have infinite measure. Thus the condition $P(K) = 1$ is omitted, and 1 is replaced by

(1)' $P(A) \in [0, \infty]$.

By these definitions, probability measures are probability functions which are also measures. I will describe a little of the history that introduced these subjects, in part to answer the question which must surely have occurred to you: why not have the probabilities defined for *every* subset of the total set K? Why this fiddling around with fields and such?

Measure theory began with some rather tentative and skeptical attempts to use Cantor's set theory in analysis.[8] These surface in the second edition of Camille Jordan's *Cours d'analyse* (Paris: Gauthier-Villars, 1893), where only finite additivity is noted as a defining requirement for measure.

Countable additivity is made part of the definition in Emile Borel's mono-graph *Leçons sur la théorie des fonctions* (Paris: Gauthier-Villars, 1898). The measure which Borel defined on the unit interval $[0, 1]$ – which we now call Lebèsgue measure – is not defined for all sets of real numbers in that inter-val. The definition runs as follows: the measure of an interval of length s has measure s; a countable union of disjoint sets with measures $s_1, \ldots, s_n, \ldots$ has measure $\sum s_i$; and if $E \subseteq E'$ have measures s_1 and s_2 then $E' - E$ has measure $s_2 - s_1$. The sets encompassed by these clauses he called *measurable*. What Borel calls the measurable subsets of $[0, 1]$ we now call the *Borel sets* on that interval. It is clear that they form a Borel field, so here we see the origin of our terminology.

It was Henri Lebèsgue who posed the explicit problem of defining a measure on all the subsets. That some such measures exist is trivial: assign 1 to any subset of $[0, 1]$ that includes the number zero, and assign 0 to every other subset. Then you have a probability measure, but a trivial one. The question is not whether it can be done at all, but whether the requirement of including every subset in the domain would eliminate important or interesting functions. Here is the passage in which Lebèsgue introduces his *Measure Problem*:

We propose to assign to each set [in n-dimensional space] a non-negative number, which we shall call its measure, satisfying the following conditions:

(i)   There is a set whose measure is not zero.

(ii)   Congruent sets have equal measure.

(iii) The measure of the sum [union] of a finite or denumerable infinity of disjoint sets is the sum of the measures of these sets.

(Lebèsgue "Intégrale, longueur, aire" (1902), p. 236)

Congruence is the relation between sets which can be transformed into each other by symmetries of the space. In the Euclidean case, one of the symmetry transformations is translation, and so we see at once that the measure asked for is not a probability measure. For suppose a given cube has measure 1. By translation we turn it into infinitely many disjoint cubes congruent to it. The measure of their union – and hence of the whole space – is therefore ∞. But more than this we can also see that the measure Lebèsgue calls for is already uniquely determined on all the Borel sets. In an n-dimensional space, not the intervals but the generalized rectangles – such as $\{(x_1, \ldots, x_n): a_1 \leq x_1 \leq b_1, \ldots, a_n \leq x_n \leq b_n\}$ – are the beginning. The class of Borel sets is the smallest class that contains all these and is closed under countable union and set-difference. All the Borel sets can be approximated by choosing as beginning instead all the generalized rectangles congruent to a single very small generalized cube R. The approximation gets progressively better as we take R smaller and smaller. But it is obvious that if the unit cube has measure 1, it can be divided into n disjoint cubes of dimensions $1/n$ which must (by finite additivity) all receive measure $1/n$. Thus the measure of all little cubes is uniquely determined, and hence by continuity, the measure of all the Borel sets.

It is well worth emphasizing this point: *only* this natural generalization of the usual (length–area–volume) measure is invariant under translation. This measure, so defined on the Borel sets, we now call *Lebèsgue measure*. The problem Lebèsgue posed, and did not solve, in his dissertation, is whether this measure can be extended to all the other subsets of the space as well.

Enter the Axiom of Choice. Borel was one of its staunchest and most vocal opponents. Lebèsgue was more moderate in his philosophical opposition, but rejected it. This was in the years 1904–1905. In his own proofs, Lebèsgue had apparently tacitly relied on the Axiom of Choice. But if the Axiom of Choice is true, it turns out that Lebèsgue's 'Measure Problem' has no solution.

In 1905 Giuseppe Vitali, already relying on the Axiom of Choice, constructed a set which was not measurable in Lebèsgue's sense. He used this to show that Lebèsgue's Measure Problem has no solution for the real line. Lebèsgue himself gave a further such example, but expressed doubts about the Axiom. Felix Hausdorff, who accepted the Axiom, used it to show in addition that even if the requirement of countable additivity is weakened to that of finite additivity, the Measure Problem has no solution for Euclidean spaces of dimension greater than 2.

Vitali had used the requirement of translation invariance which, as we saw, entails that the whole space has infinite measure. However, Hausdorff's proof that the Measure Problem has no solution for Euclidean spaces of dimension greater than 2, used rotational invariance. He discussed measure on a sphere and showed that a measure defined on every subset could not be rotationally invariant. This does just as well for a measure which assigns 1 to the sphere itself, and thus tells us at once there cannot be such a probability measure on the sphere.[9]

It will now be quite clear, therefore, that the requirement to have probability defined everywhere, would be unacceptable. We must accept as genuine probability measures also those which cannot be extended to measures on all subsets of their domain. Fields and Borel fields are their natural habitat.

Are geometric probabilities really as logically central as this story made it look? The answer is that the geometry-oriented intuitions of Lebèsgue, Borel, and Hausdorff were quite right. This is what I meant when I said that the

Muddy Venn Diagram is in a certain sense a general model.

The results that establish this are the deepest in the foundations of probability theory, for they lead us to a classification of all possible probability measures, and an understanding of all of them in terms of Lebèsgue measure.[10]

Therefore the focus will be on the paradigm probability space, consisting of the unit interval $[0, 1]$ of real numbers, the family B_0 of Borel sets on this interval, and Lebèsgue measure m defined on these Borel sets. The 'points' are the real numbers $0 \leq x \leq 1$. These are funny things: in some sense – namely by uncountable union – the unit sets $\{x\}$ generate the space, but each of them has measure zero, so their measures do not add up to the measure 1 of the whole space. We switch therefore to a more algebraic viewpoint, by identifying sets which differ only by measure zero. To be precise:

Let $S = < K, F, P >$ be a probability space with P a probability measure defined on the Borel field F of subsets of K. Call A and B *equivalent* if $P(A - B) \cup (B - A)) = 0$ and define the quotient F/P to be the following algebra:

- its elements are the sets $[A] = \{B : B \in F$ and B is equivalent to $A\}$, for A in F
- $[A] \wedge [B] = [A \cap B]$, $[A] - [B] = [A - B]$
- $[A] \leq [B]$ exactly if $A \subseteq C$ for some C in $[B]$.

Then $< F/P, p >$ is a *probability algebra* (with $p[A] = P(A)$, in terms of measure P on Borel field F).

The special case in which S is the paradigmatic probability space, yields in this way the probability algebra B_0/m with Lebèsgue measure transposed to it as in the above definition.

Define an *atom* of such an algebra to be an element y such that there is nothing between it and the zero element $0 = [\Lambda]$. That is:

y is an *atom* exactly if for any x, if $0 \leq x \leq y$ then $x = 0$ or $x = y$.

Clearly 0 is the *only* element that receives zero measure! For all sets A such that $P(A) = 0$ belong to $[\Lambda]$. So the atoms all have positive measures. We also see that the atoms cannot overlap: if y and z are distinct atoms, $y \wedge z = 0$. So we deduce at once that there can only be at most countably many atoms. For suppose we have any family of events with positive measure. There are at most 2 with measure $\geq 1/2$ because their sum cannot be greater than 1. Similarly there are at most 3 with measure $\geq 1/3$, at most 4 with measure $\geq 1/4$, and so forth. But every one of them has a measure $\geq 1/n$ for some n, so there are at most $2 + 3 + 4 + \cdots$ of them, which is countable infinity.

Therefore we now distinguish three sorts of probability algebra:

(1)   the algebra has no atoms (*atomless*)
(2)   the algebra has finitely or countably many atoms, whose measures add up to 1 (*atomistic*)
(3)   the algebra has finitely or countably many atoms, whose measure add up to some number $x < 1$ (*mixed*).

The representation theory for mixed cases is of course complicated, but consists in showing that they are all constructions out of pure (atomless and atomistic) cases.

The second, atomistic case is quite easily represented in this algebra by a Muddy Venn Diagram: divide the square into countably many distinct parts, each of which represents one atom, and put a mass of mud on each proportional to the measure of that atom. In terms of our paradigmatic probability space we can do the same thing: to represent the algebra $< F/P, p >$ divide the unit interval into a countable partition $\{A(y) \; : \; y \text{ an atom of } F/P\}$. Do it so that each set $A(y)$ has positive Lèbesgue measure. Define the function $m' \; : \; m'(A(y)) = p(y)$. This function m' represents the probability. Of course, it is itself the beginning, so to say, of a probability measure. We can extend m' to the whole Borel field just by insisting on countable additivity. Obviously then we have reached our goal here: in the atomistic case, the probability can be represented by means of a measure defined in terms of Lebèsgue measure.

The paradigmatic probability algebra B_0/m is itself an example of the atomless case. For suppose A is a Borel subset of $[0, 1]$ and $m(A) = x > 0$. Could $[A]$ be an atom? No, because there will be a smaller Borel set B with $m(B) = x/2$; and so forth. It will not come as a surprise that in set theory it is possible to find probability spaces whose Borel fields have cardinalities incredibly much hiugher than the relatively small infinities we deal with in the case of real numbers or Euclidean spaces. But what about the algebras they give rise to?

On the probability algebra $< F/P, p >$ we can define a metric: $d(x, y) = p(x \vee y) - p(x \wedge y)$, where $\vee$ is defined from $\wedge$ by De Morgan's law. Thus we can apply metric concepts, and we have the theorem:

(Birkhoff) *Any atomless probability algebra which has a countable dense subset is isometrically isomorphic to the probability algebra generated by Lebèsgue measure on the unit interval.*

I shall end this exposition here; it will be clear now that, even from a strictly logical point of view, probability theory is a subject which stays close to the earth of its geometric history.

4. SYMMETRY: AN ARGUMENT FOR JEFFREY CONDITIONALIZATION

In 1965, Richard Jeffrey created the new subject of *probability kinematics*. He did this by describing different sorts of changes in opinion, asking for a theoretical description, and proposing a rule that generalizes simple conditionalization. This new rule has since been generally known as *Jeffrey Conditionalization*. It has simple Conditionalization as a special case (and the result of this section has that in Section 2 as corollary).

Here is an example: I walk through a room, in which I glimpse roses on a table lit by candlelight. I had expected the scene, had a prior opinion that the roses would, as likely as not, be red. Now I am more inclined to think they are red. ... it now seems twice as likely to me as not that they are red. How should I adjust my total state of opinion 'accordingly'?

A bit later, or at roughly the same time as Jeffrey was thinking about this, I have been told, Wade Edwards and his collaborators were discussing a *prima facie* different problem. The rule they came up with is formally the same as Jeffrey's. Their problem was this: we have a spy in Iran, and he sends us regular reports, but he is non-reliable, he tends to lie a little. Reading his reports, what constraints shall we place on our posterior opinion? There is an obvious analogy with Jeffrey's case, a suggestion that we think of our senses as Plato did, as lying spies in the garden of earthly delights. Perhaps my eyes were saying 'red' when I walked through that room, but they did not speak with the voice of an angel, and my response was only to become more inclined toward, *not* certain of, the proposition they spoke.

Let us restate the problem in an even more schizophrenic mode. There are two spies, Cain and Abel; they are twin brothers, and their reports are always each other's negation. But we consider Cain twice as likely to speak the truth as Abel. Now we can say this: if I accept Cain's reports totally, I shall conditionalize on them; if I accept Abel's totally, I shall conditionalize on the negations of Cain's.

Since neither of these brothers speaks like an angel, I shall make my own

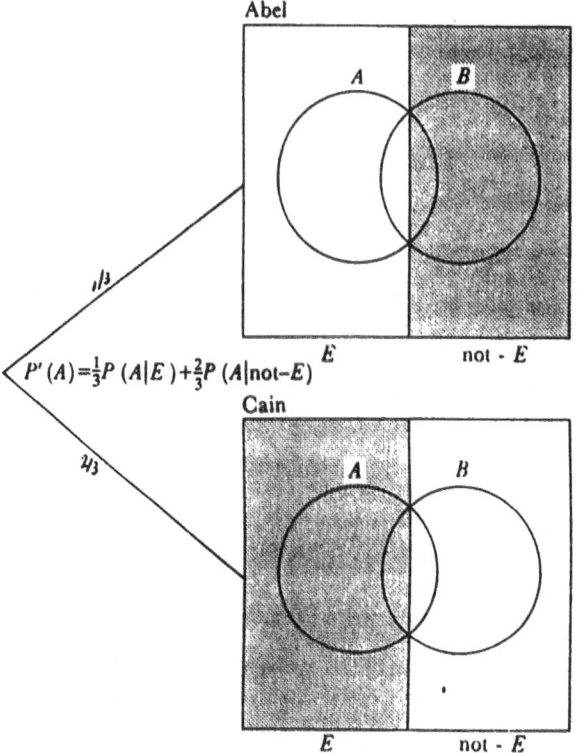

Fig. 1. The lying spies.

opinion a proportional mixture of theirs.

This idea of a *mixture* of two states of opinion works perfectly in the alternative epistemology. In the dogmatist oversimplification, the best one could do would be to take what is implied by the both Cain's and Abel's reports – i.e. tautologies only.

This is the complete description of the rule of Jeffrey Conditionalization. Can we justify this rule?

We can regard it as concerning a special case, among the following possible constraints on the posterior probability:

C_1. the new probability for B equals 1
C_2. the new probability for B equals 0.7
C_3. the new conditional probability for A given B equals 0.6
C_4. the new odds of A to B equal 6 : 3
C_5. the new expectation value of parameter x equals 4.5.

(Expectation is probability mean value: $\sum rP(x$ has a value $r)$ with the sum over all possible values of r.)

Each of these sorts of constraint requires that the probabilities – initial and final – be defined at least on the events mentioned (B for C_1 and C_2, A and B for C_3 and C_4 and the events [x has value r] for C_5). Thus if we consider the problem starting with initial probability P_1, and transform it into a different, essentially similar, problem, with initial probability P_2, we have here already one aspect which must be left intact. For the problem to make sense, P_1 had to be defined on certain events – the corresponding P_2 must be defined on corresponding events, and subject to a corresponding constraint. And of course, our solution must transform the corresponding *priors* into corresponding *posteriors* (final probability functions). We will have the usual closed diagram

[DIAG]

$$
\begin{array}{ccc}
 & \text{Rule} & \\
P_1, C_1 & \longrightarrow & P_1' \\
\downarrow & & \downarrow \\
 & \text{Rule} & \\
P_2, C_2 & \longrightarrow & P_2'
\end{array}
$$

We must now implement these intuitive ideas about symmetry requirements, in the formal framework established in the preceding section.[11]

First we isolate the significant structure which can be present in the problem situation. This is dictated by probability theory: prior and posterior probabilities must be defined on a suitable field of sets. The total structure of probability space $< F, p >$ consists in the character of field F and measure p, and the structure all probability spaces have is that F is a field or Borel field and p a measure thereon. Transformation of one problem situation into another one, which may be called 'essentially the same' should therefore preserve exactly this structure.

Let us call a *measure embedding* any one-to-one map g of $< F, p >$ into $< F', p' >$ such that g is an isomorphism as far as the set-theoretic operations are concerned, and also preserves measure, i.e. $p'(gA) = p(A)$ for all A in F. Clearly g has an inverse, and we may restate this either as $p'g = p$ or $p' = pg^{-1}$

(the first for the domain of g and the second for its range, which will generally be only part of the domain of p').

If CC is a set of constraints on probabilities to be assigned to elements of X, which is part of F, we have an equivalent set of constraints cc' imposed on the images $g(A) \; : \; A \in X$, applicable to probabilities defined on the target field F'. The requirement upon our general solution is therefore that cc', the function we present for imposing CC' on priors defined on F', should be related to cc in the following way:

SYMMETRY. If g is a measure embedding of $< F, p >$ into $< F, p' >$ then $ccp(A) = cc'p'(gA)$ for all A in F.

In other words, when p is just $p'g$ then ccp should be $cc'p'g$.

The practical effect of following this principle is that, when we try to identify cc, we are always allowed to switch our attention to a more tractable 'equivalent' probability space (either the domain or the range of a measure embedding relating it to the one we had).

The constraints CC are stated as conditions on the posterior probabilities to be assigned to a certain set X of elements of F. Without loss of generality, we can take X to be a subfield; alternatively so long as X is countable we can, without loss, take it to be a partition. I shall henceforth assume the latter.

As a working hypothesis, let us assume cc to exist, and when applied to p, to give the posterior probabilities $ccp(B) = r_B$ to members B of partition X.

It is clear that $p'(A) = \sum\{p'(A \cap B) \; : \; B \in X\}$ for any function p', so we will have identified ccp if we can determine what values it gives to subsets of members B of X. Therefore we begin with:

LEMMA 1. *Let* E, E' *be subsets of member* B *of* X *and* $p(E) = p(E')$. *Then* $ccp(E) = ccp(E')$. (See *Proofs and Illustrations*.)

Because of this lemma, we now know that for a subset E of a member of the partition, the sole relevant factor is its prior probability. So there exists for a given number B of X a function f such that $ccp(E) = f(p(E))$ when $E \subseteq B$. What is this function like? Here it is more convenient to embed our problem in a context where real analysis can apply.

A probability space $< F, p >$ is *full* when for each element A of F, p takes every value in $[0, p(A)]$ on the elements of F which are subsets of A.

LEMMA 2. *There exists a measure embedding* $*$ *of* $< F, p >$ *in a full* space $< F^*, p^* >$.

This embedding is easily constructed with $< F^*, p^* >$ the product of $< F, p >$ with $< [0,1], m >$ where m is Lebèsgue measure. Then F^* is the family $\{A \times Q : A$ in F and Q a Borel set on the unit interval$\}$, $p^*(A \times Q) = p(A) \, m(Q)$, and $A^* = A \times [0,1]$.

To continue the main argument, we now apply Lemma 1 to our thinking about this full space (see *Proofs and Illustrations*), and arrive at:

SYMMETRY THEOREM. *If there exists a function* cc *corresponding to constraints CC on the posterior probabilities for members of partition X, then for each probability function p to which* cc *is applicable and such that p is positive on all members of X, we have*

$$\mathbf{cc}p(-|B) = p(-|B) \quad for \ all \ B \ in \ X \ .$$

It was assumed, of course, that any such function must satisfy the SYMMETRY principle listed above. As an immediate corollary we have the theorem:

1ST COROLLARY (Simple Conditionalization Rule). *If $p(B) \neq 0$ and CC is the constraint that the posterior probability for B equal to one, then* $\mathbf{cc}p(-) = p(-|B)$.

Richard Jeffrey proposed, for the constraint that the posterior probability for B equal r, the rule $p'(-) = rp(-|B) + (1 - r) \, p(-|\overline{B})$. We deduce similarly the uniqueness of his rule for this constraint, in the general form:

2ND COROLLARY (Jeffrey Conditionalization Rule). *If X is a countable partition with $p(B) \neq 0$ for all B in X, and CC the constraint that a posterior probability for B equal r_B, from all B in X, then*

$$\mathbf{cc}p(-) = \sum \{r_B p(-|B) : B \in X\} \ .$$

In these cases, therefore, our first theorem already singles out a unique way of imposing the constraint.

We may sum up our results as follows: whatever the constraints CC (stated with reference to partition X) are, the effect of the function cc is equivalent to an operation that determines posterior probabilities on the partition X, *followed by* the operation of Jeffrey conditionalization with those posterior probabilities. The task that remains is to investigate the first operation with greater generality.

There are no equally satisfying results that go beyond this point. Jaynes and his collaborators have explored the rule to minimize relative information (equivalently, to maximize relative entropy) as a solution to this general problem. Of course that rule agrees with the Simple and Jeffrey Conditionalization rules – otherwise it would violate the basic symmetries of the general problem. But I shall now leave this topic, except for references, and a brief discussion of Jaynes' rule below.[12]

Proofs and Illustrations

To prove Lemma 1, we look at the subfield of F generated by $X \cup \{E, E'\}$; call it F_0. Let p_0 be p restricted to F_0. The identity function is then a measure embedding of $< F_0, p_0 >$ into $< F, p >$. We now construct a measure embedding of the former onto itself as follows:

$$g(E - E') = E' - E$$

$$g(E' - E) = E - E'$$

g is the identity function on $X \cup \{E \cap E', B - (E \cup E')\}$

$$g(A \cup A') = g(A) \cup g(A') \text{ when } A, A' \text{ are disjoint.}$$

A Venn diagram suffices to depict this subfield and this automorphism; because $p(E) = p(E')$ it follows that $p(E - E') = p(E' - E)$ so it is a measure embedding.

We can now apply SYMMETRY to this measure embedding g of $< F_0, p_0 >$ onto itself to deduce that $cc_0 p_0(E) = cc_0 p_0(gE) = cc_0 p_0(E')$. But secondly, the identity map is a measure embedding of $< F_0, p_0 >$ into $< F, p >$ and so we also deduce that $ccp(E') = cc_0 p_0(E') = cc_0 p_0(E) = ccp(E)$ as required.

To prove the Symmetry Theorem, we apply Lemma 1 to the full space described in Lemma 2. In that full space $< F^*, p^* >$, we see now that for any element B^* there exists a numerical function f such that $cc^* p^*(E) = f(p^*(E))$ when $E \subseteq B^*$. There the function is perceived to be a map of $[0, p(B)] = [0, p^*(B^*)]$ onto $[0, r_B]$. It must be additive, for if E, E' are disjoint parts of B^* we have

$$f(p^*(E)) + f(p^*(E')) = cc^* p^*(E) + cc^* p^*(E')$$
$$= cc^* p^*(E \cup E')$$
$$= f(p^*(E \cup E'))$$
$$= f(p^*(E) + p^*(E'))$$

and because the space is full we have disjoint parts E, E' with probabilities r, s respectively, whenever $r + s \leq p(B)$.

A theorem of the calculus[13] implies that such an additive function has a constant derivative, thus

$$f(x) = kx + m .$$

Looking at $x = p(\Lambda)$ and $x = p(B)$, respectively, we deduce that $f(0) = 0$ so $m = 0$, and $k = r_B/p(B)$

$$f(x) = r_B x/p(B) .$$

Hence, by SYMMETRY, the function cc, like cc*, is given by the equation:

$$\text{cc}p(E) = r_B \frac{p(E)}{p(B)} \quad \text{for } E \subseteq B \in X$$

$$= r_B\, p(E|B) .$$

5. LEVI'S OBJECTION: A SIMULATED HORSE RACE

Also writing in 1967, in a review of Jeffrey's book, Isaac Levi raised certain objections to the idea of probability kinematics. One of these has variants for all continuations of the program as well. Surely one's opinion, formed in response to a succession of experiences, should not depend on the order in which these experiences occur? That is obvious if the experience fits the Revelation Model. Simple Conditionalization successively on $E1$ and $E2$ is the same as doing it on ($E1$ and $E2$) – so there order does not matter. But Jeffrey Conditionalizing, with successive constraints $[E1 = 0.7]!$, $[E2 = 0.8]!$ for example, clearly does depend on the order. (Think of the limiting case $1 = 2$.) And so it should; but if we have here a correct description of adjustment of opinion in response to relevant experience, we have a bit of a paradox. Two persons, who have the same relevant experiences on the same day, but in a different order, will not agree in the evening even if they had exactly the same opinions in the morning. Does this not make nonsense of the idea of learning from experience?

I shall approach this challenge in two ways. First, let us ask whether experience can contradict itself? Surely not: what my experience is at two different times, is *a priori* unrestricted – what happens to me is up to nature, so to say, and there are no *a priori* bounds on nature. (This is an empiricist speaking.) But what about responses to successive experiences? Do we have any reason to suspect irrationality if those are incompatible? I think not; else why ever look twice? It is along these lines that I understand Jeffrey's own response. That is that a person who wishes to raise the probability of $E1$ to 0.7 and a bit later that of $E2$ to 0.8, has the choice of imposing the joint constraint

$[E1 = 0.7$ and $E2 = 0.8]$! at the later time, provided the two constraints are logically compatible. And if they are not compatible, he clearly *wants* to discard the previous judgement.

This response has two problems. First, Jeffrey's rule does not tell you how to impose the joint constraint. This requires a rule going beyond his; and such rules raise questions of their own, which I shall discuss below. The second problem is that the incompatibility is not a matter of simply discarding past judgements. Suppose $E1$ and $E2$ are mutually incompatible. Then they cannot have 0.7 and 0.8 at the same time; and which you impose later makes a difference to what everything gets. As an example suppose I have (Table I):

TABLE I

Successive Jeffrey Conditionalizations

	E_1	E_2	not $(E_1$ or $E_2)$	
P	0.1	0.8	0.1	
				impose $[E_1 = 0.7]$!
P'	0.7	$\frac{8}{9} \cdot \frac{3}{10}$	$\frac{1}{9} \cdot \frac{3}{10}$	
				impose $[E_2 = 0.8]$!
P''	$\frac{21}{22} \cdot \frac{2}{10}$	0.8	$\frac{1}{22} \cdot \frac{2}{10}$	

Here you see clearly that going back to the original value for $E2$, for instance, does not get you back to the original state of opinion. If the constraints had been imposed in opposite order, the first would have effected no change at all, and P' would have been the end result.

So we need a more sophisticated approach. I think we should momentarily forget the technicalities, and ask: What does happen to learning from experience? Has this become impossible? Suppose a person *learns*, in the sense that he successively accepts constraints, each of which correctly represent the real, objective probabilities. Will his opinion become more correct, or not?

This question could be investigated experimentally, to some extent. I decided to simulate an experiment on a computer. A man wishes to bet on a horse race; on each day preceding the race he is allowed to study the horses. Suppose that on each day Dn he updates his probability that a certain horse Hn will win to Pn, using Jeffrey's rule. Suppose also that the correct or objective probabilities are indeed $P1, \ldots, Pn, \ldots$. Should we expect his opinions on the day of the race to be closer to the correct probabilities than they were initially? For simplicity I used three horses, and asked the question in this form:

TABLE II

The horse-race: learning by Jeffrey Conditionalization.

	Prior	Target	\multicolumn{4}{c}{Degree of Approximation}			
			$1/100$	$1/10^4$	$1/10^8$	$10/10^{10}$
(A)	1, 1, 1	$1, 10^3, 10^6$	2	2	2	3
(A)	1, 1, 1	1, 10, 100	2	4	6	8
(C)	1, 1, 1	1, 1, 20	2	5	8	11
(D)	3, 3, 4	1, 2, 7	3	5	11	13
(E)	7, 2, 1	1, 2, 7	5	7	11	14
(F)	5, 6, 7	6, 7, 8	1	5	11	15
(G)	5, 6, 7	7, 8, 9	1	5	12	15
(H)	1, 5, 25	1, 2, 4	3	6	11	17

how many days does it take for his subjective probabilities to come within a present amount of the objective ones? Table II shows some results (stated in terms of odds rather than probabilities).

There is no doubt that this person is learning, and quickly. He gets to within one percent of the right probabilities in a few days, and to within an astronomical degree of approximation within about two weeks. Not only that, there is a definite convergence, and it is not true that successive experiences 'undo' the preceding learning process to a radical extent. Of course I assumed that the *input* – the imposed constraints – were exactly correct. But while such acute perceptions may be lacking in reality, any lack of overall success will not be due to the rule being followed.[14]

6. GENERAL PROBABILITY KINEMATICS AND ENTROPY

In 1965 Jeffrey had formulated a simple, new problem: in the roses-by-candlelight case, where Simply Conditionalization is not applicable, what should one do? And as we saw, his proffered solution has a very special status, for it can be justified *a priori* by a symmetry argument. There may be cases in which this result still gives us no guidance. This can be for one of two reasons:

(a) We may want our posterior opinion to depend on other factors besides the prior opinion and the given, simple constraint;

(b)   We may wish to impose another constraint, of more complex form.

Reason (a) is always with us, and rightly keeps us from conceiving epistemology as in principle a special sort of arithmetic or other mechanical procedure. But we should also consider the case – presumably 'normal' in the sense that deliberation and theoretical innovation and creativity must be the exception rather than the norm in our daily life – in which reason (a) is absent.

This brings to the problem of *general probability kinematics*: find rules that transform priors, subject to given constraints, and for which the prior and constraint are the sole relevant factors. The symmetry argument of Section 4 really demonstrated the following general result. Suppose that we are given constraints on the posterior probabilities $P'(A1), \ldots, P'(An)$ defined with reference to a partition $\{A1, \ldots, An\}$ of the space of possibilities. Then any rule that transforms a prior P into such a posterior P' must in effect proceed in two steps:

(i)   Assign exact new probabilities to $A1, \ldots, An$.
(ii)   Treat the outcome of step (i) as input for a Jeffrey Conditionalization.

This result is very helpful, because it means that if we are given a constraint of a new and strange sort, like:

Change your opinion so that the probabilities for rain and snow become equal.

we know that we need not worry about finer subdivisions like (rain and cereal for breakfast), (rain and scrambled eggs for breakfast), etc. For step (ii) must always be carried out in the same way. With this in hand, let us begin to investigate constraints of more sophisticated form.

If we want to go on piecemeal, here are three special instances of the general problem, which we have not yet covered:

Impose constraints of form:

(a)   change simultaneously both the probabilities of A and of B, to 0.3 and 0.7 respectively,
(b)   change the odds of A to B, to $3 : 1$
(c)   change the expectation values of quantities $X, Y, Z, \ldots$, to 4.1, 4.2, 4.3, ... respectively.

Let us look at each of these in turn. In (a), if $A = B$ we have the special case of Jeffrey Conditionalization. But the general case with A distinct from B is not handled just because, as we saw above, the order of successive Jeffrey Conditionalizations matters.

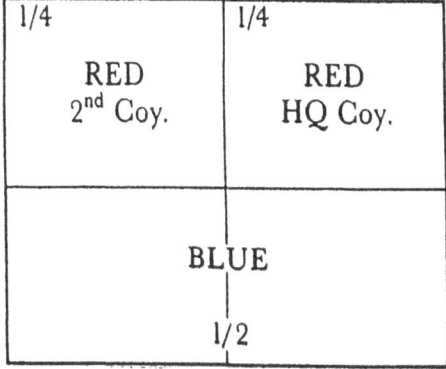

Fig. 2. The Judy Benjamin problem.

In (b) there is also a special case which is at least somewhat more tractable: when A implies B. Then the odds is a ratio of type $P(A \cap B) / P(B)$, and so the constraint is to change a conditional probability. The problem of finding a rule for this case I have elsewhere called the Judy Benjamin problem.[15] It derives from the movie *Private Benjamin*, in wich Goldie Hawn, playing the title character, joins the Army. She and her platoon, participating in war games on the side of the 'Blue Army', are dropped in the wilderness, to scout the opposition ('Red Army'). They are soon lost. Learning the movie script now, suppose the area to be divided in two halves, Blue and Red territory, while each territory is divided into Headquarters Company area and Second Company area. They were dropped more or less at the center, and therefore feel it is equally likely that they are now located in one area as in another. This gives us the following muddy Venn diagram (diagram 2) drawn as a map of the area.

They have some difficulty contacting their own HQ by radio, but finally succeed and describe what they can see around them. After a while, the officer at HQ radios: "I can't be sure where you are. If you are in Red Territory, the odds are 3 : 1 that you are in HQ Coy area ...". At this point the radio gives out. As in the movie, the platoon goes on to capture enemy headquarters.

We must now consider how Judy Benjamin should adjust her opinions, if she accepts this radio message as the correct and sole constraint to impose. The question on which we should focus is: what does her new input do to the probability that they are in friendly Blue Territory? Does it increase, or decrease, or stay at its present level of $1/2$? I originally proposed this question in a seminar in the fall of 1980, and have raised it invarious seminars, classes and conference audiences since. The 'intuitive' or unreflective response has always been, overwhelmingly, that after the radio message about Red Territory, it should seem neither more or less likely to Judy Benjamin that she is still in Blue.

Peter Williams brought out the main flaw in that sort of response. Suppose the radio officer had not said '3 : 1' but, in more general form: "The probability that you are in HQ area, given that you are in Red Territory, equals q for some value q" – 3/4 in our other example – which must lie between 0 and 1. Our task is to find a general rule to cover all. But if the radio officer has said '1' – if q had equalled 1 – he would have told her, in effect, 'You are *not* in Red-2 Co area'. In that case she could have used Simple Conditionalization, and wiped the mud off the top-left square in the diagram. This would have the result of giving Blue Territory 2/3 of the remaining mud. Hence the probability of Blue would go up; and so we cannot make it a general rule that it stays the same.

In the papers cited, three distinct rules are found which appear to satisfy all symmetry requirements, as well as some others, equally well. This raises the possibility that the uniqueness result for Jeffrey Conditionalization will not extend to more broadly applicable rules in general probability kinematics. In that case rationality will not always dictate epistemic procedure uniquely even when we decide that it shall be rule governed.

Problem (c) is the most general we have touched on yet. Indeed, (a) and (b) are special cases of (c). So a rule that handles (c) handles all the problems we have come across so far.

Does this mean we shoul simply concentrate on (c), i.e. Expectation constraints? There is a good deal of literature on this in connection with the rule INFOMIN (also called MAXENT), which I shall explain below; it is indeed applicable to such constraints. The physicist Edwin T. Jaynes proposed this rule in statistical mechanics, and had considerable success with it. The rule has also had successful engineering applications in data analysis, specifically photographic image enhancement. Lately there have appeared new deductions of the rule from much weaker premises, including interesting symmetry arguments. All of this suggests an affirmative answer to our question.[16]

But the prominence of INFOMIN is due in great part to its being the only rule that is known and has been investigated at length, which can handle expectation value constraints. Mathematically speaking, other such rules exist, but we don't know them in the same way.[17] The symmetry arguments mentioned above should rule out these others from our consideration, but symmetry arguments are often based in part on desiderata or assumptions which are not totally incontrovertible, as we know.

Apart from that, should we rule out from consideration, say, any rule that solves the Judy Benjamin problem, but cannot be extended to Expectation constraints? The putative general principle behind such a decision – *require such rules to be extendible to other sorts of constraints* – would make us discard INFOMIN as well. For it cannot be used to impose such a constraint as, for example,

(d)   change your probabilities so as to make A and B statistically independent

i.e. so that the posterior P' has $P'(A \text{ and } B) = P'(A) \cdot P'(B)$. Therefore we cannot advocate that putative principle.

If the Judy Benjamin problem has no rationally compelled *unique* answer, then – mathematically and logically speaking – it can still be a special instance of a more general problem, which does have a unique solution. (After all, solutions take the form "here is a rule that covers all the cases in your target class".) But that does not mean automatically that the Judy Benjamin problem has a unique solution. And for someone who asks: "Am I rationally compelled to change my opinions in such and such a fashion?" it is not a complete answer to be told "If you decide to adopt a rule for solving all such problems with constraints definable in terms of posterior expectation values, then *yes*".

As an example for which the more general sort of rule is needed, consider the constraint on the expectation value for tosses yielding an even number, namely

1. $E(x|x \text{ is even}) = 2\text{Prob}(x = 2|x \text{ is even}) + 4\text{Prob}(x = 4|x \text{ is even})$

    $+ 6\text{Prob}(x = 6|x \text{ is even})$

that it should equal a certain number, say 3. That implies that the number 2 is more likely to come up than the number 6; but how much more likely?

If the Principle of Indifference could do its job, it would suffice to imply a unique answer here. Since that is definitely not so, one could make a proposal. A model was indeed proposed which recommends itself by its simplicity, plausibility, and advantages in application. The proposal made by Jaynes, and since explored in many contexts, was in effects as follows.[18] We need a measure of *information*, or *informativeness*, and choose the probability function which is the least informative among those which satisfy the imposed constraints. Such measures of information are available; the best known is Shannon's, also called *negative entropy*. It is defined by:

2. $I(P) = -\sum P(x) \log P(x)$

where log is again the natural logarithm, and x ranges over the set of basic alternatives – in our case the six faces of the die. This function has the following properties:

3. $I(P) \geq 0$

4. $I(P) = \infty$ if $P(x) = 0$ for any x

5.   If x ranges over N alternatives, $I(P)$ has its minimum value, namely log N, for the measure P which assigns $1/N$ to each alternative.

This information measure also behaves very well when various probability measures are combined. In Equation 1, we really see four probability measures altogether. Besides the defined P' on set $\{1,2,3,4,5,6\}$ we have:

P1:   P1(Even) = 0.75

       P1(Odd) = 0.25

P2:   P2 is defined on the set Even = $\{2,4,6\}$

P3:   P3 is defined on the set Odd = $\{1,3,6\}$

P': $P'(x) = $ P1(Even) P2(x) if x is even

       $P'(x) = $ P1(Odd) P3(x) if x is odd

and a little calculation shows us that:

6. $I(P') = 0.75 I(\text{P2}) + 0.25 I(\text{P3}) + I(\text{P1})$.

The constraint that Even was to be three times as likely as Odd, fixes the last term; therefore to minimize quantity 6, we must minimize the other two terms on the right, i.e. choose P2 and P3 to be uniform on their domains. This gives us back the equation 2 for the correct choice of P'.

These are nice consequences. But isn't the choice of information measure I, as defined by 2, itself arbitrary? There do exist elegant deductions of the uniqueness of I, as defined, from desirable properties such as 3–6. Of course the justification for requiring such features as the additivity in Equation 6 is only something like mathematical convenience or ease of calculation. Perhaps we should be happy with this, however, once we give up the false ideal of Indifference.

What about Bertrand's paradoxes? Proposition 5 above shows clearly how the division into basic alternatives is the initial step that determines the model. When different such divisions are available, as in our paradoxical examples, these will lead to different calculations of informativeness, and contradictory solutions. But Jaynes recognizes this clearly, and considers measure I as appropriate only for finitary and countable cases.[19] When the set of alternatives is defined, in terms of a continuous parameter, we need to choose an 'informationless' prior in some other way. Then we can find the right solution for a problem that imposes further constraints by shifting from that prior to some other measure, in a way that minimizes the *relative information*. In the countable case this is defined by

7. $I(P'; P) = \sum P'(x) \log (P'(x)/P(x))$

which in the limit of a continuous parameter x becomes

8. $I(P'; P) = \int P'(x) \log (P'(x)/P(x)) dx$.

In a Bertrand paradox we encounter a function g which transforms x into $y = g(x)$. Expression 8, unlike 2, is invariant under this transformation, provided it is monotone.

The rule to choose that posterior probability, among those which satisfy the given constraint, which has the least information relative to the prior, i.e. the rule to minimize relative information (or maximize cross-entropy), I call INFOMIN. It is not universally applicable, but there are nice theorems to delineate when it is.[20]

Proofs and Illustrations

The relative information $I(P'; P)$ and its companion $I(P; P')$ are often called the *directed divergences*, and their sum the *divergence* between P' and P. The family of their linear combinations was uniquely characterized most elegantly by Rodney Johnson.[21] Let p, q be strictly positive probability densities on a parameter x, and F the functional defined by

$$F(p, q) = \int f(p(x), q(x)) dx$$

for some measurable function f on the real numbers. The following requirements then entail that F belongs to that family:

Finiteness: $F(p, p)$ is finite

Additivity: $F(p, q) = F(p', q') + F(p'', q'')$ when
$p(x', x'') = p'(x') p''(x'')$ and
$q(x', x'') = q'(x') q''(x'')$

Positivity: $F(p, q)$ is non-negative, equalling zero only
when $p = q$.

Johnson proves along the way that the first two properties already entail that the functional remains invariant when the domain of p or q is transformed by a non-singular linear transformation.

7. CONCLUSION: NORMAL RULE FOLLOWING

Elsewhere I have argued that rationality does not require rule following.[22] Rational opinion change need not be change in accordance with some rule or recipe. It does require that we do not sabotage ourselves by our own lights – for example by committing ourselves to a rule for opinion change which makes us incoherent. Besides the conclusion that it is rational not to follow rules, there was a second: if one does follow a rule, it must be Conditionalization.

In this chapter we have seen why that is. But we have also seen the limits of that point: Conditionalization is the sole rule if the input (deliverances of experience) is propositions, and the sole relevant factors are that input and the prior opinion.

The exploration of probability kinematics subject to other constraints was based on the premise that not all responses to experience (need) consist in taking propositions as newly certain. That seems a weak enough premise (Richard Jeffrey has urged that a rational agent gives probability *one* to tautologies only), and yet it has been strongly resisted in the literature. Friedman and Shimony began the resistance in an article directed at Jaynes, to the effect that INFOMIN conflicts with conditionalization.[23] There has been a good deal of literature since. In my opinion, the criticism of INFOMIN and other such rules are tacitly based on the assumption that what I called the Revelation model is the correct model of experience. In addition, there are assumptions involved about exactly what the propositions constituting the new certainties must be. None of these assumptions are demonstrated. Nor do I see how they could be.[24]

One example may perhaps suffice. Suppose I respond to my experience by accepting as (total) constraint that my posterior opinion must involve a probability of 90% that it will rain tomorrow. Then perhaps, if I am very consciously introspective, I shall also be aware that this is so i.e. that I'm accepting this constraint. Could it be that I am simply conditionalizing on the latter, autobiographical proposition? Well it is *possible*. But personally I am so aware of the unreliability of introspection that I would not take that proposition to be certainly true. Perhaps I would give it 90% probability with 10% for its opposite. And as a result my probability for rain tomorrow would become 81% instead of 90%. However, I doubt that such autobiographical commentary is normally involved anyway. It may be postulated of course. More recherché evidence taking may also be postulated. This brings us back to the argument in Section 1, that it is trivially possible to reconstruct everyone as a conditionalizer. But not fruitful.

We have now come to the end of this exploration of symmetry. It is rather gratifying to note that even this section and the last contain a number of disputed points and unsolved problems. There are, as far as our discussion is concerned, large uncharted areas that remain. To an empiricist it must necessarily be so, for whatever insight symmetry brings us for theory and model construction, scientific progress must always rest on contingent theoretical assumptions. Any *a priori* certainty it can enjoy is at best conditional.

Department of Philosophy,
Princeton University

NOTES

[1] With minor revision, this essay is identical with Chapter 13 of my *Laws and Symmetry* (Oxford University Press, 1989).

[2] 'A note on Jeffrey conditionalization', *Philosophy of Science* 45 (1978), 361–367.

[3] See my 'Rational belief and probability kinematics', *Philosophy of Science* 47 (1980), 165–187.

[4] Hacking, I., 'Slightly more realistic personal probability', *Philosophy of Science* 34 (1967), 311–325.

[5] Cf. F.P. Ramsey, 'Truth and probability', reprinted pp. 23–52 in H.E. Kyburg Jr. and H.E. Smokler (eds.) *Studies in Subjective Probability* (Huntingdon, N.Y.: Krieger, 1980); Section 3, p. 40.

[6] See my 'Rationality does not require conditionalization', in E. Ullman-Margalit (ed.) *The Israel Colloquium*, Vol. 5 (Dordrecht: Kluwer, forthcoming).

[7] I think of time here as discrete, but the unit can of course be chosen as small as you like. The proposition $E(t)$ can be identified as follows: it is the logically strongest proposition X in the domain of P such that $P(t)(X) = 1$. Note that we must take into account also the case of someone who gives positive probability to a proposition which he gave probability *zero* before; that is why we must embed P in function P' with larger domain. That case is left aside here but discussed in 'Rationality does not require conditionalization', where it is shown that nothing very advanced is needed to substantiate the assertion that such a person as here described can always be simulated by a perfect conditionalizer.

[8] For this history, I am especially indebted to G.H. Moore, 'The origins of Zermelo's axiomatization of set theory', *Journal of Philosophical Logic* 7 (1978), 307–329; and 'Lebèsgue's measure problem and Zermelo's axiom of choice: The mathematical effects of a philosophical dispute', *Annals of the New York Academy of Sciences* (1983), 129–154.

[9] We may note in passing that Banach proved in 1923 that the measure problem does have solutions for dimensions 1 and 2 provided we settle for finite additivity. But he also proved a much stronger negative result for measures properly speaking: quite aside from requirements of congruence, there cannot be a measure defined on all subsets of $[0, 1]$ which (like Lebèsgue measure) gives zero to each point (i.e. to each unit set $\{x\}$).

[10] The crucial results appealed to in the following are theorems of Kuratowski, Birkhoff, Horn, Tarski, and Maharam (see G. Birkhoff, *Lattice Theory* 3rd ed. (AMS: Providence, Rhode Island, 1967); and D.A. Kappos: *Probability Algebras and Stochastic Spaces* (New York, 1969), Ch. II.4 and III.3. For a more extensive discussion focusing on the relation between probabilities and

frequencies, see my 'Foundations of probability: a modal frequency interpretation' in G. Toraldo di Francia (ed.) *Problems in the Foundations of Physics* (Amsterdam: North-Holland, 1979), esp. pp. 345–365.

[11] The following argument was first given in a different setting, see my 'A demonstration of the Jeffrey conditionalization rule', *Erkenntnis* **24** (1986), 17–24, 'Symmetry arguments in probability kinematics' (with R.I.G. Hughes), in P. Kitcher and P. Asquith (eds.) *PSA 1984*, Vol. 2, pp. 851–869 (East Lansing, MI: Philosophy of Science Association, 1985), and 'Symmetries in personal probability kinematics' in N. Rescher (ed.) *Scientific Inquiry Perspective* (Lanham, MD: University Press of America, 1987).

[12] See further my papers cited earlier in this chapter, and also my 'Discussion: a problem for relative information minimizers', *British Journal for the Philosophy of Science* **32** (1981), 375–379 and 'A problem for relative information minimizers in probability kinematics, continued' (with R.I.G. Hughes and G. Harman), *British Journal for the Philosophy of Science* **37** (1986), 453–475.

[13] At this point the argument follows that of P. Teller and A. Fine, 'A characterization of conditional probability', *Mathematical Magazine* **48** (1975), 267–270.

[14] For the proof of convergence, see John Collins' Appendix to my 'Symmetries in personal probability kinematics' in N. Rescher (ed.) *Scientific Inquiry in Philosophical Perspective* (Lanham, MD: University Press of America, 1987).

[15] See van Fraassen 1981 'A problem for relative information minimizers'; van Fraassen, Hughes, and Harman, 1986; van Fraassen 'Symmetries of personal probability kinematics', all cited above. These articles also contain the calculations omitted here.

[16] Levine, R.D. and M. Tribus, *The Maximum Entropy Formalism*, Cambridge, Mass.: MIT Press, 1979; Shore, J.E. and R.W. Johnson, 'Axiomatic derivation of the principle of maximum cross-entropy', *IEEE Trans. Information Theory* IT-26 (1980), 26–37; Skilling, J., 'The maximum entropy method', *Nature* **309** (June 28, 1984), 748–749; Tikochinsky, Y., N.Z. Tishby, and R.D. Levine, 'Consistent inference of probabilities for reproducible experiments', *Physical Review Letters* **52** (1984), 1357–1360.

[17] Diaconis, P. and S. Zabell, 'Updating subjective probability', *Journal American Statist. Assoc.* **77** (1982), 822–830.

[18] Levine, R.D. and M. Tribus, *The Maximum Entropy Formalism*, Cambridge, Mass.: MIT Press, 1979.

[19] For this problem, see A. Hobson, *Concepts in Statistical Mechanics* (N.Y.: Gordon and Breach, 1971), pp. 36, 42, 49.

[20] See P.M. Williams 'Bayesian conditionalization and the principle of minimum information', *British Journal for Philosophy of Science* **31** (1980), 131–144.

[21] R.W. Johnson, 'Axiomatic characterization of the directed divergences and their linear combinations', *IEEE Trans. Info. Theory* IT-25 (1979), 709–716.

[22] See my *Laws and Symmetry* (Oxford University Press, 1989), Ch. Seven.

[23] K. Friedman and A. Shimony, 'Jaynes' maximum entropy prescription and probability theory', *Journal of Statistical Physics* **3** (1971), 381–384. See also A. Shimony, 'Comment on the interpretation of inductive probabilities', *ibid.* **9** (1973), 187–191.

[24] For the most sensitive treatment so far, see B. Skyrms, 'Maximum entropy as a special case of conditionalization', *Synthese* **636** (1985), 55–74, and 'Updating, supposing and MAXENT', *Theory and Decision* **22** (1987), 225–246.

INDEX OF NAMES

315

Boston Studies in the Philosophy of Science

Editor: Robert S. Cohen, *Boston University*

1. M.W. Wartofsky (ed.): *Proceedings of the Boston Colloquium for the Philosophy of Science, 1961/1962.* [Synthese Library 6] 1963
ISBN 90-277-0021-4
2. R.S. Cohen and M.W. Wartofsky (eds.): *Proceedings of the Boston Colloquium for the Philosophy of Science, 1962/1964.* In Honor of P. Frank. [Synthese Library 10] 1965 ISBN 90-277-9004-0
3. R.S. Cohen and M.W. Wartofsky (eds.): *Proceedings of the Boston Colloquium for the Philosophy of Science, 1964/1966.* In Memory of Norwood Russell Hanson. [Synthese Library 14] 1967 ISBN 90-277-0013-3
4. R.S. Cohen and M.W. Wartofsky (eds.): *Proceedings of the Boston Colloquium for the Philosophy of Science, 1966/1968.* [Synthese Library 18] 1969
ISBN 90-277-0014-1
5. R.S. Cohen and M.W. Wartofsky (eds.): *Proceedings of the Boston Colloquium for the Philosophy of Science, 1966/1968.* [Synthese Library 19] 1969
ISBN 90-277-0015-X
6. R.S. Cohen and R.J. Seeger (eds.): *Ernst Mach, Physicist and Philosopher.* [Synthese Library 27] 1970 ISBN 90-277-0016-8
7. M. Čapek: *Bergson and Modern Physics.* A Reinterpretation and Re-evaluation. [Synthese Library 37] 1971 ISBN 90-277-0186-5
8. R.C. Buck and R.S. Cohen (eds.): *PSA 1970.* Proceedings of the 2nd Biennial Meeting of the Philosophy and Science Association (Boston, Fall 1970). In Memory of Rudolf Carnap. [Synthese Library 39] 1971
ISBN 90-277-0187-3; Pb 90-277-0309-4
9. A.A. Zinov'ev: *Foundations of the Logical Theory of Scientific Knowledge (Complex Logic).* Translated from Russian. Revised and enlarged English Edition, with an Appendix by G.A. Smirnov, E.A. Sidorenko, A.M. Fedina and L.A. Bobrova. [Synthese Library 46] 1973
ISBN 90-277-0193-8; Pb 90-277-0324-8
10. L. Tondl: *Scientific Procedures.* A Contribution Concerning the Methodological Problems of Scientific Concepts and Scientific Explanation.Translated from Czech. [Synthese Library 47] 1973 ISBN 90-277-0147-4; Pb 90-277-0323-X
11. R.J. Seeger and R.S. Cohen (eds.): *Philosophical Foundations of Science.* Proceedings of Section L, 1969, American Association for the Advancement of Science. [Synthese Library 58] 1974 ISBN 90-277-0390-6; Pb 90-277-0376-0
12. A. Grünbaum: *Philosophical Problems of Space and Times.* 2nd enlarged ed. [Synthese Library 55] 1973 ISBN 90-277-0357-4; Pb 90-277-0358-2
13. R.S. Cohen and M.W. Wartofsky (eds.): *Logical and Epistemological Studies in Contemporary Physics.* Proceedings of the Boston Colloquium for the Philosophy of Science, 1969/72, Part I. [Synthese Library 59] 1974
ISBN 90-277-0391-4; Pb 90-277-0377-9

Boston Studies in the Philosophy of Science

14. R.S. Cohen and M.W. Wartofsky (eds.): *Methodological and Historical Essays in the Natural and Social Sciences*. Proceedings of the Boston Colloquium for the Philosophy of Science, 1969/72, Part II. [Synthese Library 60] 1974
ISBN 90-277-0392-2; Pb 90-277-0378-7

15. R.S. Cohen, J.J. Stachel and M.W. Wartofsky (eds.): *For Dirk Struik*. Scientific, Historical and Political Essays in Honor of Dirk J. Struik. [Synthese Library 61] 1974
ISBN 90-277-0393-0; Pb 90-277-0379-5

16. N. Geschwind: *Selected Papers on Language and the Brains*. [Synthese Library 68] 1974
ISBN 90-277-0262-4; Pb 90-277-0263-2

17. B.G. Kuznetsov: *Reason and Being*. Translated from Russian. Edited by C.R. Fawcett and R.S. Cohen. 1987
ISBN 90-277-2181-5

18. P. Mittelstaedt: *Philosophical Problems of Modern Physics*. Translated from the revised 4th German edition by W. Riemer and edited by R.S. Cohen. [Synthese Library 95] 1976
ISBN 90-277-0285-3; Pb 90-277-0506-2

19. H. Mehlberg: *Time, Causality, and the Quantum Theory*. Studies in the Philosophy of Science. Vol. I: *Essay on the Causal Theory of Time*. Vol. II: *Time in a Quantized Universe*. Translated from French. Edited by R.S. Cohen. 1980
Vol. I: ISBN 90-277-0721-9; Pb 90-277-1074-0
Vol. II: ISBN 90-277-1075-9; Pb 90-277-1076-7

20. K.F. Schaffner and R.S. Cohen (eds.): *PSA 1972*. Proceedings of the 3rd Biennial Meeting of the Philosophy of Science Association (Lansing, Michigan, Fall 1972). [Synthese Library 64] 1974
ISBN 90-277-0408-2; Pb 90-277-0409-0

21. R.S. Cohen and J.J. Stachel (eds.): *Selected Papers of Léon Rosenfeld*. [Synthese Library 100] 1979
ISBN 90-277-0651-4; Pb 90-277-0652-2

22. M. Čapek (ed.): *The Concepts of Space and Time*. Their Structure and Their Development. [Synthese Library 74] 1976
ISBN 90-277-0355-8; Pb 90-277-0375-2

23. M. Grene: *The Understanding of Nature*. Essays in the Philosophy of Biology. [Synthese Library 66] 1974
ISBN 90-277-0462-7; Pb 90-277-0463-5

24. D. Ihde: *Technics and Praxis*. A Philosophy of Technology. [Synthese Library 130] 1979
ISBN 90-277-0953-X; Pb 90-277-0954-8

25. J. Hintikka and U. Remes: *The Method of Analysis*. Its Geometrical Origin and Its General Significance. [Synthese Library 75] 1974
ISBN 90-277-0532-1; Pb 90-277-0543-7

26. J.E. Murdoch and E.D. Sylla (eds.): *The Cultural Context of Medieval Learning*. Proceedings of the First International Colloquium on Philosophy, Science, and Theology in the Middle Ages, 1973. [Synthese Library 76] 1975
ISBN 90-277-0560-7; Pb 90-277-0587-9

27. M. Grene and E. Mendelsohn (eds.): *Topics in the Philosophy of Biology*. [Synthese Library 84] 1976
ISBN 90-277-0595-X; Pb 90-277-0596-8

28. J. Agassi: *Science in Flux*. [Synthese Library 80] 1975
ISBN 90-277-0584-4; Pb 90-277-0612-3

Boston Studies in the Philosophy of Science

Boston Studies in the Philosophy of Science

Boston Studies in the Philosophy of Science

Boston Studies in the Philosophy of Science

84. R.S. Cohen and M.W. Wartofsky (eds.): *Methodology, Metaphysics and the History of Science*. In Memory of Benjamin Nelson. 1984 ISBN 90-277-1711-7
85. G. Tamás: *The Logic of Categories*. Translated from Hungarian. Edited by R.S. Cohen. 1986 ISBN 90-277-1742-7
86. S.L. de C. Fernandes: *Foundations of Objective Knowledge*. The Relations of Popper's Theory of Knowledge to That of Kant. 1985 ISBN 90-277-1809-1
87. R.S. Cohen and T. Schnelle (eds.): *Cognition and Fact*. Materials on Ludwik Fleck. 1986 ISBN 90-277-1902-0
88. G. Freudenthal: *Atom and Individual in the Age of Newton*. On the Genesis of the Mechanistic World View. Translated from German. 1986
ISBN 90-277-1905-5
89. A. Donagan, A.N. Perovich Jr and M.V. Wedin (eds.): *Human Nature and Natural Knowledge*. Essays presented to Marjorie Grene on the Occasion of Her 75th Birthday. 1986 ISBN 90-277-1974-8
90. C. Mitcham and A. Hunning (eds.): *Philosophy and Technology II*. Information Technology and Computers in Theory and Practice. [*Also* Philosophy and Technology Series, Vol. 2] 1986 ISBN 90-277-1975-6
91. M. Grene and D. Nails (eds.): *Spinoza and the Sciences*. 1986
ISBN 90-277-1976-4
92. S.P. Turner: *The Search for a Methodology of Social Science*. Durkheim, Weber, and the 19th-Century Problem of Cause, Probability, and Action. 1986.
ISBN 90-277-2067-3
93. I.C. Jarvie: *Thinking about Society*. Theory and Practice. 1986
ISBN 90-277-2068-1
94. E. Ullmann-Margalit (ed.): *The Kaleidoscope of Science*. The Israel Colloquium: Studies in History, Philosophy, and Sociology of Science, Vol. I. 1986
ISBN 90-277-2158-0; Pb 90-277-2159-9
95. E. Ullmann-Margalit (ed.): *The Prism of Science*. The Israel Colloquium: Studies in History, Philosophy, and Sociology of Science, Vol. II. 1986
ISBN 90-277-2160-2; Pb 90-277-2161-0
96. G. Márkus: *Language and Production*. A Critique of the Paradigms. Translated from French. 1986 ISBN 90-277-2169-6
97. F. Amrine, F.J. Zucker and H. Wheeler (eds.): *Goethe and the Sciences: A Reappraisal*. 1987 ISBN 90-277-2265-X; Pb 90-277-2400-8
98. J.C. Pitt and M. Pera (eds.): *Rational Changes in Science*. Essays on Scientific Reasoning. Translated from Italian. 1987 ISBN 90-277-2417-2
99. O. Costa de Beauregard: *Time, the Physical Magnitude*. 1987
ISBN 90-277-2444-X
100. A. Shimony and D. Nails (eds.): *Naturalistic Epistemology*. A Symposium of Two Decades. 1987 ISBN 90-277-2337-0
101. N. Rotenstreich: *Time and Meaning in History*. 1987 ISBN 90-277-2467-9
102. D.B. Zilberman: *The Birth of Meaning in Hindu Thought*. Edited by R.S. Cohen. 1988 ISBN 90-277-2497-0

Boston Studies in the Philosophy of Science

103. T.F. Glick (ed.): *The Comparative Reception of Relativity*. 1987
ISBN 90-277-2498-9
104. Z. Harris, M. Gottfried, T. Ryckman, P. Mattick Jr, A. Daladier, T.N. Harris
and S. Harris: *The Form of Information in Science*. Analysis of an Immunology
Sublanguage. With a Preface by Hilary Putnam. 1989   ISBN 90-277-2516-0
105. F. Burwick (ed.): *Approaches to Organic Form*. Permutations in Science and
Culture. 1987   ISBN 90-277-2541-1
106. M. Almási: *The Philosophy of Appearances*. Translated from Hungarian. 1989
ISBN 90-277-2150-5
107. S. Hook, W.L. O'Neill and R. O'Toole (eds.): *Philosophy, History and Social
Action*. Essays in Honor of Lewis Feuer. With an Autobiographical Essay by L.
Feuer. 1988   ISBN 90-277-2644-2
108. I. Hronszky, M. Fehér and B. Dajka: *Scientific Knowledge Socialized*. Selected
Proceedings of the 5th Joint International Conference on the History and
Philosophy of Science organized by the IUHPS (Veszprém, Hungary, 1984).
1988   ISBN 90-277-2284-6
109. P. Tillers and E.D. Green (eds.): *Probability and Inference in the Law of
Evidence*. The Uses and Limits of Bayesianism. 1988   ISBN 90-277-2689-2
110. E. Ullmann-Margalit (ed.): *Science in Reflection*. The Israel Colloquium:
Studies in History, Philosophy, and Sociology of Science, Vol. III. 1988
ISBN 90-277-2712-0; Pb 90-277-2713-9
111. K. Gavroglu, Y. Goudaroulis and P. Nicolacopoulos (eds.): *Imre Lakatos and
Theories of Scientific Change*. 1989   ISBN 90-277-2766-X
112. B. Glassner and J.D. Moreno (eds.): *The Qualitative-Quantitative Distinction in
the Social Sciences*. 1989   ISBN 90-277-2829-1
113. K. Arens: *Structures of Knowing*. Psychologies of the 19th Century. 1989
ISBN 0-7923-0009-2
114. A. Janik: *Style, Politics and the Future of Philosophy*. 1989
ISBN 0-7923-0056-4
115. F. Amrine (ed.): *Literature and Science as Modes of Expression*. With an
Introduction by S. Weininger. 1989   ISBN 0-7923-0133-1
116. J.R. Brown and J. Mittelstrass (eds.): *An Intimate Relation*. Studies in the
History and Philosophy of Science. Presented to Robert E. Butts on His 60th
Birthday. 1989   ISBN 0-7923-0169-2
117. F. D'Agostino and I.C. Jarvie (eds.): *Freedom and Rationality*. Essays in Honor
of John Watkins. 1989   ISBN 0-7923-0264-8
118. D. Zolo: *Reflexive Epistemology*. The Philosophical Legacy of Otto Neurath.
1989   ISBN 0-7923-0320-2
119. M. Kearn, B.S. Philips and R.S. Cohen (eds.): *Georg Simmel and Contemporary Sociology*. 1989   ISBN 0-7923-0407-1
120. T.H. Levere and W.R. Shea (eds.): *Nature, Experiment and the Science*. Essays
on Galileo and the Nature of Science. In Honour of Stillman Drake. 1989
ISBN 0-7923-0420-9

Boston Studies in the Philosophy of Science

121. P. Nicolacopoulos (ed.): *Greek Studies in the Philosophy and History of Science*. 1990   ISBN 0-7923-0717-8
122. R. Cooke and D. Costantini (eds.): *Statistics in Science*. The Foundations of Statistical Methods in Biology, Physics and Economics. 1990
                                        ISBN 0-7923-0797-6
123. P. Duhem: *The Origins of Statics*. Translated from French by G.F. Leneaux, V.N. Vagliente and G.H. Wagner. With an Introduction by S.L. Jaki. 1991
                                        ISBN 0-7923-0898-0
124. H. Kamerlingh Onnes: *Through Measurement to Knowledge*. The Selected Papers, 1853-1926. Edited and with an Introduction by K. Gavroglu and Y. Goudaroulis. 1991   ISBN 0-7923-0825-5
125. M. Čapek: *The New Aspects of Time: Its Continuity and Novelties*. Selected Papers in the Philosophy of Science. 1991   ISBN 0-7923-0911-1
126. S. Unguru (ed.): *Physics, Cosmology and Astronomy, 1300-1700*. Tension and Accommodation. 1991   ISBN 0-7923-1022-5
127. Z. Bechler: *Newton's Physics on the Conceptual Structure of the Scientific Revolution*. 1991   ISBN 0-7923-1054-3
128. É. Meyerson: *Explanation in the Sciences*. Translated from French by M-A. Siple and D.A. Siple. 1991   ISBN 0-7923-1129-9
129. A.I. Tauber (ed.): *Organism and the Origins of Self*. 1991
                                        ISBN 0-7923-1185-X
130. F.J. Varela and J-P. Dupuy (eds.): *Understanding Origins*. Contemporary Views on the Origin of Life, Mind and Society. 1992   ISBN 0-7923-1251-1
131. G.L. Pandit: *Methodological Variance*. Essays in Epistemological Ontology and the Methodology of Science. 1991   ISBN 0-7923-1263-5
132. G. Munévar (ed.): *Beyond Reason*. Essays on the Philosophy of Paul Feyerabend. 1991   ISBN 0-7923-1272-4
133. T.E. Uebel (ed.): *Rediscovering the Forgotten Vienna Circle*. Austrian Studies on Otto Neurath and the Vienna Circle. Partly translated from German. 1991
                                        ISBN 0-7923-1276-7
134. W.R. Woodward and R.S. Cohen (eds.): *World Views and Scientific Discipline Formation*. Science Studies in the [former] German Democratic Republic. Partly translated from German by W.R. Woodward. 1991
                                        ISBN 0-7923-1286-4
135. P. Zambelli: *The Speculum Astronomiae and Its Enigma*. Astrology, Theology and Science in Albertus Magnus and His Contemporaries. 1992
                                        ISBN 0-7923-1380-1
136. P. Petitjean, C. Jami and A.M. Moulin (eds.): *Science and Empires*. Historical Studies about Scientific Development and European Expansion.
                                        ISBN 0-7923-1518-9
137. W.A. Wallace: *Galileo's Logic of Discovery and Proof*. The Background, Content, and Use of His Appropriated Treatises on Aristotle's *Posterior Analytics*. 1992   ISBN 0-7923-1577-4

Boston Studies in the Philosophy of Science

138. W.A. Wallace: *Galileo's Logical Treatises*. A Translation, with Notes and Commentary, of His Appropriated Latin Questions on Aristotle's *Posterior Analytics*. 1992
ISBN 0-7923-1578-2
Set (137 + 138) ISBN 0-7923-1579-0

139. M.J. Nye, J.L. Richards and R.H. Stuewer (eds.): *The Invention of Physical Science*. Intersections of Mathematics, Theology and Natural Philosophy since the Seventeenth Century. Essays in Honor of Erwin N. Hiebert. 1992
ISBN 0-7923-1753-X

140. G. Corsi, M.L. dalla Chiara and G.C. Ghirardi (eds.): *Bridging the Gap: Philosophy, Mathematics and Physics*. Lectures on the Foundations of Science. 1992
ISBN 0-7923-1761-0

141. C.-H. Lin and D. Fu (eds.): *Philosophy and Conceptual History of Science in Taiwan*. 1992
ISBN 0-7923-1766-1

142. S. Sarkar (ed.): *The Founders of Evolutionary Genetics*. A Centenary Reappraisal. 1992
ISBN 0-7923-1777-7

143. J. Blackmore (ed.): *Ernst Mach – A Deeper Look*. Documents and New Perspectives. 1992
ISBN 0-7923-1853-6

144. P. Kroes and M. Bakker (eds.): *Technological Development and Science in the Industrial Age*. New Perspectives on the Science–Technology Relationship. 1992
ISBN 0-7923-1898-6

145. S. Amsterdamski: *Between History and Method*. Disputes about the Rationality of Science. 1992
ISBN 0-7923-1941-9

146. E. Ullmann-Margalit (ed.): *The Scientific Enterprise*. The Bar-Hillel Colloquium: Studies in History, Philosophy, and Sociology of Science, Volume 4. 1992
ISBN 0-7923-1992-3

147. L. Embree (ed.): *Metaarchaeology*. Reflections by Archaeologists and Philosophers. 1992
ISBN 0-7923-2023-9

Also of interest:
R.S. Cohen and M.W. Wartofsky (eds.): *A Portrait of Twenty-Five Years Boston Colloquia for the Philosophy of Science, 1960-1985*. 1985   ISBN Pb 90-277-1971-3

Previous volumes are still available.

KLUWER ACADEMIC PUBLISHERS – DORDRECHT / BOSTON / LONDON

The manufacturer's authorised representative in the EU is Springer
Nature Customer Service Centre GmbH, Europaplatz 3, 69115 Heidelberg,
Germany. If you have any concerns regarding our products, please
contact ProductSafety@springernature.com

Printed and bound by CPI Group (UK) Ltd, Croydon, CR0 4YY
23/04/2026
02095625-0001